FRANK GIBNEY first came to Japan at the end of World War II. He lived there for many years, mastering the language and successfully engaging in publishing and other business activities, before returning to the United States. His book, *Five Gentlemen of Japan*, has been considered a classic since its publication in 1953.

JAPAN
THE FRAGILE SUPER POWER

REVISED EDITION

FRANK GIBNEY

A MERIDIAN BOOK
NEW AMERICAN LIBRARY
TIMES MIRROR
NEW YORK AND SCARBOROUGH, ONTARIO

NAL BOOKS ARE AVAILABLE AT QUANTITY DISCOUNTS
WHEN USED TO PROMOTE PRODUCTS OR SERVICES. FOR
INFORMATION PLEASE WRITE TO PREMIUM MARKETING
DIVISION, THE NEW AMERICAN LIBRARY, INC.,
1633 BROADWAY, NEW YORK, NEW YORK 10019.

A portion of *The High Cost of Kick-backs* appeared in different
form in the *Atlantic Monthly.*

Library of Congress Cataloging in Publication Data

Gibney, Frank, 1924–
Japan, the fragile superpower.

Includes bibliographical references and index.
1. Japan—Civilization.   2. National characteristics,
Japanese.   3. Japan—History—1945–   I. Title.
[DS821.G513   1979b]   952.04   79–25182
ISBN 0–452–00593–0

For James, Thomas, and Elise

Contents

The first Japanese I ever talked to, in his own language, was a Superior Private who had survived the virtual annihilation of the Japanese Army unit defending one of the Marshall Islands in 1944. We sat in a thin wooden interrogation shack inside the remote reservation near Pearl Harbor that housed a prisoner-of-war facility, facing each other across a table with one well-used map on it. He was very frightened at the encounter and so was I. If I learned little about what military secrets he possessed, it was still no trick to realize that he was an average citizen like myself, who had gone to school, looked forward to vacation times, fallen in love with a girl, and not liked it much when he was drafted—one man among a nation of individuals, who were variously bad, good, indifferent, truthful, lying, cowardly, brave, bright, dumb.

Through the war years and immediately after, I met hundreds of others like him. It was an odd introduction to a country and its people, but it led to much. My unexpected wartime career as a navy intelligence officer not only changed my own occupational existence, from a Greek major at Yale College with ambitions to study law, to a journalist who has lately been editing a Japanese-language encyclopaedia. It also rescued me, without realizing it at the time, from the tendency of most of my fellow countrymen to see the Japanese either in terms of heedless human waves, or cloyingly friendly reception committees, alternately part of an "inscrutable" yellow peril or the animated furnishings of an Oriental curio shop. In short, as masses, not people. This tendency was understandable in the heat of World War II. It is discouraging, to put it mildly, to find it still present three decades later.

Since World War II our country has stumbled a long way toward realizing some of the biblical truth about the brotherhood of man. Europeans are our brothers. We do business with them, marry them, know their politicians, study their vintage years, and share their recipes and tourist traps. In the course of either supporting or arguing with most of Europe's nations over the postwar period, we and they have achieved rather a high level of mutual tolerance. At home in our own country, over the same period, the tidal struggles for racial equality finally broke on the beach. However difficult solutions still may seem, we have advanced much further than we were before.

About the Asians, as people, we have learned hardly anything. This is despite the fact that the sharpest crises of American foreign policy have occurred in and about Asia—Hiroshima and Nagasaki, the China involvement, the Korean War, Vietnam. Perhaps the concurrent bitterness and shallowness of the debates on these subjects over the years have betrayed a certain lack of sympathy with the peoples involved. This is nowhere so true as in the case of Japan and the Japanese, a fascinating, able, and—toward Americans—historically friendly people with whom we seem perversely interested in keeping up either an adversary relationship or one of studied aloofness.

Over the past thirty years, I have met and known many thousands of Japanese: friends, enemies, celebrities, no-accounts, bosses, subordinates, teachers, pupils, higher-ups, and lower-downs. It has been my fortune to have lived in that country with my family for the past eight years, reading their papers, sharing their business worries, complaining about their traffic and laughing at their TV comedians. I like and admire them, although proximity has made me more aware than I might have been of their failings. If I can convey something of their hopes, their potential, their humanity to my fellow Americans, this exercise will have been abundantly worth the effort. It may be useful, also, for the Japanese themselves to have this appraisal from one foreigner in their midst. Few peoples in the world are so given to national self-analysis, so curious to hear foreign comments on their situation.

The book has been written for two purposes. The first is to give one man's interpretation of the peculiar characteristics and qualities that have made the Japanese such a modern success, and which at the same time distinguish them, sometimes to the point of isolation, from the rest of us on this planet. My second purpose is to show how the Japanese—successes, virtues, and failings—relate to Americans.

Throughout the past ten years, I have been struck by the sim-

ilarities between the Japanese and the American experience in coping with the problems of the modern technological world: the
basics of energy, communication, the dispersal of resources, and
the control of invention which are dwarfing the old problems of
wars and territories as the central questions of man's future. In
the process, a complex, highly interlocking relationship has been
intensifying between the United States and Japan. It is one which,
considered positively, has much to give both parties and which,
considered negatively, would be very hard to dissolve.

It naturally depresses me, as an American, that my countrymen
realize so dimly the importance of this relationship and the relevance of the Japanese to our own present and future. That is why
so much of this book is devoted to a description of basic Japanese
attitudes and Japanese ways of doing things, as contrasted to our
own.

More than twenty years ago, when I wrote *Five Gentlemen of
Japan*, I described the civilization of the Japanese as an extraordinary society of interlocking loyalties, uniquely homogeneous, its
every segment bound by a "great steel web of contract and commitments," a society in which "historically, the individual had had
no real existence outside of his group."

When I set down those words, my description of the "web society," although stated objectively, contained an implicit criticism.
Here was a system conspicuously lacking in absolute or universal
values, yet possessing social machinery so finely meshed as to make
it a dangerous instrument indeed, if the chief mechanics chose to
turn their national consensus to ill purpose. I would continue to
make this criticism. Yet the difference of a quarter century has
altered its smugness, not to mention the security of my vantage
point. The American generation which returned from war in 1945,
one to which I belonged, represented the last flowering of a certain
tradition of liberal certitude in the West. We had fought a victorious
war, in which we had, for that time, sufficient assurance that tested
principles of liberty and justice were at stake. We had been raised
in the Anglo-Saxon tradition of the rule of law. We felt that rule
triumphant, to the extent of organizing war crimes trials to restate
it. Our society was the self-admitted inheritor both of the Christianity which gave the West its purpose and the Industrial Revolution which took that purpose and put it, so to speak, on wheels. It
was a proud polity.

Within less than one generation, however, this West has been hit as with a boomerang by the problems, reactions and counter-actions of the Scientific Revolution which we ourselves engen-dered. The feeling of superiority which Western man, *genus Americanus*, exuded in 1945 was no less present for our merging it with a wide variety of humanitarian motives and brotherly im-pulses. Barely a quarter of a century later, on both sides of the Atlantic, the converse can almost be said: A feeling of foreboding and disillusionment has accompanied the apparent failure of the West's classic faiths and ideals. The pious Marxist who believed in the absolute magic of dialectic suffers as seriously from this as the pious Catholic who believed in the absolute magic of miracles.

Where the West founded its hope on individualism and univer-sals, Japanese society, by contrast, has based its hope over the cen-turies on the fact that certain social unities of race, situation and background can be converted into a unity of purpose and achieve-ment. In the Japanese mind, consensus and collectiveness are more than a virtue. They have almost the quality of a religion.

It would be wrong to say that the impact of technology, mass communication, and their own recent age of affluence have left the Japanese unscathed. How many youthful products of the world's best-educated meritocracy, famed for its social discipline, are beating up classmates with lead pipes in vaguely justified riots, insulting the old folks, or aimlessly listening to electric guitars after surfing on the Shonan beaches, talking in almost incoherently bad Japanese about the chances for a cheap escape to Hong Kong or Hawaii or wherever? How many pillars of the work-ethic economy of Japan, Inc., are working out personnel rationalization plans for the factory on sliding scales borrowed third-hand from the Harvard Business School, sabotaging production in frivolous half-strikes or merely crowding the elevators at six minutes before five, because their civilization now takes abundance for granted. How many members of the national family society are divorced, discontented, or rebellious at the thought of any group effort that would impinge on their highly selective notion of individual rights and duties?

I do not wish to exaggerate the speed or the extent of value shifts and changes. But no one attempting to distill some definable characteristics from modern Japanese society can help being con-scious of the social ground boiling and heaving beneath him, like disturbed volcanic soil, as released energies and new frustrations of the present try to break away from a demanding past.

The fact remains, however, that, compared with other societies of the late twentieth century, the insular family world of Japan is remarkably stable. The rebel may be more apparent than he was before, but he has yet to affect his society much. People in Japan are still moving, working, and thinking together, without coercion, in a way that other free societies would still find most extraordinary. The problems that Western man is now desperately trying to solve—how to preserve one's individuality in a society taken over by varieties of "group-think," how to develop a social ethic out of mechanistic progress—the Japanese have been unself-consciously working out over past centuries. Their present international success owes much to their past development in isolation.

To deal with a broad range of subject matter, I have divided this book into four parts. The first two chapters set forth the Japanese-American relationship, its problems and its relevance. If this had been a book about France, say, I would then have gone into more specifics about the relationship itself. Being a book about Japan, however, I felt it necessary to say something in the next five chapters about the Japanese themselves, the differences between their society and ours, and the interesting ways in which they seem to cope with the same kind of modern living, better or worse.

In the section following, I have devoted two chapters specifically to Business Japan, its extraordinary success and no less prodigious problems. In the next three chapters I have talked about the idea-makers, the communicators and the political people of this economic society. The final three chapters discuss the crisis of Japan's affluent society and the wider future that the Japanese might expect. I hope I shall be forgiven for the fact that in these later chapters—not to mention the earlier descriptive portions of the book—the images, language, and statistics of business also keep bobbing up. Through circumstance and choice, modern Japan has come to make its impact on the world through its trade and commerce. It can justly be called the new international business society.

Sandwiched between the formal chapters, I have written relatively short essays about things and people and moods in Japan which I feel are important. I have tried to present these in the form of impressions and personal sketches, on the theory that one descriptive paragraph may be worth a thousand statistics. Throughout the book I have consciously kept in mind the similarities and contrasts between the Japanese and the Americans, which serves as the vantage point of this exposition.

The wording of the book's title, although hardly novel, seemed the best way to state the magnitude of Japan's postwar accomplishment, as well as my concern about its permanence. As to the book's point of view, I have tried to escape my own milieu, as a publisher in Japan, working for a specific company and with specific people, insofar as this is humanly possible. If a certain affection for the Japanese shows through at times, I hope the net effect of my comment does not seem uncritical.

Acknowledgments

I would like to thank Professor Herbert Passin, who was kind enough to read almost all of this manuscript and offered extremely valuable suggestions on it. I also appreciate the active and invaluable counsel and advice of Shigeki Hijino, who worked with me on this project at the beginning and contributed extremely constructive comments and suggestions as it progressed. Edgar Snow of Tokyo was good enough to take a look at early drafts of the manuscript as it came along; for his comments and advice I am greatly indebted.

It goes without saying that none of these mentioned should share any responsibility for whatever imbalances or inaccuracies here contained, which are my own responsibility.

Mrs. Maureen Luber was extremely helpful in copy-reading and typing the manuscript and in addition contributed some excellent editorial suggestions.

Throughout the writing I was fortunate to have the constant help and commentary of my wife, Hiroko, who was not only able to contribute comment and criticism, but corrected many errors that might otherwise have seen the light of day.

Above all, my thanks to the encouragement and patience of my editor, Evan Thomas, who is responsible for these thoughts seeing the light of day and to Mary Shuford who gave the manuscript its final editing.

Frank Gibney

JAPAN
THE FRAGILE SUPER POWER

Chapter I

The View from There

Our Occident has much to learn from this remote civilization, not only in matters of art and taste, but likewise in matters of economy and utility.

—Lafcadio Hearn

Through a rare combination of ability, conviction, and happy circumstance, it fell to the Japanese, historically a nation of politicians and warriors, to become the world's first unarmed superpower. For *unarmed* most would substitute the adjective *economic*. The spectacular rise of "Japan, Inc.," in twenty-five years from postwar destitution to a gross national product of more than $600 billion,[1] third after the United States and the Soviet Union, will go down as the world's fastest-moving, if not its greatest, economic success story. No less impressive is the fact that this has been done without any of the traditional bases of power: no conquests, no colonies, no missiles, no raw materials. Japan's principal source of energy has been the ingenuity of its extraordinarily talented, disciplined, and highly motivated people, slightly over 100 million strong. Japan has made itself factory and trader to the world, within the context of a free society whose people's democratic rights have been broadly guaranteed and scrupulously enforced.

In achieving its new goals Japan has by no means abandoned its old traditions. It is still a country of tea ceremonies and poetry contests and ritualized courtesy, a country whose surviving great craftsmen are reverently designated "living national treasures." But the picturesque country villages that nurtured these traditions

1. Until the end of 1976, all dollar-yen equivalents had been calculated at the rate of ¥300 to the dollar. The GNP figure in yen was ¥164,447,000 million at the beginning of 1976, for example, which comes to $548 billion at this rate. In 1977 the yen strengthened to the 270 level, reaching 220 by mid-1978. GNP for 1977 equalled $660 billion, at the new rate.

survive mostly in picture postcards. The 40 per cent of Japanese living in cities prewar has grown to almost 75 per cent today. The average city dweller has become accustomed to tenement-style living, in high rises whose erection was not thought possible in the Japan of twenty years ago. The Tokyo of 1955 was still a city of houses; it is now an apartment town. It is becoming one vast scale model of the modern belt megalopolis, the barely interrupted stretch of city, factory, and suburb that already marks so much of the United States. When the mayor of New York visits the governor of Tokyo he can only admire wistfully the extent of the police network, the inner-city building, and the way the Japanese manage to keep their small neighborhoods intact under the sinuous elevated highways.

A country that still prides itself on the beauty of its surviving terraced fields, scalloped by patient hand from the mountainsides, now has serious arguments about people's rights to sunlight in cities and the unrestricted building of golf courses in the countryside.[2] Children's knowledge of Chinese characters is conditioned by the characters they most frequently see written out in television commentaries and serials—the average Japanese spends almost four hours a day in front of the magic tube. A nation once famed for indifference to animals now packages diet dog food and advertises luxury dog motels. In 1950 almost everyone either walked or rode bicycles; there were only 1.5 million cars and trucks in Japan. Now there are 30 million. (In 1973 there were more than 3 million new vehicles registered.)

Urbanization, high-pressure technologies, and sudden wealth have, of course, brought with them as many problems as they have solved. Industrial pollution wracks the world's most concentrated mill and factory areas; it has the status of a national emergency. Land prices have passed out of sight. Inflation has only lately been brought under control. (The wholesale price index set at 100 in 1970 reached 169 late in 1977.) A rising number of Japan's citizens are demanding net economic welfare instead of gross national product. And the energy crisis of late 1973, brought on by the Arab oil

2. There are now almost 1,000 golf courses in Japan, most of which have been built in the past decade. In 1945 there were less than 50. Further building has been banned or restricted by some local governments. Member-ships, which are mostly saleable, run from $5,000 to $100,000 apiece. Until recently appreciating at between 10 and 30 per cent a year, they are classed as a profitable form of investment.

producers, for once made the Japanese, as a people, realize how fragile the prosperity of an economic superpower can be.

The United States has played a massive role in Japan's postwar history; on no other country, outside of brief war relationships, have we cast so wide a shadow. Although American influence on Japan is hardly a tale of unalloyed good, it has on the whole worked well for both countries.

The extent of the relationship is far wider than most imagine. Japan is, after Canada, the principal United States trading partner and America's biggest market for raw materials. In 1976 the Japanese bought more than $4 billion worth of American agricultural products—17 per cent of American agricultural exports. Fully one-third of Japan's food imports come from America.

In 1976 the Japanese bought about $3 billion worth of United States manufactures and about $3 billion worth of industrial raw materials. In addition, Japan now harbors some billions of dollars worth of American investment in plant, patents, and joint ventures; and U.S. royalties from Japan continue to swell. As a consumer market, Japan is comparable only to the United States, both in the extent of its demand and its capacity to pay. Whether American businessmen have availed themselves of their selling opportunities in Japan is debatable.

Japan is the other big consumer economy in the world. Although exports have occupied a significant margin of the increases in Japan's gross national product, they have rarely risen above the level of 12 per cent. The bulk of the cars and the color televisions made in Japan are designed for Japanese, not foreign homes and roads.[3] The export drive in electronics, for example, only grew intense after 1968, when the Japanese home market was saturated. Contrary to considerable American impression, Japan did not modernize solely to get orders in the United States.

There is, however, no debate at all about how hard Japanese businessmen have pushed their way into the American market. The United States is not only the major source of Japan's raw material imports, in 1976, Japan sold $15 billion worth of manufactures and other products to the United States, despite self-imposed export

3. The famous Sony Company, which began by concentrating on exports, was an exception to this rule. One of the reasons for this concentration was the company's weakness in domestic dealer networks in Japan, now corrected.

quotas, currency revaluations, and American restrictions on Japanese goods. More than 25 per cent of Japan's total export volume went to the United States.

Besides these economic ties, which could not be easily broken by either party, hangs the tattered but still serviceable nuclear 'umbrella', the formal guarantee of United States protection that stands behind Japan's own unarmed neutrality. It is more important than many people pretend. The United States remains the ultimate guarantor against intimidation by Japan's two neighbors, China and the Soviet Union, whose peaceful intentions are often unclear. The terms of the mortgage, however—military personnel still stationed on Japanese soil and intermittent U.S. pressures for the Japanese to assume part of America's military commitments in Asia —the Japanese have found occasionally irksome.

Yet the Japanese-American relationship has deeper roots than military bases or trade statistics. Japan is the one world power, outside the West European complex, which shares with the United States common social aspirations, a working commitment to the democratic principle, and a strikingly similar urban technological way of life. It is a technological powerhouse, whose inventive loans to U.S. industry, incidentally, can be expected to rise as its traditional borrowings decrease. It is, needless to say, the only strong and effective Asian democracy.

Ideally, Japan and the United States, as nations and civilizations, should reinforce, enrich, and complement each other. If their economies cannot be harmonized, in a healthy tension, there is very little hope for the survival of free economies in a future world. They both have much to gain by developing a sense of political and social community. For Japan, the fragile superstate, the value of the American ties are obvious. For the United States, the value of the Japanese relationship is less obvious, but nonetheless real. The American free-enterprise democracy, traditionally the youngest of countries, looks oddly old and tired as we approach another century. The Japanese, fresh and confident pupils, have a lot to tell their old teacher.

For all this, for all our dealings of the past three decades—one war, a surprisingly successful occupation, innumerable government conferences, summit meetings, books, seminars, guided tours, student exchanges, and a $25 billion yearly trade relationship—the perspective of Americans on Japan has been extraordinarily bad. The roots of the two peoples' relationship may be quite deep; it is fully a century and a quarter since Commodore Matthew Perry's black ships first anchored off Uraga, but as seen from America, the Japanese display the eccentricities of images reflected in the flawed glass and distorting mirrors of a fun house.

To roughly 200 million Americans (with few exceptions) the Japanese appears as a short, bowing character whose motives are as impenetrable as his contrived smile. He is regarded as an exotic or an adversary: either the little-man-in-a-kimono standing before the emperor's palace in an attitude of respectful submission or on his way to drink sake at a vaguely improper geisha house or the purposeful businessman on his way to sell ever-more cut-rate cars or transistors to unsuspecting Americans, special discount cards in his pocket, his neat white shirt and pinstripe suit barely hiding the old *kamikaze* pilot's scarf that we remember from the TV Late Show. His ability is established, but still rather surprising. It is all too readily attributable to low wages or human-wave sales tactics. His smile we project into a leer (and everybody knows how badly they treat their women). His general politeness is magnified into either subservience or subterfuge. "Clever, yes; but can you really trust them?"

Americans, over the years, have become relatively sophisticated about the changing situations of Britons, Italians, Russians, for instance. The French, however much they insult us, are generally treated with courtesy. Even the Germans find that their stereotypes from World War II are fading. The Chinese "hordes" who slaughtered Americans south of the Yalu in 1951 have had their image transformed, in part, to a nation of admirable puritans, skilled chefs, and sporting ping-pong players. If the Maoist ideologues had a sinister look, the Chinese people we like.

Yet in Japan's case, America's historical memory is long and inflexible. People and policies are one. The image of the bayonet-wielding automaton, killing Chinese babies in Nanking and Shanghai, was ground into the American consciousness by the save China rallies of the thirties and the stored-up anti-Japanese feeling from an earlier era. Then there was Pearl Harbor, which Americans seem to remember with extraordinary tenacity.

Another image was brought back home by the United States troops of the postwar occupation: smiling, polite little people who are friendly and hard-working, but curious, living in a lovely country with a big snow-capped mountain, delicate temples, inept taxi drivers, and lots of nice girls. That image persisted for quite a while. But as Japan became an economic threat to the United States, through the late sixties and early seventies, the adversary view re-emerged. It managed to be both hostile and patronizing, at the same time.

This adversary note was intensified during the presidency of Richard Nixon. From the clamor about excess Japanese textile exports in 1969 through successive yen-dollar problems, the surcharge on Japanese exports, and the Connolly trade offensive, the United States made its complaints against Japan in a public forum —and in an abrasive manner almost unprecedented in dealings with a friendly nation. This is not to say that the American position was wrong. (Japan's increasingly heavy exports to the United States, fueled by government subsidies, were only belatedly followed by any corresponding relaxation of restrictions on American investment or exports in Japan.) But the position was presented in a diplomacy of open denunciation, which will not be soon forgotten in Japan.

When not engaged in outright polemics, the Americans behaved with a rudeness that had to be calculated. The failure to inform Japan, since 1951 imprisoned at American direction in a sterile "me-too" diplomacy of hostility toward China, of the sudden mission of Henry Kissinger to Peking was monumental. Attempting to repair the slight afterward, President Nixon, addressing Premier Tanaka Kakuei, spoke of Japan as an "equal partner" of the United States, in a manner that would have been loudly resented by even a small European country.

In fact, the Japanese were almost too stunned by the new tilt in United States Far Eastern policy to be resentful. Accustomed to thinking of the United States, variously, as an elder brother, a teacher, or a cornucopia of inexhaustible talent and resources, they found it hard to believe, at the beginning of the seventies, that a certain portion of their own prosperity was in a sense coming out of America's hide, and that Americans, feeling the pinch, did not take kindly to this. The image of America in Japan, over the years, had grown larger than life. In contrast to most Americans' view of Japan, it moved, it was dynamic, and it was basically sympathetic. Fingers snapping with Yankee ingenuity, packsack full of imagined technological treasures, an idealized Uncle Sam figure lived on in Japan long after the image was quite dead everywhere else.

Even the Vietnam experience did not absolutely shatter the friendly view of America. At least before the large-scale bombings of the North, the Japanese public in general was little exercised about Vietnam, even though the press and the intelligentsia were increasingly angry about what Tokyo newspaper editorialists began to call a "racial war." Annoyance with the United States involvement there was more pragmatic than ideological ("If you had to get in there, why can't you wind the thing up quickly?"). Gradually this forbearance melted. By the end of the sixties Vietnam was as dirty a word in Japan as it was in the United States. Yet both the quality and quantity of American dissent against the war were noted, which did much to dilute the charges of imperialism made by the Socialist and Communist parties. No one ever won an election in Japan on the Vietnam issue.

Vietnam and adversary diplomacy notwithstanding, the Japanese image of Americans stayed tall. There was a good bit of distortion in it, yet the United States, as the average Japanese sees it, is at least a three-dimensional study. He sees a people whose achievements and energies spell competence. His regard for American technological skill and the essential modernity of things American verges on the superstitious. America is the place where most Japanese overseas students go for an education, where technicians go for exchanges of information. American English is Japan's basic means of communication with the outside world.

Socially, the American comes on as likeable, if often unpredictable, with an envied informality. Americans visiting Japan after residence in Europe are interested to find that their nationality is no longer a pejorative. Culturally, the kinetic modernity of today's Japan—rock songs, wide-screen movies, quiz shows, billboards, drive-ins, business management seminars, computer schools, even the majority of new phrases in its language—expresses itself in American terms. Hemingway, Faulkner, Steinbeck, John Updike, and Norman Mailer are about as real to Japanese university students as their own Kawabata, Mishima, Oe, Endo, and Abe.

The phrase "Made in America" still has a cachet. First blue jeans, then bourbon, bowling alleys, Head skis, silly putty, every variety of American women's fashions, Colonel Sanders' Kentucky Fried Chicken (100 franchises), MacDonald's hamburger emporia (127 branches), and Dipper Dan ice-cream stores (47) have marked the passage of culture across the Pacific. As a Japanese resident of some years' standing, I am by now inured to the sight of pretty girls wearing misspelled American college-type sweatshirts (Calumbia), Pepsi and Coca Cola signs plastered over the front of otherwise respectable Japanese restaurants, pirated Playboy emblems

on every other Fairlady sports car, and a Japanese *Playboy* magazine, complete with pneumatic nude blondes in the centerfold.[4]

Thanks to a largely benevolent postwar occupation and their
own peculiar reaction to the one military defeat in their history,
the Japanese rarely think of the United States as an enemy, past or
present. Even the recollection of the horrible fire-bombings of
Tokyo in 1945, which came to light in the early seventies in a
series of books, was enough overshadowed by the wave of postwar
American geniality that it hits the average Japanese consciousness
as more of a postscript to history than a separate chapter heading.

The American occupation of Japan began in September 1945
and ended on April 28, 1952, when the peace treaty signed at San
Francisco took effect. Although formally speaking an Allied occupation, for practical purposes it was directed solely by the Americans; in particular, General Douglas MacArthur and his staff in
Tokyo. Operating on the assumption that the Japanese government
and society as it existed was feudal and hopelessly disposed to rule
by authoritarians, the occupation gave Japan an overhaul almost
unprecedented in history. A new constitution was imposed and a
new civil code written in which the imperial institution was reduced to symbolic importance and ultimately its power placed in
the hands of a universal-suffrage electorate and its representatives.
The educational system was drastically changed. A sweeping reform of land ownership was put through. Existing business cartels
were dismantled and some 200,000 officials who had run the Japanese government and businesses were purged automatically from
office. Labor unions were revived and unprecedented forms of collective bargaining actively encouraged. In this period, slightly
more than $2 billion in aid funds was tendered to Japan by the
United States.

Many of the occupation directives were unwise and unsuited to
Japan. Many radical reforms were altered, under the growing pressure of the Cold War, as the United States began to regard Japan
more as a potentially useful ally than a defeated and hopefully
penitent enemy. But on balance, the effect was good. The basic
well-meaningness of the occupation was accepted as such by the

---

4. When a Japanese publishing
company produced a magazine called
Playboy in 1966, lawyers from Hugh
Hefner's far-flung girlie-book empire
sued. The Japanese judge decided
there had been no infringement because the magazine in question used
the Japanese kana syllabary characters,
pronounced literally "puray-boy" and
not the English letters spelling playboy. Ultimately, the publisher joined
forces with *Playboy* to produce an
authorized version.

great majority of Japanese. They were impressed both by the power and the justice of the American democracy; and the impression has proved an extraordinarily lasting one.

Even the editors of *Asahi*, Japan's largest newspaper, whose bias is chronically anti-American, concede in their periodic public opinion surveys that—Vietnam, Okinawa, and race problems notwithstanding—the Japanese people have consistently regarded the United States as the country whose friendship Japan needs most.

The United States is also, after Switzerland, the country Japanese most wish to visit. Although few Japanese actually make it to Switzerland, up to half a million a year visit the United States. Paris may be for intellectuals, but for the average Japanese, the ultimate in cities remains New York, or more comfortably, Los Angeles, with its Japanese hotels, banks, and shopping sections. The ideal baseball team to beat is still the U.S. World Series champions. The ideal vacation plan is to go to Hawaii, which is also regarded as a good place to put your yen. By early 1973, Japanese had bought eleven hotels there and invested more than $250 million, mostly in real estate and leisure industries. (One of the new Japanese golf courses, as might be guessed, overlooks Pearl Harbor.)

The similarities of the urban-technological way of life in Japan and America are obvious. High-rise constructions increase and multiply, traffic jams on choked superhighways are a commonplace. Along Japan's highways, as in the United States, the promise of new life through newer and faster cars beckons on every other billboard. Tokyo has a rhythm of barely controlled frenzy that evokes New York more than any other city, and vice versa.

The externals of Japan's modern technology, visitors quickly notice, are pure American or American-Japanese with no European filtering: ice in everything, rasping public address systems and canned music, a neon-devastated landscape, and outdoor vending machines (hot sake for ¥100 and warmed-up noodles for ¥150) on every other corner. The television commercials are of the intrusive American sort that Europeans would shrink from. But it is Japan's unique contribution to have trailer ads running through the bottom of the screen during the entertainment, in addition to the regular commercials.

Pushbutton service, automatic car washes, radar speed traps, supermarket carts and checkout lines have long been commonplace. The air conditioning and the computer systems in both countries function well. Mechanics and plumbers know their business (and in Japan they do not overcharge). Efficiency is widely regarded as its own reward. Nowhere else in the world has the American ideal of material progress through know-how so spectacularly taken root.

At election time, the ubiquitous sound trucks, violent electioneering, but proportionately strict election laws, testify that a commitment to the democratic principle and an urge toward a better society founded on this principle are as much a part of modern Japan as the big buildings, the bounding stock market, and the rotary-engine cars. Political demonstrations are noisy and constant. The pollsters, as in the United States, keep telling the voter how his vote is committed before he casts it. But the electoral system works. The voter in this universally literate country has as good, if not a better idea about the issues he faces as do Americans. The press is rambunctiously free, in its peculiar way. So are the courts, also in their peculiar way, although jury trials are unknown.

The American people, over the years, have developed a sense of group living and an urge for collective identity that half belie our widely announced principles of free enterprise and individualism. The ancient tribal nation of Japan has, meanwhile, come some way toward developing a sense of individual expression as well as political freedom among its citizens. We are not nearly so far apart as the distorted mirror images would suggest.

All of this is not to say that the Japanese view of Americans is entirely worshipful. It is, in fact, decidedly mixed. If America has been Japan's leading teacher, it has also been a big competitor and, in the eyes of some Japanese, an oppressor. Not the least of the motives behind the surge in Japanese productivity was the urge to catch up with and surpass the United States. The Japanese may envy American achievements; but unlike people in many European or other Asian countries, he does not want to change his nationality or his permanent address. Goethe's wistful *"America du hast es besser"* is not for him. "American-style" thinking is not necessarily a term of praise; and someone with an American-style way of dress (e.g., long hair, wide ties, colored shirts) will not go too far in most large Japanese companies. The Japanese is secure, even arrogant, about his own formidable culture, which shelters members of his homogeneous nation like a cocoon. But he is also aware constantly of the pull of things foreign, which are by and large things American. He is constantly testing, checking, comparing. He is never quite sure of himself, for this.

The ambivalence of Japanese about Western thought and ways is historic and deeply rooted. The debate is as endemic to the modern Japanese nation as our old literary arguments about native-American versus Europeanized habits and culture ("redskins" and "palefaces") were endemic to the United States. The very speed of Japan's advance into the world of industry and technology after the Meiji Restoration of 1868, and its apparent ease of achievement, obscured the vast societal changes and reactions which the reforms provoked within the Japanese.

The Meiji Restoration of 1868 was not something that happened and was over with, a two-dimensional collection of laws and decrees taken from the history books. It had been preceded by a steady evolution of Japanese society, within the protected isolation of the Tokugawa Shoguns. This created the conditions which made the Meiji reforms such a quick success: (1) a sophisticated, well-knit, and disciplined society, accustomed to working under tight political control; (2) a relatively high degree of education, which in turn stimulated the appetite for more; (3) well-developed business and political traditions, with ample talent to put them into practice. The seeds for Japan's centralized urban civilization of the twentieth century had been planted long before Commodore Perry made his expedition.

Indeed, no country or civilization—in Japan for centuries the terms were historically synonymous—had ever developed so much in isolation. Emerging early in the Christian era as a distinct nation society, out of misty warring tribalisms, the Japanese by the fourth century A.D. had the marks of a relatively unified state. Considerable immigration and cultural importation from Korea and, less directly, from Han Dynasty China were assimilated by the time of Shotoku, the princely regent of Japan's seventh-century Nara Period. Prince Shotoku, known as Japan's great law-giver, presided over the importation of Buddhist culture into Japan. His basic principles of justice and order, founded on Buddhist thought, became the warp and woof of later Japanese ethics.

By the time of the Heian Period, from the eighth to the eleventh centuries, Japan's civilization was in flower, vastly more sophisticated in its culture than anything comparable in Europe. The Emperor's nominal rule was enforced by a series of regencies, powerful hereditary families who governed on behalf of the throne. It was only after the Fujiwara regency declined in the late tenth century that Japan became a battleground of warring military clans, notably the Taira and the Minamoto whose struggles strikingly foreshadowed later civil disturbances like the Wars of the Roses in Britain. Even when military dictators established control of the country, however, they were generally content with the title of Shogun ("general") plus some high-sounding court ranks extracted from the emperor.

Through all of Japan's internal turmoil, no foreign armies, whether invaders or allies, ever got a foothold there. The only attempts at military invasion were the two assaults of the Mongol fleets in the late thirteenth century, defeated partly by the hard fighting of Japanese warriors, but mostly by a disastrous typhoon that scattered the Mongol fleets.

At the end of a second civil war, the so-called *Sengokujidai* ("era of the warring countries") of the fifteenth and sixteenth centuries, Japan did produce in succession three dictators who were virtually contemporaries. Oda Nobunaga, Toyotomi Hideyoshi, and Tokugawa Ieyasu took over in turn the government of the nation. While they lived, the ties between them were not broken. On Oda's death Hideyoshi succeeded him. After Hideyoshi died, Tokugawa seized power, had himself declared Shogun, and established his successors in what proved to be a 250-year-long regency.

The Tokugawa Shogunate, true to Japan's isolationist tradition, persecuted Christian missionaries who had come to Japan and almost totally restricted even peaceful maritime commerce with the outside world. The emperors in the old imperial city of Kyoto were almost totally eclipsed, if not imprisoned, by a dictatorship which, for all its oppressiveness, gave the Japanese people unbroken peace and a freedom from foreign interference almost unprecedented in the histories of major powers.

The enforced peace of the Tokugawa *Sakoku* ("closed country") generated discontent as well as prosperity. Pressured by the demands of European and American traders and warships to open their country, a whole generation of young samurai, both alarmed and fascinated by what they heard of Western technology, joined forces with the older anti-Tokugawa barons and restored the young Emperor Meiji to direct rule. The national emergency symbolized by the appearance of Commodore Perry's squadron destroyed the Tokugawa Shogunate. But unlike the case of China, the fall of the established order did not lead to weakness and anarchy in Japan. In something like a national consensus, the relatively low-ranking samurai who supported Meiji were given license to rule and ride roughshod over the inherited prejudices and taboos of past centuries. A hundred years before Mao Tse-tung coined the term for modern China, Japan had made its own "Great Leap Forward" into the nineteenth-century world of European nationalism and the Industrial Revolution.

The explosive changes decreed by the young Meiji era guardians of the state [5] set off a century of continuing change in Japanese thinking and institutions. A century of modern living is a long time. One year of adjustments in life and technology in this era is worth ten or twenty or a hundred years of gradual change in the past. During the century that began in 1868, the Japanese assimilated a considerable Western heritage of their own. A Japanese University student may have an imperfect idea of what Dante, Cicero, and Rousseau each contributed to the development of Western culture, but he knows something about them—and still more about Kant and Marx and Lincoln and Jefferson. His culture has been shaped, in good part, by the traditions these men represent—as has the culture of his father, his grandfather, and his great-grandfather before him.

The very way in which Japanese intellectuals sometimes champion the cause of the East has been conditioned by the repository of culture and education they share with the West. The great novelist of the Meiji era, Natsume Soseki was rightly regarded as a very Japanese writer; but it is interesting to note that, when he describes the moods and emotions of the artist protagonist in *The Three-Cornered World* (*Kusamakura*), he finds it useful to invoke thoughts and comparisons from, among others, Sterne, Wilde, Lessing, Ibsen, and DaVinci.

When a Japanese speaks of himself as an Asian, he is speaking as a very special kind of Asian who has been conditioned by one hundred years of universal literacy, Beethoven symphonies, mass-circulation newspapers, flush toilets, Scotch whiskey, high-pressure advertising, and double-entry bookkeeping. Without underrating the purely Asian influences that shaped Japan—Buddhism, Confucianism, and its own singular island culture—it is also true that

5. The Platonic phrase is used advisedly. The idea of the guardians of the state, as expressed in the *Republic* is very close to the notion of disinterested, paternalistic government which the Meiji reformers, as they became aging state councilors (*genro*) bequeathed to several generations of Japanese bureaucrats which followed them.

Japan, along with Europe and the United States, has been one of the principal legatees, not merely of Western technology, but of Western civilization as well, as it has advanced into the modern era.

The meeting of East and West, part struggle, part fusion, goes on constantly within Japanese society. Active Western influences from the outside—outside Asian influences on Japan are minimal—serve only to reinforce or punctuate a tension which has existed within every Japanese from the time, at the beginning of this century, when Soseki was warning his countrymen about uncritical acceptance of Western ideas.

To a great extent prewar and overwhelmingly after 1945 the active outside Western influences on Japan came to be embodied in the American. The public dialogue between East and West often shook down to a conversation between Japanese and American, with the American doing very little listening. While the Japanese listens with unfailing interest, he does not approve of everything that is said; and often he gets angry. Nothing is so irritating to the human spirit as picking up the melody after someone else has called the tune.

The Japanese have played this role before in history, however, and they do it well when they have to. They played second fiddle to the Chinese and Koreans in their wholesale cultural importations between the sixth and tenth centuries, and to Americans and Europeans once before, in the Meiji modernization of the late nineteenth century. In 1945, surrounded by the total ruin of their country, physically and psychically, they had no choice but to accept the options the American occupiers gave them, and learn yet a new score.

For a number of years, from the New Deal to the new leisure, there was a gap of a decade or more between trends, moods, and innovations in the two countries. At first Japan merely echoed the United States, although the timing of the echo was often ironic. In the late forties, for example, while Harry Truman was fighting for his political life against a rising tide of conservative sentiment in the United States, the young bureaucrats at SCAP [6] headquarters in Tokyo, who had cut their teeth on the New Deal's economic optimism of the thirties, were still trust busting, union organizing, and democratizing the schools in Japan. By the late fifties, after

6. For Supreme Commander for the Allied Powers, the expansive official title for General Douglas MacArthur, as head of the occupying forces in Japan.

the so-called Sputnik reappraisal, American educators were beginning to wonder in public whether courses in life adjustment and enhanced student government facilities were an adequate substitute for learning. There were agonizing reappraisals of rising mass enrollment and falling standards in university education, Meanwhile, waves of Japanese students, liberated by almost a decade of the occupation's educational reforms, were finally swamping their old elitist universities and forcing their elders to build up Japanese multiversities to supplement them.

In the middle fifties, while Japan was still working hard and worrying whether the balance of payments deficit would ever clear up, Americans were jamming the sports-car showrooms, building ski lodges, and unwrapping do-it-yourself boat kits in a new mass leisure boom. In the late sixties, the leisure boom hit Japan. The spillover in bonuses and dividends from the massive surges in GNP helped spawn a huge industry of land developers, villa salesmen, car-rally enthusiasts, skiers, tour-takers, and scuba divers. While the Japanese leisure class was polluting the waters of the Fuji lakes and darkening the Shonan beaches, a troubled America was sunk in the race problem and just beginning to sense the implications of the bog in Vietnam.

In the fifties America was full of talk about computers, cybernetics, and the world of automatic calculating, printing, education. By the mid-sixties, the information industry boom was all over Japan, with every major company planning some marriage of hardware and software. The ministries were hurrying along Japan's young computer industry, to see how quickly the American monopoly might be broken.

The American tourist was, alternately, the wonder and the terror of the world in the late fifties. By the late sixties, the Japanese group tour had superceded that status. Museum and gallery people in Europe, fresh from pushing back the invasion of the American tax-deductible picture buyers in the late sixties, braced themselves against an unexpected influx of Japanese who would pay premium prices for almost any canvas with paint on it.

In the early seventies, with Wall Street weak and liverish after years of reckless high-living, the brokers in Tokyo had just begun to feel their oats; manipulations that had been outlawed in Wall Street for a decade briefly turned *Kabutocho*, the downtown Tokyo block where the stock exchange is housed, into a shady speculators' paradise. Meanwhile Wall Street, from its sickbed, was talking reform.

As a visiting American economist remarked, after reading some recent pollution statistics and hearing a briefing on Japanese agricultural farm price supports, "Lord help us, they've taken all our bad habits and repeated them. You'd think people would learn."

The bad habits were not, of course, all conscious imitation. Japan's new affluent generation was bound to swamp the old universities, occupation or no. And the surges of Japan's growth economy would have brought on a leisure boom, land speculation, and a swinging stock market in any case. But the constant communication between the two countries did provide a feedback that intensified tendencies which might otherwise have grown more slowly.

Progress of whatever sort has a way of catching up with itself. The more successful and complicated Japan's postwar economy became, the faster Japan moved in the direction of the postindustrial society, the more the lead time between American firsts and Japanese seconds dwindled—or became a lead time in favor of the Japanese. In the hardware-software world of learning systems, VTR, cable TV, and the like, the Japanese were hard on the heels of the Americans, and often ahead of them, as, for instance, with the civilian-market transistor. They led in instructional and educational television. Beginning in the early seventies, Japanese automobiles edged significantly into the compact car market in the United States. Detroit nervously noted the past successes of Japanese electronics manufacturers in the field of calculators, transistor radios, and color television, and changed its own product mix. In productivity and, consequently, price, the utterly modern Japanese steel plants have put the U.S. in their shadow. In some areas, like shipbuilding, Japanese innovation and efficiency have eliminated any American competition.

The social side-effects of progress appeared almost simultaneously with the progress. Ralph Nader's publicity-conscious consumer movement spread to Japan faster than the credit card. By the 1970s Japanese consumer protest, thanks partly to the concentration of population, was more vocal and effective than anything in the United States. The Fair Trade Practices Commission (*Kosei torihiki iinkai*) in Tokyo has begun handing down the Japanese equivalent of cease-and-desist orders with a confidence that rivalled the storied arbitrariness of Washington's Federal Trade Commission. Student protest movements started in Japan first. The Japanese student violence of the late sixties, although it did not have the widespread social confusion of the American protest movement, was superficially far more militant and dramatic.

In the late sixties, however, while American protests against pollution were gathering, Japanese plants were still recklessly expanding. The leisure industry, helped by local governments, hacked through the forests to build more six-lane toll roads and seaside motels. No nationally organized Sierra Clubs sprang up in Japan to plead for conservation and a clean countryside. But by the early 1970s, the cumulative impact of local protests had produced strong antipollution sentiment in Japan. Agitated by a series of industrial accidents, the antipollution movements, by then led by a government agency, developed support strong enough to reverse the national priorities of growth unlimited.

The energy crisis of 1973 hit both countries full amidships; although, in the case of Japan, the popular reaction was delayed. Coming on top of pollution, inflation, and an increasing awareness that technology means confusion as well as progress, it fueled a mood of uncertainty in both countries. In America the soul-searching was at first more political than economic, in Japan it was more economic and social. But it was significant that the world's two greatest modern exponents of bigger and better were now beginning to talk seriously about the quality of life.

The similarities between American and Japanese problems, the relation between American and Japanese progress, as well as the normal political and economic interchanges, are painstakingly reported to the Japanese public. They are much discussed. The Japanese daily press is full of American news, as are the magazines. No issue of a self-respecting journal would be complete without at least one article examining (and generally deploring) the state of Japanese-American relations. Television stations run continual commentaries on American people and politics, not to mention a steady stream of United States sports events, rock songs, folk singers, and the 9 P.M. Japanese-dubbed American movies. American political campaigns are given the most intensive coverage, as is news about changes in the United States bank rate.

With a deep pride in his own national worth, the average Japanese tends to assume that his intense scrutiny of American fads, social ferment, economic indicators, and legislative processes is naturally to some extent reciprocated.

Of course, it is not. The view of Japan from the United States is cursory and casual. Its limits are accurately suggested by the relatively slight coverage of Japan in the daily and weekly United States press. To be sure, there is Economic Japan, a clouded but impressive picture of pell-mell, purposeful industry—Sony-Toyota-Honda-Tanker-Mitsubishi. But despite the relatively heavy coverage of Japanese business and U.S.-Japanese economic confrontations, despite the treatises by American economists and business school professors on the economic miracle, Japan's success story is still regarded by most of the general United States public as something of a devious riddle.

It is true that American reportage on Japan has greatly increased over the past five years. The *New York Times, Time, Newsweek,* and the *Los Angeles Times* have at least made attempts to give Tokyo something like the continuing news treatment which their readers have come to expect from datelines such as London, Paris, and Moscow. Among the bulk of American newspapers, however, and the periodical press, the coverage of Japan remains very thin. There are reasons for this. For instance, there are few crowd pleasers in Japanese factional-group politics; those politicians who do have strong local followings are rarely interesting to foreign observers. The Japanese reporter's clubs which cover most of the big government news are closed corporations, almost never open to foreigners. Of course, it is not hard to find good stories all over Japan. But when they do, most United States correspondents still have to reckon with the old tendency of their editors (and presumably their readers) to think of Japanese news, again, in terms either of the immediately obvious news story, like a huge department store fire or a Washington trade ultimatum or of the exotic. If a geisha commits suicide in a hot-springs bath, some American editor can be trusted to find out about it and make it his Japan story of the month.

There is, apart from newspaper stories, a continuing informed dialogue between Japan and America; but it is much overrated. The meetings, seminars, and symposia on relations between the two countries are generally conducted by the same group of Japan and America experts who are all too accustomed to hearing each others' voices. On the American side, especially, the representation is limited. There are probably no more than two thousand Americans who can handle spoken Japanese at all competently; less than one thousand who are capable of reading that language. Although American studies are fairly well developed in Japan, barely 5 per cent of

American universities have anything resembling a Japanese department (history or literature).

Small wonder that the picture the average American gets of Japan is hopelessly two-dimensional. About one of the world's liveliest and most sophisticated literatures, the American knows nothing, except when one of its major exponents either receives the Nobel Prize or commits suicide. He is unaware of the urban ferment in discoveries, ideas, and new designs for living going on in Japan, despite the fact that Japan's mass communication and city civilization is facing precisely the same problems of transport, pollution, and technological squeeze as his own—excluding, of course, the racial problem. The quaint Japanese may be cracking the postindustrial sound barrier right behind us, but not many Americans know about it.

It is easy enough to demonstrate that this other-end-of-the-telescope view of Japan is unavoidable, that much closer contact between Americans and Japanese is an illusory hope, that as peoples we are worlds apart. Japan, a nation of utter racial homogeneity, is the most tribal of countries. America is the least. America, proud of its past successes with democracy and the melting pot, has been, at least until recently, the most publicly didactic of nations; Japan the least. Ritual in both countries is completely different. It is not merely a matter of the Japanese pulling a saw (where Americans push it) and tending to go through doors ahead of their womenfolk. The commands and controls that the Japanese enforce in behind-the-scenes conferences and compromises tend to be accomplished in the United States by public confrontations and debate. The Japanese have traditionally prized the form of things, to the detriment of their content. The Americans tend to worship content, loudly professing annoyance with form. The Japanese dotes on enclosures, such as the tiny garden in the city house, and manufactures space with artificial illusion. The American insists on a room with a view. Americans delight in exhibiting feelings. We explain our moods, meanings, hopes, hates, and concerns to the world with the shameless exuberance of Lyndon Johnson showing reporters his operation scar. The Japanese are far more secretive. Like members of a family which has learned to live together in one room, they try to hide random emotion wherever possible. Between the extremes of silence and explosion, they find it very difficult to tell the world what is bothering them.

A Japanese tourist enjoys seeing New York or Paris as a visitor, of course, for a few days, buying up souvenirs by the gross, visiting the Louvre and the Crazy Horse with equal zeal—like the touring American innocent of the twenties, vaguely hoping for adventure ("Here we are in Paris. Where are the naked foreign women?"). The Japanese has become the most grimly determined traveler of this century. But he prefers to travel in groups and rarely likes to stay. After his two weeks or so in foreign parts, he is annoyed by the odd smells and complains about the lack of Japanese pickles. As soon as the volume of steady Japanese visitors picks up in a foreign city, someone builds a well-equipped Japanese hotel, like the Miyako in San Francisco or the Kitano in New York City.

Of all the world's peoples, the Japanese are one of the most diffident and awkward in establishing relationships with foreigners. The Japanese correspondent is the last to ask a question at an international press conference. The Japanese diplomat will generally speak only when spoken to. The Japanese businessman, although he will plunge single-mindedly into the matter of the contract which he is pursuing and fulfill all the formal demands of good fellowship which are deemed necessary to this process, will rarely attempt to go beyond. He lives in his Japanese ghetto in Düsseldorf or Bangkok or Queens by choice. It is the sort of place he assumes is fitting and proper for foreigners, like the places built for foreigners in Japan.

When the Japanese does meet the outside world, he prefers to do so under highly controlled circumstances, preferably on his own home ground. The nation will put its best face forward for a 1964 or a 1972 Olympics in Tokyo or Hokkaido or a 1970 World's Fair in Osaka. Everything will be done at such times to assure the happiness of the visitor from Oswego or Indianapolis. But presenting the Japanese case in Oswego or Indianapolis is a different matter. There is still too much visible doubt as to what role he should or could assume in a world outside, which is, after all, peopled with foreigners.

In March 1973 a blue-ribbon panel of Japanese communications experts, led by the chairman of the Fuji Bank, Iwasa Yoshizane, made a serious study of Japan's international public relations. "The Japanese people," the panel reported, "are by and large unskilled in expressing their thoughts and presenting their positions. We often display traits unintelligible to Westerners. We patiently and silently endure discomfort and hardships, dislike talkativeness, rely heavily on nonverbal communication and tacit implication. The difference in language becomes a tremendous barrier in dealing with people of other countries. As our influence grows stronger in international society, however, these barricades must be overcome and our point of view must be rationally and articulately presented to the rest of the world."

The very inarticulateness of the Japanese abroad is a by-product, in a sense, of his overprotected atmosphere at home. His failings as a visitor in foreign parts reflects the utter security he feels at home.

For Japan's *nation society* is unique. A tightly threaded, two thousand-year-old mesh of human fabric, it has an extraordinarily broadly based culture, a low common denomination of artistic tastes and living standards, and a talent for narrowing the national energies to well-defined tasks or objectives. It is a consensus democracy where people consider it bad form for the majority to outvote the minority without concessions. The words *manjo-itchi* ("unanimity") have a magic ring to them in any Japanese meeting. In fact, the Japanese have virtually achieved of their own free will the kind of society that Marxist ideologues only dream about. Their unities came naturally from long incubation in their own island society. Thus, when they borrowed the arts or inventions of other nations, they generally did so confidently and purposefully. It takes an extraordinarily self-confident culture to assimilate so many outside influences without losing its own identity.

The Japanese can wax indignant on occasion, over rising prices, bad housing, or faulty garbage disposal. Yet they have an extraordinary faculty for enduring inconveniences that would readily provoke or enrage others. A United Nations survey recently showed how, in all West European countries, the incidence of crimes of violence always increased as living standards rose; but in Japan it decreased. In adding his own appreciation of Japan's high social boiling point, a popular columnist for Tokyo's *Asahi* once wrote, "Given what we have here—the world's highest rise in commodity prices, horrible pollution and environment problems, severe housing difficulties—the people of any Western European country would probably have thrown out their government many times over."

Such patience and group consciousness have made the economic success of Japan, Inc., so spectacular. But there is another reason: the protective insulation of Japan over almost two decades, between 1945 and the beginning of the sixties. During this time the country was able to gather its energies for a leap forward, quite as it had in the last years of the Tokugawa period, before the samurai of the rebel Satsuma and Choshu clans had become the technicians of the Meiji Restoration. It might be said that the Japanese, considered as members of the world community, passed with rocket speed from the militarist thirties to the scientific sixties, without ever having really lived through the international turmoil of the late forties and fifties. In the light of the changes now being wrought by the postindustrial society and the overturn of methods and thought patterns that the forties and fifties considered modern, this asset seems considerable.

This is, of course, hardly to say that the Japanese missed the quarter-century of history which included war and the A-bomb. The atrocities perpetrated by the Japanese army on literally hundreds of thousands, bystanders as well as enemies, are ground into the memory of Asia. In turn, the American fire bombings of Japanese cities and the privations of the late wartime years, not to mention the deaths of 1,850,000 Japanese (almost 700,000 of them civilians), were enough to make any society savor the consequences of historical hybris. Nonetheless, the Japanese, as a people, were isolated during this time from the sense of living in an international community. This was due, first, to the censorship of a militarist government and, later, to the firm, avuncular guidance of MacArthur's occupation. The military character of SCAP was at constant war with its reformism. Its inevitable use of censorship and directives gave a hothouse atmosphere to the country and inevitably had an inhibiting effect on Japanese energy and initiative.

When Japan's energies were released, they were thrust immediately into the pursuit of economic gain. The late Ikeda Hayato, prime minister from 1960 to 1964, led his countrymen away from the left-right political quarrels that followed in the wake of the occupation to concentrate on his famous double-your-income policy (*shotoku baizo seisaku*). They followed with zeal. Given his relaxed political leadership and the ultimate guarantee of safety under the American nuclear umbrella, the Japanese felt free to concentrate on being economic animals. Throughout the sixties and into the seventies, they showed a conspicuous disregard for the sensibilities of both their customers and rival tradesmen in the process.

They also almost deliberately ignored the political conflicts of the ideologically divided world around them, with significant exceptions: Japan's rising capitalists made considerable profit from the economic windfall of U.S. military orders during the Korean and, later, the Vietnam wars. An almost professionally socialist intelligentsia, comfortably protected from the invasions and implosions of Communist powers, viewed the United States as the real imperialist enemy, somehow spiritually akin to the prewar Japanese militarists! But the people as a whole, having fallen into a classically Japanese situation of mass dependence during the United States occupation, tended to feel throughout most of the sixties that international politics, pro-American or no, were simply not Japan's concern. The effects of this political head-in-the-sand attitude plague Japan to this day. Aside from sudden bursts of enthusiasm, or concern, e.g., almost universal support for re-establishing relations with Peking, and almost universal concern over Arab plane hijackings and Indian nuclear testing, there has been until lately little sustained interest in international politics.

Over the last few years, however, the mood has changed somewhat. *Economic animal* has become a dirty word. A new generation is emerging: more traveled, more outspoken, less hard-working, but physically bigger than its elders (thanks largely to the spectacular changes in the Japanese diet). There is little of the revolutionary about this generation in Japan. Although young people are quicker than past generations to denounce inept bureaucrats, greedy businessmen, or excessive ceremony, they are less obviously interested in changing the structure of their society than young people elsewhere. The new generation is genuinely, if uncritically, internationalist, with an urge to be thought cosmopolitan. *Kokusaijin* (an "internationalist") is a new word with a highly positive value. Yet the youth are also vigorously proud of being Japanese.

The new impulse of the Japanese to be internationalists is wellfounded. Just a few months after the Iwasa panel made its judicious remarks, Thai students were burning and boycotting Japanese goods, in protest against the arrogant behavior of local Japanese businessmen and the market-grabbing of Japanese trading companies. Prime Minister Tanaka's visit to Southeast Asia in January 1974 brought on more of the same, car-rocking student riots in Bangkok and Djakarta and renewed protests against economic imperialism.

Being a superpower of any sort, as the Americans found out before them, can be a difficult job. The parallels in many ways are tempting. Like the Americans a few years earlier, the Japanese rushed out to seize power where they found it, without being psychologically prepared, either for the consequences of their success on others, or the responsibilities that go with the franchise. That the power was only economic made it no less real and oppressive to others.

The affluence of power, conversely, is a fragile thing. The successive shocks and buffetings given to Japan at the beginning of the 1970s only underscored its dependence on the constant flow of energy, food, and raw materials from the outside. It is an odd turn of fate for a nation which historically has prided itself on its poor, but safe island self-sufficiency. American trade confrontations, the enforced kowtowing to China, the Soviet pressure tactics about economic concessions and seized Japanese territory, the European economic boycotts, trouble in Korea, and, most spectacularly, the Arab oil cutoffs, all struck at Japan within little more than five years. Each blow emphasized how thin is the line for Japan between dynamic prosperity and the sort of chaos that can only happen to a finely tuned engine, when it runs out of gas and oil.

The View from There, as Japan looks on the rest of the world, is confusing and rather bleak. It poses unending problems. The day of the successful, single-minded trader is gone. Yet there is continuing power in Japan's wealth and achievement, which most of the world envies. If the Japanese do not wish to be perennial hostages to their neighbors, like a fat 747 awaiting random hijackers, the power must be used to play a world role. The Japanese have become internationalists by necessity.

It is shortly after noon, and I am sitting in a modest Japanese-style restaurant near the Isetan Department Store in Shinjuku, one of the huge business and transportation centers in metropolitan Tokyo. I have finally asked for some service, after waiting for seven or eight minutes unattended, not because of any hostility on the part of the serving staff, but simply because everyone was afraid to approach me first. "What does the foreigner want? We don't have an English-language menu. Does he know this is Japanese-style cooking? Who knows enough English to ask him?" The buzz of discussion behind the screen I only partly overheard and there were studiedly impassive expressions on the faces of the waitresses hanging back at the end of the room, but I am familiar enough with their concern.

When I finally order in Japanese, the service comes with customary politeness. I leave with bows and smiles, followed by a dozen sets of covert looks and unspoken questions. "Why did he come here? Does he live or work here? How does he know the Japanese language? Where is he going? Will he come here again? Does he have Japanese friends who know this place? Why did he come here?"

The situation in the restaurant is a familiar one, constantly relived. If I have exaggerated the curiosity of the staff about my presence, it is only because the foreigner living in Tokyo comes to have a sense of being constantly looked at. The farther he goes outside of Tokyo and a few other big cities, the more his curiosity value increases. There is no country in the world where the outsider feels so solitary and apart, like a stranger breaking in on a family party.

This is not to say that he is unwelcome. Quite to the contrary —especially if he comes for only two or three weeks. The Japanese make superb hosts for transients. They are solicitous and, indeed, genuinely kind. The most casual acquaintances offer trips and excursions. Passers-by on the street volunteer their help with directions. Presents arrive wrapped in a delicate sense of fitness. And the good guest flies out of Haneda on a cloud of happy memories.

But living here is another thing. Even for the foreigner with a knowledge of the language and some past experience in Japan, the sense of apartness weighs heavily. Without such a background, it can be terrifying. It is a crushing thing to be surrounded by spontaneous uniformity, especially when one does not understand any of the rubrics involved. "Why must all Japanese cars have white doilies on the headrests? Why must all Japanese senior businessmen be driven in the same black car, by the same white-gloved driver, with the doilies supplemented by white seat covers? Why do all the offices look the same?

"Why do all newlyweds look the same as they board the airplane or bullet train on their way to borrowed bliss in the hot-spring hotel, she wearing the same flowered hat and bouquet, he wearing the same nervous grin? Why do all Japanese grin when they are obviously embarrassed or unhappy? And gather in ceremonial clusters seeing off travelers or welcoming the boss back to the office, the pecking order faithfully mirrored in the angle of the compulsory bows? Why do so many men publicly relieve themselves by the roadside? Why do the women put a hand over their mouth, when they essay a smile or a nervous giggle? (And when will they learn that purple or yellow knee socks are unattractive with Western clothes). Why do the men stare fixedly ahead, purposely ignoring you, as they try to edge your car out of a lane in heavy traffic? Why are all those faces so noncommital and impassive, anyway? Can't anyone laugh or yell outside of the movies?"

The answers take a long time to formulate. It is hard enough, even assuming that the language barrier can be surmounted, to get to know people, to have the kind of continuing social conversation that ultimately gives you the sense of the society around you. Invitations to hoped-for Japanese friends go out and are responded to, but rarely reciprocated. Even in the few chinks and crannies of Japanese society where foreigners can penetrate, real friendship is not given easily and then only after long incubation. This is not because of hostility. It is only partly because Japanese are uncertain about foreign manners. It is mostly because friendship in Japan, once acquired, is not readily disengaged from.

Living in Tokyo is not like living in Paris or Rome or Buenos Aires, anymore than the Japanese language is like French or Italian or Spanish. Japan is totally other; and it is assertive. Inescapably, the resident foreigner who has been cradled in the Greek Christian culture of the West comes to sense around him the confidence of a people who have been raised in a totally different world, based on totally different premises, which has nonetheless made an equally successful transition to technological living. In an odd way he finds their confidence annoying, their success unfathomable. ("How could they have made those bullet trains themselves?").

For outgoing people, especially for Americans who like to be liked, life in Japan can be a cruelly isolating experience. The men, at least, have their work associations with the Japanese, which are on the whole successful. There is the manufactured comradeship of the golf course and occasional evening entertainment, of varying degrees of formality, to which the Japanese invite male guests so readily.

The women, once they have despaired of learning Japanese and exhausted the resources of the flower arrangement class and the ladies' club, are thrown back on their own resources. This is no world for them. Business wives, who in America have been led to believe they are part of the corporate team, are treated in Tokyo with the barest sufferance. On rare occasions when they are invited to a big Japanese reception, they find no wives of Japanese present at all. By long custom, female companionship is afforded by several platoons of hostesses from the top-flight Ginza bars and cabarets, who wend their sinuous way through the banquet halls of the Hotel Okura or Otani, pushing the hors d'oeuvres and greeting familiar customers.

Cut off even from household chores like marketing (the maid would object) and denied almost any opportunity either for work or the normal society of an American or European community, women are reduced to a state of uncomfortable dependence. It is no surprise that there are many severe psychiatric problems among them, as well as their menfolk.[1] Perhaps one should give psychological tests to resident foreigner candidates for Tokyo, like those given candidates for the navy submarine service.

Yet even for the best adjusted, the feeling of apartness is always

1. For some years the Air Force, which had some psychiatric treatment facilities at its Tachikawa base, kindly allowed the worst of the American civilian breakdown cases to exit on a hospital plane.

there. Your children come home mad when small Japanese children run after them, shouting and pointing, "*gaijin, gaijin,*" ("foreigners") in tones half-descriptive, half-pejorative. You are frequently allowed, with a smile, to precede others in queues for tickets (that is, if you are a white *gaijin,* not a Korean, Chinese, or Indonesian *gaijin*) and only rarely are ticketed for traffic violations; but you are also a moving target for every student who wants to practice English conversation phrases as well as half-tipsy young "salarymen," rocking out of a Ginza bar, who show their sophistication by shouting, "Harro, wheruh you frommmm? You Engrish speakku." That is because you are not only *gaijin* (literally, "a person from the outside") but also *tanin,* ("someone other"), which is to say that you are in no way involved with any of the family, social, or business obligations or courtesies that bind together the hundred million people around you.

As a foreigner, you are supposed to like the things foreigners like. If you happen to prefer *sashimi* ("raw fish slices") to beef or pork at a meal, the Japanese are visibly pleased, but subliminally uncomfortable. *Sashimi* is their thing, after all. You are not behaving the way a foreigner should. Similarly, if you want a Japanese-style house with a garden. The Japanese real estate man thinks in terms of houses-suitable-for-foreigners and houses or apartments which are for Japanese. Foreigners are supposed to demand lots of chrome, lots of heat, lots of bathrooms, and so forth. If particular foreigners do not like all of these, so much the worse for them. They are not living up to their image.

The apartness of the foreigner in Japan is equated to the apartness of the Japanese in the world outside their tight shores. Of course, it is uncomfortable to be away from your own. How they know it!

After a while, the Japanese view of foreigners itself becomes contagious. One of the tertiary symptoms of adjustment among foreign residents in Tokyo is to see an unfamiliar foreign face on the street and find oneself instinctively asking the question, "Why did that foreigner come here? How did he get here? Does he live or work here? Does he know the Japanese language? Will he come here again?"

Chapter II

The Tangled History of Japan and the United States

Over the Western sea higher from Niphon come,
Courteous, the swart-cheeked two-sworded envoys,
Leaning back in their open barouches, bare-headed, impassive
Ride today through Manhattan.
> —Walt Whitman, *Leaves of Grass*

On a quiet Saturday evening in April 1950, the finance minister of the sovereign state of Japan, Ikeda Hayato, and his official secretary, Miyazawa Kiichi, were relaxing in the $7.00 a day double room they shared at the Hotel Washington, as visiting guests of the United States government, enjoying some Japanese hors d'oeuvres. Ikeda had brought several bottles of sake with him in his baggage, which Miyazawa heated in the bathroom sink, so they might have something to wash down the pickled preserves which the finance minister's favorite restaurant had packaged for them in Tokyo.

They were tired. Directly from their propeller plane journey across the Pacific, they had spent several hours in a highly secret conference with Joseph Dodge, the Detroit banker. Dodge was still working in Washington, with offices in both the Pentagon and the State Department, after his largely successful 1949 mission rationalizing the economy in occupied Japan. The conference was secret, because Dodge was almost the only man in Washington they knew and, consequently, could trust.

Ikeda, Miyazawa, and Prime Minister Yoshida Shigeru, who had sent them, were all too aware that United States policy about Japan's future had three power centers: the State Department, the Pentagon and—off by itself—the headquarters of General Mac-Arthur in Tokyo. All three groups disagreed about Japan's future, which troubled Mr. Yoshida greatly. He had sent Ikeda and Miyazawa to the United States so that they might directly present

Yoshida's case for the speedy conclusion of a peace treaty and, with it, the end of the United States occupation. Their pretext for leaving occupied Japan was a study trip. This at least got them the $7.00 a day double in the Hotel Washington and subsequent meetings with Dodge and a rising Republican foreign policy expert named John Foster Dulles.

Yoshida Shigeru, an old-fashioned liberal (he consciously used the word in its nineteenth-century Gladstonian meaning) had opposed the war. He was a professional diplomat with aristocratic background and connections. One of his first assignments had been to accompany his father-in-law, Count Makino Nobuaki, who led the Japanese delegation to the Versailles Peace Conference. Retired in 1938 and kept from entering at least one prewar cabinet because of his antimilitarist views, he received a clean bill of health, so to speak, in 1945 from an American occupation intent on purging virtually any Japanese who was part of the national authority structure before and during World War II. He was also, however, a firm Japanese nationalist. He fought the occupation to keep the Japanese government apparatus intact and as independent as possible, after his accession to the premiership in 1948, for 5½ stormy years.[1] His primary objective was to secure a peace treaty for Japan, so that his country could resume the conduct of its own affairs.

Dulles and other Americans in Washington were sympathetic to Yoshida's views, as Ikeda presented them. Their sympathies were reinforced when the Soviet-trained North Korean Army invaded South Korea barely two months later. At that time Dulles was himself in Tokyo, on a trip taking soundings for what became the 1951 peace treaty. The treaty that was signed the following September reflected the American concerns of the Korean War period, as well as Yoshida's desire to get rid of the occupation, at almost any cost. Under pressure of a wartime situation, the idea of Japan paying formal reparations to the victorious allies was tabled in favor of Japanese payment in goods, while the network of United States private and government-sponsored trade with Japan was strengthened. A firm relationship was established with Chiang Kai Shek's government, which would long delay Japanese rapprochement with mainland China. The Security Treaty with the United States emerged, by which American military bases were retained in Jap-

1. It was his second term in office. He had become premier, briefly, in 1946, but was voted out by the Socialist coalition election victory in 1947.

anese territory long after the weathered English street signs (A Avenue and 40th Street, etc.) had faded away and MacArthur's writ itself had passed into memory.

Ikeda and Miyazawa played a prominent part in the 1951 treaty discussions at San Francisco, when Premier Yoshida joined Dulles, Secretary of State Dean Acheson, and other representatives of the fifty-one nations who signed the treaty. Mao Tse-tung, then engaged in a shooting war with the United States, was conspicuously unrepresented. So was the Soviet Union, despite the fact that Japan had agreed unilaterally to confirm the Russian occupation of southern Sakhalin and the Kuriles. Writing in 1971, Miyazawa, who had by then become a cabinet officer several times himself, described the scenes and situations above and justly commented, "Although it is twenty years since that time, the questions decided at that peace treaty have cast their shadow on the present. When you contemplate the problems of modern Japan you can't help but go back to the starting point, twenty years ago." [2]

The issues raised by Miyazawa in his essay have indeed continued to be critical to the Japanese-American relationship: the United States Security Treaty; the economic interrelationship between the two countries that began with extensive American raw material imports in the occupation period and efforts to encourage Japanese to export to the United States; the problems of closer Japanese ties with both China and the USSR, her two dwarfing Communist neighbors. The three problems came to a head in 1970 and 1971, along with the "Nixon shock" of United States economic restrictions on Japan.

The Security Treaty was renewed, in the most painless way possible, by allowing it to continue in force without change. Yet the United States government continued to press for a more active Japanese defense role in Asia. The acute American trade imbalance with Japan, however, posed problems for the United States which no one in San Francisco had anticipated. The dollar devaluation, the import surcharge, and the intensified quota system for Japanese imports were all marks of a suddenly revived American protectionism. The new American relationship with China (and to a lesser extent the USSR), with the implied downgrading of the Japanese alliance, was itself a shock, coming after twenty years of anti-Chinese containment.

2. "Ihojin Nippon no Ikiru Michi," *Bungei Shunju*, October 1971.

Both in content and magnitude, all three problems were enough to cause concern. But to understand the Japanese feelings of misplaced trust, disillusionment, and worse, if not the strangely casual way in which the United States handled its side of the relationship at this time, it is necessary to go back to the beginning, almost one hundred years before the furtive visit of Ikeda and Miyazawa to Washington.

The first Japanese embassy to the United States arrived in Washington on May 14, 1860, after a long voyage from Japan aboard the U.S. steam frigate *Powhattan*. Headed by Shimmi Masaoki, the Lord of Buzen, it comprised seventy-seven people and proved a great crowd-pleaser, as the samurai and their retainers wheeled up Pennsylvania Avenue to the Hotel Willard, behind several massed companies of military escort.

"We were amazed at the crowd of people that packed the streets," wrote the expedition's second in command and diarist, Muragaki Norimasa of Awaji. "It was a surging sea of faces. Every window was full and there were spectators even on the roofs of houses. . . . Occasionally flowers were showered on us from the windows and Captain Dupont kindly explained that the flowers came from ladies. This is said to be the highest expression of kindness on the part of the fair sex. How lucky we are!" [3]

The American press of the day was equally enthusiastic and more self-congratulatory. The presence of the embassy, a special Japanese supplement in *Frank Leslie's Illustrated Newspaper* began, "may be considered as proof of how much more the peaceful and yet dignified diplomacy of our Republic is adapted to such remote and bigoted peoples as the followers of Buddha, Confucious or Sintoo, then the overbearing and violent system of the Western Powers."

The purpose of the Japanese mission was to make formal ratification of the treaty signed by the United States consul general, Townsend Harris, in Edo (the present Tokyo), in 1858. This followed by five years Commodore Perry's mission to open the doors of Japan to foreign trade, after more than two centuries of determined and almost complete Japanese seclusion from the West. As

3. The Lord of Awaji was less pleased at his first exposure to a Washington grand ball: "It was, of course, with no small wonder that we watched this sight of men and bare-shouldered women hopping round the floor arm in arm. . . . It seems very funny to us, as dancing is done in our country by professional girls only and is not at all a man's pastime."

Harris's were the first treaties with a foreign power, the visit to Washington was Japan's first embassy.

The choice was not by accident. Both the British and the Russians, the two other powers involved, pounded on the Tokugawa Shogunate's shut doors. But they had aroused considerable fears of military aggression in Japan. Russian forces had already skirmished with the Japanese on the island of Hokkaido (then called Yezo), which was for that reason declared a military zone by the Shogun. The British had assaulted China in the Opium War of 1839, when Hong Kong was annexed, and with the French in the "*Arrow*" incident of 1856.[4] Although Perry's guns were as real as those of visiting British and Russian squadrons, the Japanese judged, correctly as it happened, that the United States was interested in trade, not territory. Besides showing his open gun ports, Perry at least had the nicety to bring along a few things of use and wonder to the Japanese, such as a sample working telegraph and a small, but excellently functioning souvenir locomotive.

Whether they were influenced by the locomotive demonstration, the technical assistance of the small United States Navy crew in navigating Japan's first steamship, the *Kanrin Maru* (which sailed with the *Powhattan* to San Francisco), or missionaries like Guido Verbeck or J. C. Hepburn, it was at that time that the Japanese first came to think of Americans as "people-to-learn-from," as the essayist Yoshida Kenichi put it. "Such people who teach," he wrote in 1963, "are not necessarily to be loved, as witness our schooldays; but they cannot be ignored. They have something. They represent a standard and we have to cope with them in some way or other if we are ourselves to make good."

The Meiji Restoration was activated by the menace to an unmodernized Japan of acquisitive foreign fleets off its shores. But its raw materials, as noted before, were the changes in Japanese society, the rise of an active middle class, the urge for discovery and invention, the pent-up hopes and abilities that had been fermenting through two centuries of enforced seclusion. The lower-ranking samurai who led the Meiji Restoration (actually "renovation" is the literal translation of the Japanese word) were tactical innovators, but for the most part strategical conservatives.

Politically, the leaders of the Meiji Restoration were far from

4. Following the Chinese seizure of a British registry ship, the *Arrow*, and the later murder of a French missionary, British and French warships took action and captured Canton in 1857.

revolutionaries. Oddly like the American Revolution which, for all its talk of "taxation without representation," had very few Tom Paines and was largely led by gentlemen interested in modifying the status quo rather than destroying it, the Meiji Restoration was constantly at war with its trademark. It was no sudden revolution.

Bloodshed was minimal in the transfer of power from the Shogunate to the emperor's men; and, indeed, many of the talented people on the Shogun's side, like the recalcitrant admirals, Katsu Kaishu and Enomoto Buyo, were readily assimilated into the new national government, after coming to terms with it. Nor was Meiji, as Japanese Marxist historians like to say, the conscious beginnings of a new authoritarianism. The militarists who seized power in the thirties had little, if anything, in common with the pioneers of Meiji. They may have been able to distort the revived imperial system into something not too far distant from European fascism, but they were not necessarily the logical consequence of that system.

The modern political systems most congenial to people like Ito Hirobumi, who in 1885 became Japan's first prime minister, were the semiauthoritarian parliaments of Germany, which gave a limited amount of popular representation without really jolting the interests of a vested establishment.[5] It is interesting, if slightly ironic, in view of the generally anti-American stance of Japan's contemporary Marxist Socialists, that the first Japanese socialists received their ideological education in the U.S.A. Katayama Sen, the great leader of the Second International—and the one Japanese to be ceremonially entombed in the Kremlin Wall—learned his basic socialist theory working his way through Yale College. Kotoku Shusui, the intellectual firebrand who was executed in 1911 for allegedly plotting to assassinate the Emperor Meiji, had his education in the United States. Even Nozaka Sanzo, the veteran Communist leader of today, got his start on the West Coast, pamphleteering among Japanese seamen.

For the last century the motives of the men who led Japan's Meiji Restoration of 1868 have been dissected and debated. They indeed represented a strange combination of modernizers and traditionalists. Even Confucian scholars were utilized in the debates of

5. Oddly enough, Ulysses Grant, who visited Japan after leaving the White House, stopped off to see the emperor and gave Meiji some unexpected cautions about instituting representative government too soon. "Privileges like this," he said, "can never be recalled."

the time to justify the modernizers' attempts to restore the imperial system. But their one common denominator, from Ii Naosuke, the far-sighted minister of the last Shogun, to Mori Arinari, the new government's first minister of education, was a desire to capture the riches of foreign technical and cultural achievement and bring them back to Japan. From the beginning of Meiji, the Japanese saw the need for a strengthened school and university system as the very basis of modernization. *Bunmei-Kaika* ("culture and enlightenment") was, in fact, a juster and more typical slogan of the Meiji years than *Sonno joi* ("revere the emperor and destroy the barbarians") which, over the decades, got a lot more publicity.

After the first trip to Washington, Japanese ambassadors traveled to Europe, students and professors with them. Like frantic contractors, putting up a new building from whatever materials were at hand, they reached out for different aspects of different cultures. German law fascinated the men of Meiji, as did the ideal of the German university. The French army was widely admired (before its defeat in 1870) and the customs of the British navy were imported wholesale. Admiral Togo Heihachiro, who destroyed the Russian fleet at the Tsushima Straits in 1905, went to Dartmouth as a midshipman.

It was in education that the Americans made their most lasting impact. Something in the didactic missionary urge of the American teacher struck a spark in the knowledge-seeking of the Japanese student. From the 1870s to the 1920s and early thirties, long before a United States occupation of Japan would have been conceivable, American teachers were packing their bags to go to Japan; and Japanese students were struggling to get through their first lectures at Amherst or Rutgers.

The Japanese Ministry of Education was founded in 1871. The following year, the government dispatched the Iwakura mission to tour Europe and the United States in search of new and adaptable instructional programs, among other things. The compulsory education law was also promulgated. This initially set up four years of basic education, as well as providing for a system of high schools and upper schools, and eight universities. Probably because of their zeal for a universal public education system, the Japanese took more readily to American mass education ideas than to the European elitist education systems, such as the French lycée and the British public school. (The elitist concept the Japanese saved for their universities.)

The new education minister, Mori Arinari, was a confirmed admirer of American education. He did not lack for native American interpreters. Between 1860 and 1900 roughly two hundred Americans came to Japan to teach in the new schools, most of them in positions of considerable authority. The whole system of women's education in Japan was founded by Americans, the Japanese themselves being rather remiss, until very recent times, about educating females. The women's colleges, not to mention the first women's normal schools, were begun mostly either by Protestant missionaries or by Japanese who had studied in the United States.

The first agricultural school was founded in Sapporo, Hokkaido, in 1875 and the redoubtable William Clark, then president of Massachusetts Agricultural College, came over with some of his teachers to set Japan's future scientific farmers on the right track. Clark gave the Hokkaido A. & M. students sound agricultural instruction, as well as a crushing indoctrination in Protestant Christianity, and one of the more simplistic of typically American slogans, "Boys, be ambitious."

In 1873 a mathematics professor from Rutgers named David Murray was hired as the first superintendant of educational affairs. For almost six years he remained in Japan, setting up everything from study courses to school buildings and desks, which were planned on the model of those in Philadelphia. The urge for Western education was pervasive. When the missionaries were not looking, presumably, schools were even opened in the Yoshiwara brothel quarter of Tokyo to teach the girls English.[6]

After 1890, the Meiji Rescript on education and subsequent infusions of German philosophy tempered much of the made-in-America atmosphere. But until the beginning of the militarist era in the thirties, American educational ideas continued to have some hold on the Japanese. Progressive schoolmasters flourished in the twenties. John Dewey came to lecture in Japan in 1929, as did others of similar stature. Schools like Seijo Gakuen and Jiyu Gakuen in Tokyo patterned their unfettered curricula after exemplars in Boston and New York.

American traders came to Japan almost simultaneously with the missionaries and teachers. Although they speedily encountered problems with the protectionist trend in Japan's economic direc-

6. For this as well as the more substantive facts in this section, I am indebted to Herbert Passin's indispensable work, *Society and Education in Japan* (New York: Teachers College Press, 1965).

posed on behalf of his government that a racial equality clause be inserted in the treaty. His proposal was turned down, largely because of British opposition, but the Americans unhesitatingly supported the British action. Until that time, the idealism of Woodrow Wilson had aroused much enthusiasm in Japan. The American action on the equality clause turned enthusiasm to indignation.

In 1921, the Washington Naval Conference put an end, it was hoped, to the rivalry between the great maritime powers by establishing the 5:5:3 ratio of naval strength between Britain, the United States, and Japan. Japan's civilian government leaders of the time privately welcomed the arms limitation. Their economy was in a precarious position and heavy military expeditures could have wrecked it. But in the gathering mood of nationalism, the treaty was denounced by the Japanese right as a surrender, "Japan's defeat in a bloodless naval battle."

Through the twenties and the thirties the political relationship between the United States and Japan grew steadily worse. The peace diplomacy of men like Baron Shidehara Kijuro (who later, as Japan's first postwar premier, enacted the famous antiwar clause, Article 9 of the Japanese postwar constitution) received little notice in the United States. Hostile gestures and slights, however, were widely publicized in both countries. Japanese popular propagandists dwelt on the horrors of the Exclusion Act, while the American Sunday supplements were fond of examining the possibilities of a Japanese fleet and army invading California.

At the same time, however, the image of Americans as people-to-learn-from continued strong in Japan. American jazz, American movies, American ideas on education, socialism, and child care tumbled into Japan from across the Pacific. Despite the Exclusion Act, Charlie Chaplin, Clara Bow, and Harold Lloyd became household words in Japan (horn-rimmed glasses are still called *Roido-megane*—"Lloyd-glasses"—after the old American comedian).

On September 18, 1931, the Japanese began their invasion of Manchuria, which culminated in Japan's creation of the puppet state of Manchukuo. American-Japanese differences quickly escalated to the level of public confrontation. In 1932 Secretary of State Henry Stimson publicly declared the nonrecognition principle vis-à-vis Manchukuo. This was later adopted by the League of Nations, after the report of Lord Lytton's investigatory commission. On February 24, 1933, Matsuoka Yosuke led the Japanese delegation out of the Palais des Nations at Geneva. Shortly thereafter Japan formally quit the League.[9] On February 26, 1936, young army officers in Tokyo led an abortive coup against the government, which reflected, among other things, the Japanese Army's desire for further expansion on the mainland of Asia. In 1936 Japan joined Nazi Germany in the first Anti-Comintern Pact. The China War began in July 1937. In the same month Japanese military aircraft bombed the *U.S.S. Panay* in the Yangtse River. It was the first Japanese act of war against the United States. In December 1937 the Japanese army perpetrated mass rape and slaughter in Nanking.

As the thirties came to an end, the American view of the Japanese narrowed down to a simple matter of black or white—or, admitting a not so latent racism, a case of white against yellow. The Chinese, however, were excepted from the color ban in this instance. China was a victim. Not only that, but the efforts of American missionaries, valiantly overdramatized at home, led a considerable portion of their countrymen to believe that the Chinese were an almost universally likable, wise, and friendly people, veritably on the brink of becoming good Presbyterians. As Samuel Eliot Morison later wrote in his *History of the American People*, simplistically but accurately describing the United States voter's mood at that time, "He was sorry for 'John Chinaman' and detested the Japs." Every weapon in the public relations arsenal available to the Chiang Kai Shek government and its China Lobby friends was used to foster this impression.

The realistic accounts of the arrogance, cruelty, and greed which distinguished the Japanese Army operations in China were bad enough alone. But they were reinforced by America's one-sided view of what was happening inside Japan. Moderate opinion in Japan was actually quite strong. There were dedicated people in Japan who wanted peace—bureaucrats, editors, diplomats, socialists, and not a few professional navy men as well. In fact, the balance between a militarist and a reasonably democratic Japan was even, in the mid-thirties, delicate. It was probably tipped, in the end, by the willingness of Japanese big business to partake of economic profits of the army's aggression, as well as a general national pride in the success of the Japanese armies in an area where Western armies had already done their share of land-grabbing.

9. The following year, having recently annexed Outer Mongolia and most of the Chinese province of Sinkiang, the Soviet Union was welcomed into the League.

One of the few balanced views of Japan presented by the American press during that time was an impressive special issue, "The Rising Sun of Japan," which *Fortune* magazine published in 1937. Not only did *Fortune* see Japan's China adventuring in some historical perspective, but the editors pointed out the effects of the corollary Japanese economic advance into world markets, expedited by the artificially devalued yen. Their judgment was accurate and in some ways, prophetic:

The Japanese were no longer merely dressing up like a great power and talking like a great power—they were actually doing what the great powers had long reserved the special and peculiar right to do. They were appropriating pieces of Asia. And that wasn't all. Indeed it wasn't the half. Trade statistics began to come in. The Japanese, not content with pricking the diplomatic pride of the powers, were picking their purses as well. And not only picking their purses but picking them at a time of world depression when world markets were shrinking overnight. . . . Japanese exports of textiles, up 117 per cent from 1924, had somehow, in 1933, passed the diminished British total and now led the world. In Malaya, British cotton sales were off by half in three years while Japanese sales had mysteriously almost doubled. The same thing was true in Ceylon. In India the Japanese were underselling not only the British mills but the Indian. All through Africa the story repeated itself. Six Johannesburg factories closed in 1933 as the result of unforeseen and ruinous Japanese competition. Japanese textile exports to Kenya and Uganda rose to six times the British.

. . . The Dutch found themselves yielding leadership to the Japanese in their own Javanese markets and were eventually driven to imposing quotas and restrictions which infuriated the natives and seriously diminished Dutch prestige. The Germans were violently concerned about Japanese competition not only in their South American markets but in Germany itself. America was in the same stew, with imports of Japanese goods into such countries as Ecuador 750 per cent higher in 1934 than in 1933, and with the Argentine textile market almost lost to Japanese competitors.

. . . There was no sinister explanation. . . . The great competitive superiority of the Japanese is not a superiority in natural resources of which they have few, nor in resources of capital, which are limited, nor in mechanical genius, which is still rare, but in a homogeneous, highly integrated, and beautifully adapted social organization permitting a unification of national effort not possible in any other country. . . . She acquired the industrial revolution by mail order and fitted it to her pre-regimented population like a ready-made house dress to a perfect thirty-six.

. . . At present the Japanese masses are educated in the hands and the eye and the ear but not in the mind. Their hands are educated to work in one world: their minds are left to live in another. The methods of propaganda, of censorship, of "moral education" of police terror, now in use and now effective, may retain their power for some time unless a serious war occurs. A serious war, lasting more than two years, would almost certainly exhaust the country, producing economic collapse and the beginnings of social revolution. But sooner or later war or no war, the products of Japanese industrialism must necessarily be desired by the people who make them and by the people who feed the people who make them.

The objective analysts of *Fortune* represented a tiny minority. For the most part Americans throughout the thirties shook their heads at the newsreel pictures which reinforced their one-sided view of the "Jap" aggressor. Walt Whitman's picturesque two-sworded envoys, Commodore Perry and brave Admiral Togo were forgotten. Although the Americans were free with their words, until 1939, at least, they were strongly isolationist as far as deeds were concerned. Indeed, the Japanese militarists could not have hoped for a better bogeyman. For the repeated anti-Japanese statements of U.S. congressmen, as well as President Roosevelt's famous "quarantine the aggressors" speech of 1937, were like the earlier Stimson China declarations, hostile gestures with little follow-up. They were irritants rather than deterrents. It was only in 1939, when the United States renounced the commercial treaty with Japan, that the gestures began to hurt. But by that time most of the Japanese people had already come to believe that America was their enemy. The efforts of the small, albeit highly placed minority in Japan that still argued against war were made all the more difficult.

The climax came, of course, in 1941, when Japan's advance into Indo-China resulted in the United States peremptorily cutting off oil shipments. There were people in Washington, at that time, who still believed that this action, taken with the usual amount of public rebuke, would make Japan stop its aggression. How badly they read their potential opponents is amply recorded.

Certain parallels exist between that day and this. Twice in the seventies, just as in 1940 and 1941, a confrontation of sorts had developed between the United States and Japan. At the beginning of the 1940s the military adventurism of Japan in China pointed the way to war. The confrontations in 1970–71 and later in 1977–78 arose over what might equally justly be called Japan's economic adventurism—in selling so aggressively to the United States and other countries, all the while zealously protecting its own economy against too much foreign competition. In the case of Southeast Asia, where Japanese expansion had once created the proximate cause of the Pacific War, parallels to the seventies were also tempting. There the single-minded advance of Japanese international businessmen, tempered by relatively little in the way of economic, technical, or social assistance from Japan to those countries, provoked reactions almost as indignant as if there had been a military invasion.

In all such instances, the United States countered the Japanese moves with ultimatums and rejected Japanese overtures at continuing private negotiation. In 1940 and 1941 the threat in the United States ultimatum to Japan was explicit, "Unless you clear out of Southeast Asia and wind up the China War, we shall cut off your raw materials." In the seventies the threat was clearly implied by American economic diplomacy, "Unless you cut down your large trade balance with us, we will curtail our trade with you, on which you critically depend."

Japan's response to the American ultimatum of 1941 is well known. But just because the ultimatums of 1971 and since have been met by no economic or diplomatic Pearl Harbors, we must not assume that their shock in Japan was not felt. As the United States simultaneously made its rapprochement with China, it began to dawn on the Japanese how illusory was their faith in single-minded economic expansion. However successful the business of Japan, Inc., has been, its success ultimately depends on a completely free system of international trade. This system in turn depends on the active support of the United States, widely regarded as the major force behind the international free trade system. For twenty years Japanese economic planners based their single-minded growth policy on these premises.

Japan's military leaders of the thirties and the forties were insular and, on the whole, politically illiterate men. They believed not merely that military aggression could solve problems, but, most fatally, they believed it could solve problems cheaply. They had their own illusion of America as a power preoccupied with its own pursuit of economic gain. People like Premier Tojo Hideki and the wily Foreign Minister Matsuoka Yosuke thought that an American military reaction would be either delayed in coming, or ineffective when it came.

By contrast Japan's political leadership in the seventies is genuinely inclined to peace and relatively sophisticated in dealing with the world outside their islands. They are committed to securing Japan's future as a peaceful international citizen, whatever the argument about their skill in presenting Japan's image. They and their electorate generally agree that a modern great power need not be a military power and, in Japan's case, had best not become one.

Their particular illusion, however, was to overrate the international political consciousness of the United States. They overvalued the price that a shortsighted government in a weary, post-Vietnam America would put on the long-range political, diplomatic, and, indeed, economic advantages of the Japanese alliance.

It is true that the military relationship of the two countries has been ironically reversed. Japanese commentators have dwelt for a long time on the parallel between the Japanese China Incident and the American undeclared war in Vietnam. The lesson of the China Incident—the case of "a man walking into a marsh, sinking in deeper at each step," as the last wartime Japanese Foreign Minister Shigemitsu Mamoru put it—was forcibly recalled. The war which the United States fought in Korea had been a defensive action, which the bulk of the Japanese people understood and supported. But the protracted struggle in Vietnam, with its clouded origins and its escalating American bombing, for the first time drew popular attention to the old arguments of the Japanese left that the Americans were imperialists, bent on entangling Japan in their schemes to maintain hegemony in Asia. It was also clear to the Japanese financial and business community that the billions spent by the United States on the Vietnam War were not unrelated to the sudden crisis of the seventies in the United States balance of payments.

The situation of a peaceful internationalist Japan in the 1970s, if viewed with a realistic eye, is more precarious than it was in the militarist 1940s. Japan now imports 96 per cent of its raw materials. As a modest, second-rank power, it could have profited by technological advances and attained by now at least an agricultural self-sufficiency which the economists of the forties would have envied. But the rice fields have been plowed under for other more immediately attractive economic uses. The entire country's development has been pushed in the direction of large-scale industrial and commercial expansion. So, despite urbanization and scientific advance, the economic weaknesses of the thirties have become magnified in the seventies, albeit in a different way.

Japan is utterly dependent on its trade with the outside world. There is in Japan as of this writing less than a sixty-day supply of oil. A line of eighty-thousand-ton tankers stretches from Kuwait to Yokohama constantly bringing the oil to keep the world's fastest growing economy running. With so much of Japan's food coming from overseas, the consuming public is easy to panic. Fully 96 per cent of the Japanese dietary staple soybeans, for example, comes from the United States. One can imagine the shock of the news in July 1973 of Washington's temporary soybean embargo (because of local commodity price considerations), especially after the United States government had been pressuring Japan to import more and thus become even more dependent.

Although Japan is planning a gradual switch to reliance on nuclear power, its nuclear energy programs are running in low gear. It has no military force to speak of that would count in international power politics. Its only weapon, aside from moral suasion, is the utility to others of the goods which Japanese industry and ingenuity can produce. That weapon has its uses, but they are limited.

However distasteful the reality may be, a balance of military and political strength between two ultimately antagonistic value systems is still the world's basic insurance against self-destruction. Pictures of Henry Kissinger shaking hands with Chou En Lai or Jimmy Carter and Leonid Brezhnev exchanging bear hugs may be reassuring, but there are few accompanying signs that the Communist countries are relaxing either the shape or the goal of their institutions. In security terms, many Americans now worry about the Finlandization of Europe. Assuming that the United States military presence is completely withdrawn, this term connotes a Europe effectively dominated by the physical power center of the Soviet Union. Yet compared to the Finlandization of Europe, Western sentimental ties aside, a future Finlandization of Japan, with the Japanese ultimately depending on China or the USSR or both, could be a worse American disaster. It could be worse to the extent that Japan is a more effective unitary economic, social, and potentially political force than the divided Europeans.

Such a Finlandization might occur, with the blessings of one or both partners to the present alliance. On the one hand, Americans have grown increasingly lukewarm about foreign military commitments. It is particularly irritating, in Japan's case, to reflect that an expensive nuclear umbrella is being spread over an ally who uses the money saved in his own arms spending to subsidize export products that compete with America's. From the Japanese side, too, there is some doubt. After diplomatic and economic Nixon shocks and later "export" confrontations, the Tokai Mura crisis, the denial of Alaskan oil, etc., who trusts American guarantees?

For the balance of the seventies, Japan would have, in the extreme sense, three great power options:

(1) It can continue its relationship with the United States. This course relies on American self-interest as well as good will to keep the trade flowing and minimize economic abrasion between the two countries, as well as a continuing United States military guarantee, in an unobtrusive way.

(2) It could cast its lot with one or, given an end to their quarrel, both of the two great Communist powers, thus turning into something like the workshop for Peking or Moscow, which was the nightmare of so many American commentators in the early fifties. With Peking, at least, the relationship is a conceivable one.

(3) It can play a lone, adroit diplomatic game with the other superpowers, increasing its independence and possibly, but not inevitably, developing its own brand of international armament as a deterrent against becoming a victim of big power politics, rather than a participant. Oil producers are already giving ultimatums to oil consumers. Small-country militarism is becoming potentially as dangerous as superpowerism. (France, the home of chauvinism has refined its H-bomb, and India demonstrated its own potential in 1974. What would it be like if Libya got one?) To remain a big power, Japan must play a big power game in matters other than simply buying and selling. To do this without resorting to arms will be quite a feat.

The best possible course for Japan and the United States would be some combination between alternatives (1) and (3) with probably more of (3) than the American government would be happy with. We are finished with the twenty-five-year-old postoccupation relationship between Japan and America. Had it been used and understood better, the relationship might have flourished longer. Yet an alliance that was frozen in the shape of a union of a very senior and a very junior partner, founded on the military premises of past years, was overdue for revision. Japan is no longer either a client state or a potential enemy. Nor can the United States be cast forever in the role of an understanding and always indulgent protector. The Nixon shocks of 1971 and 1972 gave Japan a dousing of cold water, both economically and politically. The 1973-74 energy crisis which followed necessarily ended an era of complacent drifting in Japan, which recalls nothing so much as the euphoria of Coolidge America ("The business of this country is business") in the twenties.

Since Japan's only wealth, further, resides in the talents and the efficiencies of its people, Japan's maneuverability as an international power depends directly on the degree to which the Japanese people are concerned, united, and motivated. "The homogeneous, highly integrated and beautifully adapted social organization" which the Fortune editors observed in the thirties is much changed after four decades. The crude censorship of that time has given way to the world's most professionally critical press; the single-minded "moral education" of prewar Japan has been replaced by various kinds of moral and ideological suasion, operating in a fairly open market. Yet the economic growth of the sixties and early seventies could never have been attained without "a unification of national effort not possible in any other country." For all its strikes, ideological battles, and economic bickering within, the innate cohesiveness of Japanese society is still the wonder of the world outside it.

Among the other superpowers, the Soviet Union and China still control their populations by a combination of mass indoctrination and armed force. In the case of the USSR, at least, the former capability is faltering; but it is difficult to imagine that any meaningful revolt of Soviet citizens is around the corner. The United States, in the middle of a national soul-searching, can protect a considerable amount of bickering and divisiveness among its citizens behind the armor of its natural resources and its industrial and military power. Japan is the fragile superpower. Its huge economic wealth can continue to work and grow only if the consensus of the Japanese people is firm behind it. Consensus can be a fragile thing, even in a country that achieves it better than almost anyone else.

On a bright fall day in 1950, the twenty-fifth year of the Emperor Hirohito's reign, a slow-thinking countryman pushed his cart out of an alley into the middle of the traffic on National Route One, just north of the city of Atami, and ran almost smack into my jeep station wagon. I jammed on the brakes, almost in time. The countryman fell, but got to his feet. He looked stunned, but did not seem to be hurt.

A policeman came out of the crowd, bowed to me and poured invective at the pedestrian for getting in the way of traffic. I asked if the man had been hurt. The policeman said he was merely stupid and asked if my car had been damaged. I said it was not, then again asked the victim if he had been hurt. He said not and started to apologize. The policeman apologized. He asked if I wanted to call the MPs. I said that would not be necessary. He looked grateful, and told the pedestrian he had better watch where he was going in the future.

That was part of the United States occupation of Japan. An ever-present sense of privilege. You rode the trains in the special, anti-septic first-class cars marked "Reserved for Occupation Forces," while the cars fore and aft of you, windows cracked or missing, were packed with the great and small of Japan, from kids to company presidents, everyone who could not afford the rare taxi or the still rarer private car. You played golf at the Kawana Hotel's immaculate course, where the only Japanese you saw were the caddies and the pro. You took the hot-springs bath at the Unzen Kanko Hotel in Kyushu or rested grandly at the Fujiya in Miyanoshita; and read in the *Stars and Stripes* about how tough inflation was among the Japanese. You ate steak at occupation messes, with vegetables specially prepared (no night-soil for fertilizer) by the occupation's hydroponic farm. You furtively took cases of Crations with you to the country inns and cartons of cigarettes to give to the girls who would always say they were geisha. Liquor from the PX was cheap and "Off-limits" signs were for fun as long as the MPs did not check that day. At work you read tons of English-language paper—press digests, opinion digests, economic reports, Civil Information and Education reports, democratization surveys—and dealt with the people around you through a protective screen of interpreters, secretaries, drivers, and supernumeraries for whom the job meant regular eating without having to go out to the country for food.

Part of the occupation was work. You worked to write the reports and you worked hard trying to make them come true, at the same time; because, although you laughed at the anachronistic old man in the Dai-Ichi Building, with the five stars and the corncob pipe, there were few people in the occupation who remained completely untouched by his sense of faith in the mission. With some it was "Onward Christian Soldiers" all the way. A military government team officer in a remote prefecture, trying to interpret the latest occupation political directive, between picking out the good guys and the bad guys at City Hall, would have little to help him except instinct, a few area study guidebooks, and his enthusiasm that democracy could really take hold here. The economic analyst back in Tokyo felt a certain sense of communion with Teddy Roosevelt, or at least Thurman Arnold, as he found a *zaibatsu* trust that had not been busted or a cartel that was yet to be cleaned up. The education expert would shake his head over the translation of prewar Japanese textbooks and wonder how fast some new material could be imported from the States and translated for use.

There were other Americans not so nice: black marketeers, military and civilian, money-grabbing foreign traders who bought up Japanese land and copyrights for almost nothing, would-be imperialists who shouted all the more at Japanese when they could not understand their loud English, people who treated every polite gesture from a Japanese as a sign of servility or treachery. But by and large, the relationship was not so bad. Certainly nothing like it had ever before been attempted. While the Soviet armies across the Japan Sea were busy looting Manchuria and North Korea of their industrial movables, the Americans made an extraordinarily honest effort to give something to Japan. We liked to be thought of as the people-to-learn-from, at least then.

The sweep of the occupation reforms, viewed from the perspective of a less optimistic era, seems presumptuous almost to the point of fantasy. They uprooted, or tried to uproot, virtually all of the prewar Japanese political structure, along with a good portion of the social and economic network, on the theory that it all had led to war. It was justly observed that, from land reform to abolition of the old civil code, most of the proscribed Communist party aims of the thirties were realized by the Americans in the forties. Everything old was wrong. The bowing was servile. The militarists were evil. The caste system was real. It should all be changed. And it could, "Because, damn it, those people are pretty smart. . . . That guy in the motor pool does the job as good as an American." It was here, of course, on the level of know-how and can-do, that a feeling of some community was reached.

On the part of the Japanese, the young were interested, the old were relieved. They had expected worse. The businessmen and the mechanics dusted themselves off and started to work again, with an idea that they might do better. The bureaucrats held fast and somehow kept the old machinery of forms and applications and administrative decisions going while the occupation rebuilders were hammering outside and in. Japan, unlike Germany, kept its emperor and its government from war to peace.

Many suffered nevertheless. The purge of government officials and company directors was made without regard to individual desserts. People who had fought against the war were removed from their jobs as readily as people who supported the war. No distinctions. "Why didn't you kill them all," a Japanese businessman once said, talking of his purged friends, "it would have been more humane. You took away their jobs, with no appeal. They had no way of supporting their families. Their world was smashed. And they were the people who wanted to help you." If it was justice, it was rough justice.

A few of the bureaucrats watched with no small amusement as the American idealism of 1946 and 1947 gave way to the Cold War concerns of 1950. When the occupation sent word to arrest the Communist party leadership in 1950, the premier of Japan was known to remark, "Americans are very interesting people. When you came here in 1945, we had all the Communists in jail. You made us let them all out. Now you tell us to put them back in jail again. That's a lot of work, you know."

By 1952 the novelty had worn off. The general with the corncob pipe had already gone back to the United States and told the Congress about old soldiers fading away, in a grossly theatrical speech which those who heard it will never forget. And there were rioters rocking the occupation cars in the broad square before the imperial palace. In the course of democratizing the schools, the Japan Teachers Union had gone Communist, but Mitsubishi was back, as capitalist as ever, papering over some of the trusts.

The battles of the occupation will be refought for a long time. The Socialists and Communists continue to denounce any attempt to revise the American constitution as part of an American and Japanese rightist, imperialist plot. The conservatives keep wondering how to repeal an antiwar clause that leaves Japan unprotected in the face of continual provocation by its Soviet Communist neighbors.

Many Japanese still think of the occupation as an excuse, a straw man to beat, for their own lack of decision, as with the businessmen who blame the occupation's labor policies for an unruly union movement, but are disinclined to fight the issue out themselves. For others, however, it remains part of a dream (good or bad), a piece of an inspiration, an irritating but stimulating experience. It was the only time in Japan's history that a foreign guest ever arrived, went past the entry hall, and stayed for dinner.

Similarly, it is doubtful that the names of MacArthur's occupation personnel will go down in Japan's annals. They had their moment. But for those who lived through that time, Japanese and American, the memory of the experience remains amusing and sad, prideful and embarrassing. In a sense the experience was too strong, too heady. It fostered in both peoples an attitude and a relation to each other which was not, in the long run, healthy, but which took a long time to die. History will probably record that, on balance, those years reflect credit on both players in the game.

On a bright fall day in 1970, the forty-fifth year of the Emperor Hirohito's reign, I was pulling out of the driveway of my house in Setagaya Ward, when I collided with a young tradesman on a bicycle. I had been going about five miles an hour, so his only casualty was a bent front wheel guard and a broken light; but he was not about to let the accident pass lightly. The house I had come out of was big and he thought he had a good claim, which he began to assert in a loud voice. "Why had I not looked? Did I have a license?"

After a few minutes a policeman walked up from the *koban* ("police box") on the corner and asked what was wrong. He sized up the situation, took the young man aside for a few minutes' conversation, then came back and asked if I were willing to have the bicycle repaired. Not only that, but I would loan him one of our bicycles while we had the job done.

"Are you satisfied," the policeman asked the victim politely, "no further claims if the foreigner does this for you?" "No further claims," he confirmed and rode away on the borrowed bike.

"That's better," said the policeman, "we could have had a report and a formal accident hearing, but with foreigners, there's all that paperwork. You get involved, we have to call the authorities, and then maybe they'll call the MPs. That's such a lot of trouble calling the MPs," he said. He was a quiet gray-haired man, close to retirement. "You remember the MPs . . . when they were here . . ."

I said I certainly did.

Chapter III

The Unlonely Crowd

If any peasant abandons his fields, either to pursue trade or to become a tradesman or laborer for hire, not only should he be punished but the entire village brought to justice with him.
—Toyotomi Hideyoshi (1591)

Sagata Mura is a small village of farmers and weavers set alongside the Ashida River, in the low mountains of Hiroshima Prefecture. It has 150 households, with a total population of about 700 (a few score of whom may well have moved to the city during this writing). Apart from the remains of the castle of the Mori clan and a sixteenth-century Buddhist temple, Sagata has little to recommend it but some peaceful scenery, two river bridges, cool forest land, and a set of complicated, interlocking village and household relationships that would keep most social anthropologists out of harm's way for months.

The village itself is subdivided into upper and lower sections, which are in turn divided into *kumi* ("groups") of ten or more households each. The *kumi* join and unjoin for various traditional work tasks, some of which have been allocated and defined centuries before. Households have their own interrelationships, with a spiderweb connection between *honke* ("principal families") and *bunke* ("subsidiary families" i.e., cadet branches) which are bound as much by economic function as by geneology. Although the fortunes of households shift with the years, a working arrangement of seniority always prevails.

By tradition and preference, whatever the people of Sagata Mura do, they do in groups. Work, play, and civic duties are performed on the basis of households. The individual, as such, is rarely considered apart from his house (*ie*) and his *kumi*. They, in turn, fit into larger groups. Even now, despite the steady erosion of people to the cities, marriages are arranged as much as possible within the village or with village people. It is extremely difficult to do the elaborate detective work that the families of the bride and groom need for mutual reassurance if one of the prospective partners comes from far away.

Status can change, of course. Houses have had their ups and downs, over the years, as the business of weaving the attractively rough *kasuri* textiles waxed and waned and the job of farming transformed itself from the prewar and immediate postwar level of subsistence to the affluence of modern Japan where, thanks to wildly inflated land prices, the members of the remotest farmers' cooperatives (*nokyo*) suddenly could afford group vacation trips to Hong Kong, Guam, or Hawaii.

There is no longer the compulsion to stay in the household, still less the village, that there was in past generations. In most surviving Japanese farm villages, much of the male population goes off for seasonal or steady work in the cities. Yet people keep up their ties because they want to, even when they go away. Although similar villages exist in Europe and remote parts of the United States, they are mostly simply survivals, out of step with the modern urban society around them. Only in Japan, the technological industrial wonder of the century, does the old group society still cast such a strong shadow.

Far from Sagata Mura and hundreds of other hamlets like it, Japanese cluster in their ungainly cities, work in the endlessly proliferating factories, and sit at corridors of desks in the service industries that lubricate the wheels of their demanding consumer civilization. The carefully structured work tasks and taboos of specific villages like Sagata Mura may be forgotten. But in most places, the Japanese manage to transmit the ideals of the village. In neighborhoods, companies, and schools, they instinctively set out to build up the same kind of hierarchical microcosm that the village mind remembers.

The city of Tokyo is the classic example of this innate group consciousness among the Japanese. It is also perhaps the world's most apparent triumph of mind over matter.

Tokyo is one of the world's oldest metropolises. Founded in the fifteenth century as a cluster of huts around a warrior's castle, it had a population of 1.2 million by 1800, 3.7 million in 1920, and about 7 million at the outset of World War II. It has never known any kind of sustained, systematic planning, neither the ancient geo-

metric patterns which the eighth-century founders of Kyoto borrowed from the Chinese nor the modern network of broad avenues
and intersections which Nagoya's city administration started developing in 1945. Tokyo has simply grown, appallingly.

Originally the political capital only, it has become Japan's postwar business, communications, and cultural capital as well. Two-
thirds of the country's business leaders and fully one-third of all
Japanese university graduates live there. The population is about
11.5 million. If one includes the seven surrounding prefectures in
what Japanese call the capital sphere, the total metropolitan area
population is more like 30 million. It has been increasing every five
years by more than 3 million souls.

Tokyo is hopelessly crowded, polluted, and ugly. Flimsy
wooden shacks still cluster around the hundreds of graceless, if
allegedly earthquake-proof, new high rises. There is frighteningly
little green space. Most of the available open land, for a twenty-
five-mile radius from the center of the city, has been used up by
multithousand-unit *danchi* housing developments. More than half
of the city lacks adequate sewage disposal facilities (a large portion
of Japan's sewage is still piped into tank trucks and carried away
thereby). There are terrifying smog problems, a strangling, unimag-
inatively planned suburban belt, incredible land prices, hopeless ed-
dies of crowds everywhere.

Public and private rapid transit facilities are packed and people
are hurled into subway cars like chattels by professional pushers.
(One of the immediate results of the energy crisis in late 1973 was
a gratifying 10 per cent decrease in highway traffic; but this in
turn necessitated hiring two thousand extra platform "pushers" for
the added crowding on Tokyo trains). The elevated belt highway
system was already obsolete before it was finished, thanks to the
annual increase in automobiles. Water is periodically in scarce supply; the supply of fresh vegetables is growing precarious. Garbage
disposal plants are overflowing. How can Tokyo go on?

This description of Tokyo's physical problems is not exaggerated and the problems are continually getting worse. Tokyo seems
the logical candidate, whenever city planners meet and the talk
comes around to that awful possibility: When will a modern urban
center simply stop, choked by its own swill? Yet Tokyo, possibly
alone of the world's great cities, preserves through the crowding
a certain sense of spiritual buoyancy, a vitality that has gone from
London and is ebbing from New York, a kinetic energy that is

somehow guided usefully. Tokyo's surges of people are amazingly diked and channeled. They have yet to explode in violence or confusion or, worst of all, the apathy and the sense of alienation that afflicts city-dwellers in the West.

The crime rate is incredibly low. Tokyo averages about 2 robberies a day, as against about 6 in London and 200 in New York.[1] People find purses in the street and return them. Shopkeepers let customers come back and pay later. A promise does not have to be backed up by a threat or a written contract. People religiously observe queues for taxis and buses. They wait patiently at street crossings for the signal. Protestors demonstrate against the government, carrying banners advocating total destruction of authority, while patiently observing the directions of policemen steering the *demo* past traffic. The people of Tokyo complain, argue, buy earthquake survival kits, read best-selling science fiction thrillers about the total destruction of the Japanese islands, and join consumer boycotts against high-priced retailers. They may get angry. But they are not alienated.

There is some irony in the fact that the Western civilization which was premised on the Greek city-state of Athens, which developed its democracy through the independent spirit of city-dwellers in London and Paris and stamped the modern megalopolis in the image of America, can demonstrably no longer handle its own artifact. It took the Japanese to make the modern city half-bearable by transcending it and imposing their own illusion on the urban reality.

The Japanese city keeps its sanity, against most Western rules of logic and order, because its people are, paradoxically, villagers. In their hearts they remain villagers. In this sense Tokyo is not a city at all. It is a modular assembly of hundreds of villages clustering around and radiating from the big transport centers like Ikebukuro, Shibuya, Shinagawa, or Shinjuku. People live everywhere in Tokyo. Except for a few pockets, such as the financial-district skyscrapers of Nihonbashi and the palace area, there are houses and apartments in almost every district. There are few Park Avenues; people of greatly varying income levels live side by side. Each group has its neighborhood stores, its schools, its shrine, its own

1. Even allowing for differences in statistics-gathering, the contrast is formidable. In 1973, for example, there were 196 murders in Tokyo, 1,680 in New York. New Yorkers reported 3,735 cases of rape; in Tokyo there were 426. Against 82,731 car thefts in New York, Tokyo had 3,550.

little Ginza, where the shopkeepers put out paper flowers at sale time and push the *omikoshi* ("portable shrine") on the feast days of the local neighborhood dieties, as if they were dwelling in a small mountain hamlet (*buraku.*)

Most American cities have systematically eliminated their neighborhoods. The children go to school in buses and the mom-and-pop stores have all but vanished under the slide rules of housing developers and supermarket chain planners. In Japanese cities, the neighborhood store abides. The local noodle shop delivers indiscriminately to fancy "mansion" suites and to tiny twelve-mat [2] apartments in the public-housing *danchi*. People leave their dishes outside, and dishes and money can be collected in the morning.

The small machine shop and the lean-to cleaners and shoemakers stay in business—and are still protected from real estate enterpreneurs trying to engulf them. (In Japanese custom and law the squatter has strong rights.) The officer at the neighborhood police box is part of the local scene. He is rooted in it, not an alien presence who rides past in a patrol car. The children walk to the public school, which almost all attend. Every neighborhood in Tokyo keeps its human core. Each has its small, precious bit of distinction, enjoyed by the group that lives there.

The Tokyoite's consciousness of being a villager is often literally true. Far more than Americans or even Europeans, the Japanese city-dweller keeps up strong ties with the actual place in the country from which his family came. If you ask a Japanese where he is living, he will tell you that he resides in the Ebisu district of Shibuya ward. But if you ask him where he is from, in the American manner, he is more apt to give the name of his grandfather's village in a far-off prefecture, a place which he may have visited only once or twice on holidays or to pay respects to the family graves. The idea of the national citizen, as such, was late in coming to Japan; Japanese scholars themselves are the first to point this out.

In his fascinating study, *The Ways of Thinking of Eastern Peoples*, Professor Nakamura Hajime has written, "Japanese cities were rather nothing more than densely populated areas controlled by warriors. Even after the Meiji Restoration of 1868, when cities

---

2. One mat (*jo*) is the standard-size woven straw floor covering in the Japanese house, measuring six by three feet. Whether actual mats are used or not, the mat is still taken as a measure of space in Japanese housing and, by extension, other matters, e.g., a four-and-a-half-mat mind is a nice way of saying small-minded.

expanded rapidly, their citizens did not come to possess the self-consciousness of European citizenry. And particularly noteworthy here was the fact that there was a constant flow of farming population in and out of the cities. This farming population continued to be bound by blood relation and economy to the farming village during its residence in the city and was free to go back to the country if subsistence became difficult. This situation formed an obstacle to the growth of a general public morality as well as an ethic for the individual in Japan."

Be that as it may, the same people who delayed the development of public morality in building the nineteenth-century city seem remarkably well equipped to face the fissionary pressures of the modern city as it struggles into the twenty-first. In adapting to the megalopolis, the Japanese have simply used their old facility for turning work or political situations within a village into family-type relationships [3] and have magnified it. Almost unconsciously the Japanese have broken up their cities into manageable pieces and imposed village relationships on them. The strong sense of neighborhood solidarity is partly self-perpetuating. As local ties develop, people feel a sense of obligation to the neighborhood tradesmen, who in turn save their best goods for their regular customers. The tradesmen also buy their customers' wares, insofar as possible.

These loyalties and a sense of community are highly particularized and have their negative side, too.[4] A respectable businessman who is a model of probity within his own neighborhood will regularly carouse at bars and dubious "Turkish baths" in the next ward. School children who scrupulously help keep their own street clean turn into heedless litterbugs when they enter an adjoining neighborhood. This spirit was epitomized in the so-called garbage war of 1972, when one Tokyo ward refused to accept the neighboring ward's refuse at its garbage dump. Citizens and ward officials joined to bar entrance to trucks from the outside.

As streets of houses turn into high-rise rabbit warrens of apartments and the Japanese supermarket becomes more and more a fact

---

3. The so-called *dozoku* ("single-family") principle by which, for example, a branch of a shop is entrusted to a relative or former servant or, as still happens, a promising employee is adopted into the family, his name changed, and he is married to the boss's daughter.

4. The old proverb, A man away from home need feel no shame ("*tabi no haji wa kakisute*") has its counterpart in all countries; e.g., the Grand Rapids furniture man whooping it up at the Chicago convention; but nowhere else is it so literally true.

of life, the traditional bonds of the Japanese neighborhood inevitably loosen. But for many, the sense of living in a village is supplemented, if not replaced by their businesses. The corporation becomes the village. Within it, as with the old villages, there are the departments and the divisions with their own inner sense of people who belong, the families with their sense of obligation.

Each of the transplanted villagers in a Japanese metropolis goes through life, for the most part, as a secure member of a group, whether business or cultural or geographical. It is a group organized, like life in the village, vertically rather than horizontally, part of what the sociologist Nakane Chie calls the *Tate Shakai* ("vertical society") a congeries of universities and corporations; trade unions and farming cooperatives; golf clubs, group tours, and cabarets, where almost everybody has his niche and his rank. The professor from Tokyo University is introduced before the professor from Hokkaido University; the banker from Sumitomo headquarters in Osaka ahead of the branch manager from Kyushu; the high school principal before the primary school principal.

The *honke-bunke* relationship of the households in Sagata Mura is mirrored in the relations between the Japanese company and its subsidiary. (The average Japanese businessman loathes leaving his head office for a subsidiary company. He is only partially comforted for the loss in group status by the higher rank which the subsidiary may give him.) Or in the storied relationships between the professor and his *deshi* ("apprentice") which have continued not for years, but for decades, until one or the other dies. The apprentice, himself a professor of distinction, in turn develops an apprentice network of his own.

The same is true of student relationships. When an American talks about someone with whom he went to college, he would say, "Oh yes, there's old Stillman. He was at Yale when I was." This means he might have been a class or two ahead of the speaker, or a class or two behind him, or maybe in his class. It does not really matter much. With the Japanese, however, the wording is always precise. One can never simply be at college "at the same time as I was." One has to be either a classmate or a *senpai*, or a *kohai*, one who has "gone ahead" of me or one who was "behind" me. The relationship never changes all through life. One is always either the *senpai* or the *kohai*. Even in the rare case where the *kohai* becomes a superior, there is a certain dignity accorded to the senior man in terms of years or prior experience, always.

When an American is talking about a sibling, he will generally say, "my brother." He will rarely say, "my older brother, Harry," or "my younger sister, Jane." A Japanese will always specify an older or younger brother or sister. Indeed, there is no simple word for brother in the Japanese language other than *kyodai*, which can also mean sister. These relationships also abide. Younger sisters can marry and have families, but there is always some tie back to the older brother. He is still consulted on family decisions, however much distance militates against it.

The American is taught to chafe at hierarchy. It is our paradox that, although we accept the book of the law, we are conditioned to challenge the word of personal authority. "What does he have that makes him better than me?" is a familiar American reaction to any structured situation. The Japanese does not have this strong reaction. He may not like the situation, but he is conditioned to accept it. He acknowledges that the other fellow was born before he was, or is assumed to be senior in some other way—he may have come into the business first. "*Akiramemasu*," he says, in the constantly used phrase, "I am resigned to the situation. I accept it." In this resignation the Japanese have a built-in stabilizer in their society, one which our society lacks. This makes them a very difficult people in one sense: structured, embarrassingly polite, and overly conscious of protocol. And to be resigned often leads to complacency or loss of initiative. It is also a powerful force for security, however, especially when there is no other stabilizer around.

It has been this way for centuries. Given this sense of seniority and protocol, one wonders why Japanese society did not long ago congeal, as did China by the time of the Manchus or feudal-aristocratic Europe, with no one allowed to go beyond his rank and station.

Various types of frozen seniority systems have, indeed, existed in Japan. The most conspicuous was the imperial aristocracy of Nara and Kyoto, who kept their world quite satisfactorily closed until about the eleventh century. It was a world of poetry, protocol, and considerable high-level promiscuity. (The reader of the *Tale of Genji* is led to speculate how the prince could find any time for his official duties, with so many literary ladies to keep entertained.) But the growing military rivalries of the time put an end to that class society.

After Tokugawa Ieyasu won the battle of Sekigahara in 1600 and established his Shogunate, he and his successors closed the country and again set out to enforce a class system, but this time based on the supremacy of the feudal leaders and their samurai. Borrowing freely from Confucian China, the Tokugawas divided Japanese society into four levels, the famous *shi-no-ko-sho*: warrior nobility, farmers, artisans, and, last of all, merchants. The scholar class, which had been at the top of the Chinese social pyramid, they ignored—or implicitly included with the warriors.

Two hundred and fifty years of enforced peace in turn fermented considerable changes within this structure. By the 1850s there were a great many poor samurai and rich merchants, farmers who had once been warriors and commoners who had the right to carry swords. When the Meiji reformers restored the imperial house to the center of Japanese society in 1868, the Tokugawa aristocracy was irreparably shaken.

The desperate national impulse to master Western learning subsequently brought new people into the governing class. Scholars were valued as highly as warriors. The newly established system of universal public school and public university education began a leveling process which has continued in Japanese society to this day.

In many ways, Japan is a true meritocracy. One can move up, if one obeys the rules, and get quite far through sheer competence. The railroad motorman's son from far-off Akita prefecture who excels in the exams and enters Tokyo University will find himself, by that very fact, elevated to a pre-eminence in the Japanese community that his family might not otherwise have achieved for generations. He will be accepted into a government office or large corporation on the basis of his school affiliation (as long as he does well in the corporation's testing process) and handled as a member of the elite. But he is an elitist the way the Chinese mandarins were elitists, by right of scholarship and merit (plus possibly some help from a few acquaintances). He does not think of himself as part of an elite social class. And, unlike people who climb from humble circumstances in America, he rarely turns his back on his antecedents. Family is family. Akita is still home. And, after all, everybody is Japanese.

There are no society pages in Japanese newspapers, no beautiful people to read about, almost no social clubs to aspire after. There are far fewer social distinctions, reinforced by education and upbringing, than in the democratic United States. Japanese tend to live within their own neighborhood or work groups. There is still little pretense of a society founded on social classes. For example, businessmen and professors will not mix at a party, even if they share a common educational background and general style of life. Within their particular village structures, the Japanese are content. The stockroom boy is just as assured of his position in society as the president of his corporation; and he feels as much a part of it.

A certain cultural and educational togetherness, too, has made the gulf between an intellectual, hard-working elite and the rest of the population less offensive and obvious than it is becoming in most other countries. The very language, its nuances, the time lavished on learning it, and the constant need for altering its expressions to fit different situations, in rather rigid patterns, is an adhesive rather than a disruptive factor. The Harvard professor and the truck driver may speak a quite different level of English, but the Tokyo University professor and the truck driver use the same Japanese, where the occasion demands it. Gradations of tone and politeness are part of their common linguistic heritage constantly used.

Foreign travelers never cease to note that the manner of Japanese living differs surprisingly little as one moves from high to low in the society. As there is a certain common denominator of Japanese artistic taste, so there is a common denominator of living standards. The contrast between the dwelling appointments of an American corporation president and his lowest employee is vast. But the similar contrast in Japan is often more a matter of size and quality of furnishings than of sharp differences in living appointments.

The Japanese house imposes a certain commonality on all who live in it; hence the relatively slight difference between the houses of the very rich and those of the poor or middle class—as compared with the class-conscious housing of Americans. Barring a few gaudy conspicuous consumption monuments erected by high-livers in Tokyo and Osaka, even city housing manages to retain a generally common design.

The respect for beauty in the institution of the *toko no ma* [5] has not gone out of Japanese life, nor has the corollary love for captive nature expressed in the garden or the small flower or bonsai grow-

5. The slightly recessed alcove in a Japanese living room which gener- ally contains a treasured scroll, flower arrangement, or other art object.

ing inside the house. These national trademarks still are being carried over into changing modes of mechanized household existence.
In the matter of culture, the Japanese tend to resist both the
American desire to lower the common denominators of culture as
well as the concurrent world tendency—demonstrated in Communist and democratic countries alike—to have two parallel cultures,
one for the intellectual patricians and another for the huge, mass
market plebs. Books of a high intellectual level can easily become
best-sellers in Japan. A treasury of the world's great writing can
sell more than 200,000 copies per volume, despite the fact that it is
not composed of easy abridgements, but is a complete and authoritative set in which the reader must digest such writers as Lenin,
Schopenhauer, and Nietzsche. Poetry collections do astonishingly
well.

"In the beginning was the Word," say the Scriptures of Western
society; the Word not merely of St. John's Gospel, but of the ideals
of truth and justice and all the abstract virtues which lay their
weight on the individual conscience and idealized the single soul,
or the single reason, alone with its logic, or its God. Japan has
different spiritual roots. In the beginning were the People, is the
thought running through the scriptures of the Japanese. For them,
reality is the sense of the group living within itself, lower than
some groups, higher than others, but always contiguous to some of
them. Considering the later codes and manners of the Japanese, it
was totally characteristic that the first laws of Japanese mythology
were not tablets handed down on a sacred mountain, but the result
of a discussion among the gods meeting in a riverbed.

While Americans, firm in the tradition of the Greeks, the Romans, and the Jews, place our greatest hopes in a government of
laws, the Japanese will opt for the government of men. John Adams' pious hopes for the Commonwealth of Massachusetts [6] would
find no response in the village of Sagata, and not much in the cities
of Japan either.

The law, as the West understands it, has descended directly from
the opposition of the apparent reality and the universal ideal, between things-as-they-are and things-as-they-should-be. If a given
state of affairs is running contrary to the law, we feel, in the West,

6. "To the end that it might be a
government of laws and not of Men," he wrote, working under "the Great
Legislator of the Universe."

that it must be brought into conformity with the law. If not, the law must be repealed or reinterpreted. But the law, bend it, twist it, or interpret it as one may, must remain the norm.

To the Japanese the law is not a norm, but a framework for discussion. The good Japanese judge is the man who can arrange and settle the most compromises out of court. When an American calls his lawyer, he is confident and happy to rely on the strength of his whole social system, the rule of law. When a Japanese calls his lawyer, he is sadly admitting that, in this case, his social system has broken down. A system of personal commitments has failed. The American, if the case goes to court, looks for victory and a judgment for his side. The Japanese, if the case goes to court, looks to a happy compromise—in or out of court—which will favor his side as much as is feasible.

In a perceptive essay, "The Japanese Notion of Law," Tokuyama Jiro of the Nomura Research Institute has written, "[Japanese] people or even judges and police do not regard the law as determinate. Not distinguishable from reality, the legal norm is forced to compromise with the reality of social life to a great extent. The Japanese regard with disfavor the strict application and execution of the letter of the law; as they feel it to be too unrealistic and too rigorous.

"In Japan law is often compared to a treasured sword handed down from generation to generation in a family. The sword is not for killing; it is rather an item used for decoration or as a prestige symbol. Likewise law is not for controlling social life; it is a mere decoration."

Whatever are our contemporary Western boasts about the sacredness of the public weal and the supremacy of purely human values, our history and the ancient creeds that formed it have conditioned us to revere the law as something more than human. It is the Ten Commandments reaffirmed, God's sanction of man's codes. Our idea of the individual and the individual conscience draws its strength from the same source as does the law. Western man tends to depend on the sanction of universal truths that are no more and no less believable than the Christian afterlife. The Japanese, in contrast, draws sanction for his acts—if not, indeed, the reason for his existence—from the approval of the community, his participation in it, and from his ancestors. It has always been this way. There were no Japanese Ten Commandments. Nor was there any Japanese Grotius to explain how the natural law regulated intercourse among nations.[7]

Historically, the Japanese has tended to reject the world of universals in favor of the world around him. He lives in the here and now. His scriptures are the laws of the community. He is, first, last, and always, a villager. Whether his identity symbol is his place in the village family register, a country fireman's coat, the tatoo of the *yakuza* gangster in Osaka, or the neat button announcing to the world that he is a member of the Mitsui Company, the Japanese's identity is greatly dependent on his group.

Ostracism—the so-called *mura hachibu*—has been, down to modern times, the worst possible fate for a Japanese villager; just as to be a *ronin* ("masterless samurai") was a sad fate in the days of feudal Japan, signifying as it did a man who had lost his feudal leader. *Ronin* is, incidentally, a term of some opprobrium in modern Japan. It refers to a student who has repeatedly failed college entrance examinations, but is still trying, searching for a university which can give him the group identity on which most of his future life's pattern—the kind of job, the kind of marriage, the kind of friends—will depend.

Japanese intellectuals worry about their society of colleagues ("*nakama shakai*") and the inability of their countrymen to think of themselves apart from their particular group. As Nagai Michio wrote some years ago in a wise book, *Several Varieties of Heterodoxy (Ishoku no Ningenzoku)*, "I regret to say that virtually all organizations in present-day Japan, whether conservative or progressive, are much too solidly organized to permit single persons to express their individuality. Most people have abandoned their efforts to express themselves as individuals."

They have a point, just as Professor Nakamura does in his historical comments about the growth of Japanese cities, considered as a modern negative. When a country, facing a crisis, has difficulty bringing together people from business, government, and the intelligentsia even for the purposes of discussion, there is cause for concern (see Chapters X to XIII). The constraints and discipline of

7. One of the most puzzling problems the Meiji reformers of Japan faced in the late nineteenth century was the workings of Western law. It was a total mystery to them. Yet they had to learn it, to secure abolition of the extraterritorial privileges of Western countries in Japan, to make war properly on China in 1894, and to secure favorable treaty rights of their own.

the unlonely crowd in Japan can be daunting to live with. Even Japanese who are born into it chafe under its unwritten bonds and codes. The long-time foreign resident claustrophobically longs for the occasional vacation to Hong Kong or Manila, where people argue and shout in the streets. There is something quietly terrifying about a society where, as a Japanese friend of mine once spoke out in frustration, "We are like people driving high-powered cars down a six-lane superhighway, quite capable of passing others and going sixty or seventy miles an hour, but condemned to stay at a safe forty because this is the way everyone is driving today."

Yet it is only the rare Japanese who makes this objection. The majority in the unlonely crowd find their group existence not at all disturbing. Whatever its failings, their society has a unique cushion for coping with the practical problems of this era. The villagers of Sagata Mura, and of Tokyo, are the least alienated of peoples in a world where technological pressures make citizenship a demanding role and individualism an increasingly expensive luxury.

Chapter IV

Sacraments in Banks and Department Stores

Nowhere is there a shadow in which a god does not reside. Peaks, ridges, pines, cryptomerias, mountains, rivers, seas, villages, plains, and fields, everywhere there is a god. We can receive the constant and intimate help of these spirits in our tasks.

—Shinto Shrine Chant

"*Irasshaimase, irasshaimase, mina sama iraaasshaiii.*" The man waving the new automatic rice scoop stands coatless, and soon to be lungless, at the juncture of two counters in the packed housewares section of the Tokyu Plaza Department Store, trying, with ritualized desperation, to draw some of the passing customers to his particular cash register. "Step right up, ladies and gentlemen. We have here for a mere 750 yen one of the newest and most useful discoveries. Just take a look, ladies and gentlemen."

Throughout the weekly Saturday boom, at this and other counters, the barker's spiel continues. A few buy. Most pass on. But the salesman continues to shout his message, secure in the knowledge that he, along with the smiling garage attendants, the patient demonstrators in the toy department (Japanese children treat display toys as roughly as they treat their own), and the uniformed girls wiping imaginary specks of dust from the escalator handrails at every floor, is an essential part of a total effort at institutional—or, better, family—salesmanship.

Customers in American department stores may be familiar with part of this scene, but the totality of it is unique to Japan. Every other pillar seems to conceal a new information desk. The masses of girl elevator starters and operators, dressed in costumes worthy of a Busby Berkeley musical in the Hollywood thirties, change shifts like the guards at Buckingham Palace. Clerks stand attentively at cashier counters and perform package-wrapping rituals of unbelievable complexity. Everywhere, from the squads of ground-floor doormen to the countermen at the roof restaurants, one hears the chorus of "Welcome," "*Irasshaimase*."

Japanese customers expect this kind of thing. They expect the counter demonstrations, the food-for-tasting, the tiny souvenirs on holidays, and the thousands of hand-lettered "sale" and "bargain" signs dangling from the ceilings in women's wear or draped across the half-acre of electrically heated quilts in the dry goods department. Most American department store controllers would at best look askance at such obvious pieces of nonpaying frippery. But in Tokyo, whether at the Tokyu Plaza or more senior establishments such as the Nihonbashi Takashimaya, Shinjuku's Isetan, or branch stores of the three-hundred-year old Mitsukoshi firm, these extras are not thought of as extras. They are part of the seller's relationship to the buyer and, as such, not lightly discarded.

In a Japanese store even the act of packaging is important in itself. The simplest purchase is dignified by careful wrapping, not to mention the effort given to the many-splendored paper surrounding gifts. A bill should be presented in an envelope and on a tray, just as a gift should be presented in its proper setting. Bills and gifts are to be considered and contemplated before receipt, payment, or proper expressions of admiration are allowed to enter the picture. They are all essential forms, without which business would be something less than business. And in case anyone should think that such niceties are not profitable, he might take a look at the astounding rise in Japanese department store sales, from $3.5 billion (¥1,042.7 billion) in 1966 to $8.9 billion (¥2.669.0 billion) in 1972. While the elaborate wrapping and escalator handrail wipers multiplied, profits multiplied with them.

Such stress on the form of things occurs in every section of Japanese society. On the same platform where the *oshiya-san* ("pushers") shove subway passengers' arms and legs inside the cars so doors can close efficiently on the rush-hour commuter trains, an assistant station master, wearing white gloves, stands at attention and salutes the arrivals and departures of major intercity expresses. In the middle of a deadly national traffic problem, where fatalities from drunk or reckless driving threaten to rival World War II infantry losses, drivers still blow their horns to acknowledge routine road courtesies. The most reckless weaver never fails to raise his hand apologetically (and sometimes with a weak smile) as his careening Mazda sports car cuts in front of your sedan. Japanese

bankers, who come closer than most bankers to owning their customers in this high-debt society, religiously make their round of
ceremonial courtesy calls on the humblest of corporate clients.

There are in Tokyo roughly 100,000 restaurants and eating
places. Most of them are only a few hundred square feet of floor,
with a counter, a few tables and chairs (or *tatami* mats), an out-
sized refrigerator, and a precariously flammable gas range. They
are tucked in shacks, basements, and arcades of huge buildings or
excess space under expressways. All restaurants the world over try
to provide some kind of illusion for their eaters, be it the ambience
of a nineteenth-century French bistro, a Renaissance German rath-
skeller, or an American Western saloon. Nowhere is this done so
relentlessly and so successfuly as in Japan. Wood, pictures, green-
ery, paneled boards are brought in to make the dank concrete par-
tition in the third subbasement of the Dai Ni Station Hotel Building
a corner of old Edo. Daily the waitresses exchange their pants suits
and chewing gum, in the back room, for a delicate looking kimono
and, with the kimono, as many courtesies of manner as their educations and recollections permit. The boys behind the counter, if
it is a *sushi* (raw fish) establishment particularly, tie around their
heads a *hachimaki* cloth like that favored by fighting samurai (although they roll theirs differently) and greet the customers with
traditionally loud shouts of welcome. Even if only for the duration
of the 800 yen *teishoku* ("table d'hôte") meal, old Edo manages
briefly to reappear.

Whenever a Japanese house is built, whether of concrete, plastic,
steel, or wood, somewhere the sake is brought out for the *muneage*
ceremonies, just as it always was when the builders actually put up
a real wooden ridgepole. For sports victories, business successes
and family parties, the *banzai* cheer and the ritualized burst of
handclapping are insisted upon. The city apartment dweller dotes
on the few plants or dwarf trees he has bought for the living room.
The Japanese letter writer begins the most routine communication with the obligatory comments on the state of the weather,
the recipient's health and the blossoming of seasonal flora.

Festivals too go on: Boys Day, the Feast of the Dolls, Coming of
Age Day, Spring Equinox Day when you hurl beans out of the
door to cast away the equinoctial devils. They are celebrated with
the comforting consistency of the old Christian ecclesiastical year.
People with neither belief in nor concern about the afterlife build

and repair thousands of shrines and temples every year as community efforts. Drums beat in the hearts of cities for the local neighborhood festivals.

Countless acres of valuable floor space in Japanese corporation offices are taken up by Reception Rooms. Few Japanese businessmen would think of welcoming guests of any importance to their own offices if a Reception Room were unavailable. This is despite the fact that the Hollywood-type office, with its out-sized desk, deep sofas, television set, and handy bar is not exactly an unknown commodity in Japan. But a private office is for work. To receive guests, a special place is preferable, and special forms. In crowded or humbler quarters, a great point is made of the host leaving his desk and joining his guests at a tea table three feet away. Even when visiting relatively small fry in a company or government office, the guest waits in the small reception enclosure on the office floor until the host polishes up his glasses, shuts his ledgers, and approaches, signaling for tea. Although his desk may be next to the visitor, it is poor form to recognize him until he arrives in his host posture, sitting across the reception table on one of the white-doilied reception armchairs.

In my own experience, when I have had occasion to visit the head office of our leading bank in Tokyo, our financial staff prepares for the excursion with a concern worthy of a papal audience. Indeed, walking in lock step with members of my peer group through the high-ceilinged halls and slowly opening Reception Room doors, I have thought of suggesting to the president that a few Swiss Guards with halberds might not be inappropriate. There is the nervous wait, with subsidiary bank officials. At last, the moment comes when the members of the bank's main office curia enter. They are arranged in order of sacerdotal rank, as are we. Everyone sits in the same straight reception chairs, set precisely 90 degrees to the table. Each talks in order of his rank, through many servings of tea, coffee, and/or soft drinks, until the audience is over—which fact is communicated by half-telepathic exchange of signals among lower-ranking members of the group, who then proceed to marshal the bishops back into the sacristy, while corporate acolytes scurry off to alert elevator operators, door openers, and the drivers waiting below.

The same ritual is observed in visits to Mitsubishi, Mitsui, Sumitomo, Japan Steel, Fuji Film, or the latest chrome-plated P.R. agency on the Ginza. Always there is the Reception Room. Always the introductions. Always the exchanges, which supplement the normal business conversations. Ritual is not a waste of time to the Japanese, whatever foreigners may think about it. It is their way of reminding all parties concerned that nothing is perfunctory or automatic, that they are people as well as corporations, and that people have faces and meet and exchange greetings as well as reports and balance sheets. The meeting is, in its way, a business sacrament. Along with the department store demonstration the illusory wood paneling in the restaurant, and the stationmaster with white gloves, it is a part of Japan's national rubric, an almost religious ceremonial.

This insistence on the forms of life as a sacrament can have its exasperating side, as anyone will confirm who has had to stand in line at a Japanese reception (either as a guest or host), participate in the almost endless gift-exchanging rituals that mark the New Year and the June gift-giving (*chugen*) season in Japanese corporate life (for years I wondered whether there was any significance in the fact that our advertising agency always gave me small bath towels on these occasions), or attempt to cash a traveler's check in a provincial Japanese bank. Many young Japanese, it is true, are slack in their observances and, by classic standards, unbelievably rude in their speech. But as a nation, the Japanese continue to accept and defend the importance of ceremony-as-art in keeping a civilization together.

A certain amount of ceremony infuses even the simplest rituals of the home. A member of the family leaves the house with the words "*Itte mairimasu*," (literally, "I'm going now and I'll be back"). The person remaining at home will say, "*Itte irrasshai*" (literally, "Go and come back," or "I'll be waiting for you"). Few Japanese would think of beginning to eat without the words "*Itadakimasu*" (literally, "I accept the food") which was, is, and will continue to be a kind of ethical-culture version of the old Christian grace before meals. "Bless us, O Lord, and these Thy gifts." It has much the same meaning behind it.

The house itself can also be a part of ritual. Living in an old Japanese house, with its daily ceremonies of opening the wooden shutters in the morning and closing them at night, airing the house, clearing the bedclothes in a *tatami* mat room, and spreading them again in the evening—along with the traditional creaks and groans from shutters, floors, and sliding panels—gives one a sensation like living aboard ship, with the same kind of enclosed security.

It is, of course, true that the paper-and-wood houses are rapidly being replaced by high-rise apartments and municipal housing developments, but many of the so-called nonfunctional elements are stubbornly preserved nevertheless. New lamps are built to look like old wooden lanterns. *Shoji* sliding panels are preferred to Venetian blinds. Where everything is steel or plastic or concrete, wooden panels or plastic made to look like wood are included in the furnishings, to keep up the illusion. It is no accident that some of the most successful new Japanese buildings, like the National Theater in Tokyo, have succeeded because their architects could reproduce in metal and concrete the age-old wooden forms of structures long ago burned and gone.

Tange Kenzo, besides being one of the world's great architects, is one of the few people who has tried to conceptualize this feeling for nonfunctional forms as a necessity of living. The city, he writes, "cannot be thought of simply in terms of 'housing' or construction. For, among other things, the revolution in thought and communication has forced us to think of the city in terms of art and symbol as well." For Tange, the need for art to combine with function is not merely a modern thing, but a feeling rooted in the Japanese past. He is fond of showing the importance of absolutely nonfunctional beams and pillars and other pieces of construction in producing the Grand Shrine of Ise, which along with its counterpart at Izumo, comes as close as one can imagine to the physical idea of what is Japanese. In Tange's view, the nonfunctional elements were as essential as the actual foundations and props and pillars to the success of a building that was trying to express a thought, as well as to give shelter to certain religious observances.

A Japanese cathedral builder would not insist that a gargoyle conceal a waterspout, or that decoration had to be supplied through the effigies of saints. Nor would he resort to the acres of delicate filigree and deliberate ornamentation of the Muslim architect who wanted to decorate, even though the Koran forbade him the luxury of depicting human face or form. What others might have thought ornament, the Japanese regarded as close to essence. The temple would say what it had to say through the way it was built. The form counted.

Who else but a nation of stubborn formalists would have persisted, quite unlike the Chinese and Koreans, in defying the cold and snowy winters of Northern Asia with houses built of sliding paper and flimsy wood, just because they embodied some national memory of the eternal summer in the south, where the ancestors, some say, came from.

To understand this emphasis on form is to understand a basic difference of philosophy between Japanese and Americans (not to mention Europeans) which has survived several generations of intensive technological coexistence. The Japanese stress on ceremony in everyday life expresses far more than a conflict between Japanese formalists and American functionalists. Japan is the world where how-it-is-done seems virtually all-important (or so it appears to one brought up, as I was, in a tradition which held that the important questions to ask were "Why is-it-done?" or "Should-it-be-done?").

Examples of this difference are met constantly. There are, for example, the frequent stand-offs in the Japanese Diet over matters as diverse as the school curriculum and the yen-dollar crisis. A vote of confidence in the Diet is rarely involved. The four opposition parties simply refuse to attend the Diet sessions in protest against the government's not assuming public responsibility, as, for instance, in the latter-named case, for handling the yen re-evaluation better. It is a common tactic. The majority Liberal-Democratic party mediators in this instance proposed a declaration by which the premier admited to "keenly realizing" his government's concern in the matter. The opposition debated whether this wording was sufficiently repentant to gain their acceptance, allowing them to return to the chamber.

One day over lunch, a foreign executive in a joint-venture company confided in me the details of a terrible blow-up with his Japanese partners. When he handed them a letter from his parent company in New York rejecting the Japanese company's proposal, his Japanese opposite numbers were furious—not because the terms of their proposal were turned down, but because the president of the American concern stated his "perplexity" at receiving this kind of proposal. That sort of language represented to them a deep loss of confidence, which was going to take much explaining to restore. It was a bad gesture.

A Japanese schoolmaster I know was in the eighteenth month of protracted discussions in a law suit for damages by a crooked real estate firm. The suit would have been laughed out of court in almost any other country. The real estate "brokers" were trying to get a wholly undeserved commission. In Japan, however, the law on the subject is quite vague. The judge, admitting this fact, has repeatedly asked the plaintiff and defendant to submit to arbitration and compromise their differences, even though the plaintiff has very thin grounds on which to argue. A mediation effort (*wakai*) would be wound up in a matter of weeks. But no one could agree on the right wording of a compromise statement. If my friend trusted to justice and wanted court proceedings, he would probably get it, but it might take several years. In the end, whatever the actual rights and wrongs, some kind of compromise will be ordered.

If a driver injures someone in a traffic accident, one of the principal factors in determining his sentence, if not his innocence or guilt, is the alacrity with which he visits the victim in the hospital to offer gifts and apologies. Japanese judges set great store by such things. The rite of apology is almost as important in the late twentieth century as it was in the early twelfth.

The accent on form is also reflected in clothing and vacation habits. Japanese like to look the part they are playing. If the occasion is a funeral, a wedding, or a formal dedication, the gentlemen will appear en masse in their own or rented cutaways and striped pants outfits. (An irreverent visiting Englishman once suggested that the introduction of a Moss Brothers' cutaway rental outlet in Tokyo might have dramatically rectified the trade balance problems between Britain and Japan.) The ladies will wear *de rigueur* their black kimono with the family crests adorning them. A new Japanese skier will buy out as much of the sporting goods store as his wallet permits, before he ever gets near the slopes. The traffic in golf clubs and accessories shows the intensity of what has become a national craze: In 1972 more than $100 million was spent on golf equipment, domestic and imported, by Japan's six million-odd golfers. In some quarters of Japanese business society, a man's handicap (pronounced "*hahndee*") is almost as necessary a part of the introductory process as his calling card. Yet form is also important. Whether it is a golf swing, a high dive, or the type of pro-style climbing boots used in mountaineering, the ultimate tribute from one's fellows is, "*Kakko ii ne*" ("good form!").

The nineteenth-century British public school product brought the notion of good form to its artificial peak in Western society. In Japanese society, on every level, a basic feeling for good form has been endemic since at least the seventh century and need hardly be taught. The appearance of doing a thing well counts, in some ways, as much as the actual execution or the result. There is a right way of doing things, there is a right time, and a right place.

When planning vacations—and choosing the times to leave the city and return to it—the Japanese conform to the same standard with unconscious precision. Foreigners in Japan are unceasingly surprised at the way beaches are virtually empty before and after the prescribed July-August vacation time and hopelessly crowded during those months. This is not a matter of climate. September and June weather may be balmy and warm. There is just a powerful sense of fitness and tradition at work in Japan's pursuit of leisure, which has little to do with actual enjoyment.

The giving and receiving of gifts, although not now as rigidly prescribed as it was before, is still so widespread a custom that visitors to Japan are amazed, not to say embarrassed by it. ("What a lovely cloisonné vase that Mr. Yamamoto from the sales department brought over. But Bill, don't we have to give *them* something?") Gifts between people in the same company, gifts from one company to another, gifts to members of the family, gifts on coming back from a trip are part of the rite. Many Japanese tourists spend a considerable portion of their shopping time overseas trying to find adequate presents to hand out to their friends and well-wishers, who will of course be waiting at the airport when the plane lands. The higher one's position in Japanese life, the more gift obligations one has. Well-to-do executives in modern Japanese business society, caught in ascending circles of presents, sponsorships [1] and entertainments, sometimes offer a small-scale parallel to the feudal lords of the Tokugawa era, who were kept in a state of semibankruptcy by the necessity of lavishly entertaining the Shogun and his officials, not to mention their own family retainers.

At the apex of these social sacraments stands the institution of

1. Among sponsorships the most common is that of go-between (*nako-do*) at company marriages. The go-between in a Japanese wedding performs a function that combines elements of the American best man, as well as father-of-the-bride. He is traditionally a man of substance in the local community, acquainted with both families. In the old arranged-marriage tradition, he often had something to do with getting the bride and groom together. In modern Japan he is more apt to be a business superior of the groom. He is traditionally available for consultation long after the marriage and keeps a benign eye on the couple or couples he has honored with this commitment. One Tokyo executive, who served as a go-between for almost twenty marriages has to devote a considerable portion of his and his family's New Year's vacation to receiving visits (and presents) from the happy couples.

the tea ceremony. Almost with the solemnity of Christians celebrating the Divine Eucharist, the Japanese for many centuries have practiced the rite of *cha-no-yu*. In it the simple act of preparing tea, serving it to a guest, and drinking it is stylized almost to the point of absurdity. (Among some contemplatives, the tea ceremony is performed and enjoyed alone.) The ritual gestures, studied and acted out like a gavotte in slow motion are designed to foster self-discipline as well as a feeling for beauty. They serve to show, in the most extreme way, how a simple daily action can become of itself a work of contemplation, almost a sacred thing. There are, indeed, strong overtones of Zen Buddhism in this rite. But as normally practiced, it is more aesthetic than religious, a way to exalt the daily life of man without the need to call on God.

Throughout Japanese history, poets and generals, priests and merchants have practiced this cult. Some of the loveliest specimens of classic Japanese architecture are the small, one-room houses (*chashitsu*) constructed solely for the tea ceremony. One of Kawabata Yasunari's great modern novels, *Thousand Cranes*, is built upon the tea ceremony, with the unchangeable rites and artifacts of *cha-no-yu* contrasted to the volatile emotions of some of its devotees. Tourists visiting Japan have to make a positive effort to escape *cha-no-yu*, for the Japanese like to demonstrate its importance to their culture. At the Expo 70 World's Fair in Osaka, the Matsushita Pavilion thoughtfully staged a daily mass tea ceremony for important visitors, in which a platoon of young ladies stirred, bowed, and poured in unison.

There have been some magnificent dissenters. In the early nineteen hundreds Natsume Soseki wrote, "There is nobody as ostentatious or as persuaded of his own refinement of taste as the man who performs the tea ceremony. He deliberately reduces the wide world of poetry to the most cramped and limiting proportions." But dissenters have always been in a minority. The tea masters now teach their rarefied art to the millions. Every department store has a large section given over entirely to the expensive pottery, trays, and bamboo stirrers used in its ceremonies. It is Japan's *agape*, adapted to a mass culture.

Are the formal Japanese trying to tell us something with their devotion to rites and gestures? If so, it is not apparent. In fact, when the average Japanese thinks of American society in contrast to his own—he is apt to envy the lack of restriction he sees, the apparent directness of the American, the American's freedom from the kind of convention which surrounds each Japanese like a cara-pace, from his first days in primary school to the complex social ceremonies attending his demise. The Japanese may, at the same time, look down on what he feels is an American national lack of manners. But he would hesitate proposing to install any of his own codes as a substitute. Probably not least of the reasons for the abiding popularity of Westerns with the Japanese is that the cow-boy heroes can dispatch the bad guy with a six-gun pronto, with-out the obligatory ceremony that Mifune Toshiro must initiate before he can carve up his enemy with his samurai sword.

Conversely, the American generally reacts to Japan's formalism with less appreciation than impatience: "They *are* awfully polite. But it takes a long time to do business with them. Conferences, conferences, cards, courtesy calls; and did the people in the store have to take all that time to wrap a single screwdriver?"

There was a time, it might be argued, when ritual played a large part in American life. Family sleigh bells and the reading of Dick-ens's *Christmas Carol;* the train conductor with the authoritative pocket watch whose, "All aboard" was as formal as the Japanese station master's white gloves; long Catholic Lenten fasts; and the stylized greetings to newcomers at First Methodist. Buying a new car had its own rubrics and so did the long shoeshine on the court-house steps. Social introductions took some time and there was even a certain obligatory ritual to business meetings. Amusement parks were still places of fun, before the pitchmen moved to Wall Street and Washington. The newsstand proprietor folded his papers with a flourish and the conductor on the old double-decker Fifth Ave-nue buses in New York made quite a ceremony of it when he pressed the bell for the driver to move on.

Perhaps such memories are no more than sugar-coated nostalgia. They may have been no more real in their time than the precise white houses on the *Saturday Evening Post* covers. But there is no doubt that what passed for ceremony and amenity in American daily life is increasingly hard to track down in the United States today, a casualty, perhaps an inevitable one, to the stronger pull of the American exaltation of function over form. This headlong progress, broken only intermittently in our history, has ended in a society whose incidence of alienation and spiritual or social con-fusion is almost unrivaled. Early in 1972 the public television net-work produced a program featuring an assertedly typical American family in Santa Barbara, California; its members were democratic, free and easy, unbound by tradition, but, as recorded in their daily life on the screen, conspicuously incapable of holding serious conversations with each other. Extrapolating from this one disturbing example, people made the logical comment on the irony of a society in which the systematic neglect of formalities or conventions, in the priority of ease and communication, paradoxically made the simplest communication more difficult.

Most Japanese families, by contrast, still communicate rather well. Is it because the sum of a thousand little shared formalities in their daily lives keeps up the fabric of shared communal existence, so that disagreements may be endured and worked out, without necessarily swamping the whole family boat? For all their problems—which are immense—most Japanese schools still function as places of study and learning. The school system may be stuffy and overly structured, but it has yet to be transformed into a series of community work projects. For all his own physical and gathering spiritual inconveniences, the Japanese city-dweller, as I have suggested, has a sense of constant participation in his society.

Are the ceremonial hand clappings, the attentive department store attendants, and the basement restaurants disguised to give the illusion of old Edo simply pieces of window dressing? Or do they add up to an impression of belonging—high or low, rich or poor —belonging? Is the ceremony which the Japanese incorporate into their daily lives, and keep there against such strong technological pressures, the human equivalent of the nonfunctional beams and pillars which, according to Tange Kenzo give the Grand Shrine of Ise its peculiar character, like an enlightened sixth sense infusing an animal nature?

A recent popular book, the *Japanese and the Jews*, is, despite its title, actually a shrewd look at what makes the Japanese community tick, by a Japanese disguised as a foreigner.[2] The Jews were chosen as a counterpart to the Japanese because they, too, represent a separate and vigorous cultural tradition which is equally hard to describe. In this book and a sequel, the author coins the word *Japanism* as, in one sense, almost a counterpart of Judaism. By Japanism, he means a kind of overmastering tribal ethic, with

2. First published in Japanese in 1970 as *Nihonjin to Yudaijin*, it has since been translated into English. The author, who calls himself Isaiah Ben Dasan, at first refused to reveal his identity. Although his knowledge of Judaism is evidently more scholarly than experiential, he does know a lot about Japan. In his real-life existence as Yamamoto Shichihei, he has since written several best-sellers.

strong religious overtones, that imparts to most elements of Japanese life a cohesion not generally found elsewhere. In the sense he uses it, the ceremonies and rituals of Japanese life are indeed part of a secular religion. It is, however, religion determinedly non-eschatalogical, a creed designed solely to make one feel useful and well occupied during one's allotted span. In its stress on ceremony and visible ritual, Japanism echoes the formal Shinto worship of land and ancestors which has now largely retreated within its shrines. The gods are gone from Japanism. But without bothering to speculate on ultimate purposes or final destinies, Japanism is capable of dignifying the simplest acts of daily living with a ritual that helps justify them.

Chapter V

The Religion of Giri-Ninjo

"Father," said the Lord of Chikugo, "you and the other missionaries do not seem to know Japan."
—*Silence* (*Chinmoku*) by Endo Shusaku

In the summer of 1549, the first Christian missionaries landed in
Japan. Others soon followed. They were Portugese, Spanish, and
Italian Jesuits, led by St. Francis Xavier, one of the founders of the
order; their zeal and resourcefulness were immense. They made
converts quickly, aided by the corruption of Buddhism at the time
and the generally troubled atmosphere of Japan's anarchic *Sengokujidai* ("the age of the warring country"). At that time, particularly, a passport to heaven seemed more secure than most forms
of sublunary allegiances. At its height, barely a half-century later,
Catholicism numbered about three hundred thousand believers out
of a total Japanese population of twenty million. Although their
knowledge of Christian theology was slim, the Japanese said their
Latin and Portuguese prayers faithfully, built their churches, and
defended them with typical stubbornness. More than two thousand
Japanese martyrs were recorded in less than four decades.

Within a century, organized Christianity was dead, its priests
exiled, hunted, killed, and apostasized.[1] Its last large group of followers, the Catholic samurai and peasants of Shimabara, were
slaughtered after the fall of their castle stronghold in 1637. Their
heroic story lives on in legend. Their faded armor and banners,
ornate Japanese helmets with crosses, and war flags with wispy
portraits of the Virgin Mary are exhibited to endlessly fascinated

1. Endo Shusaku's best-selling novel
of recent years, the subject of a wide-
screen technicolor movie, was based
on the apostasy of Father Christian
Ferreira, the Jesuit vice-provincial at
Nagasaki in 1633.

Japanese tourists at Shimabara, a small coastal town in Kyushu conveniently close to the hot-springs resort at Unzen.

When Christianity was reintroduced in Japan at the time of the Meiji Restoration, the missionaries' hopes were high. But unlike steam turbines, high-speed printing presses, and machine guns, Christianity never really took hold again. At present some six hundred thousand Japanese are professed Christians, evenly divided between Catholics and Protestants. Most of them are quite fervent.

There have been some distinguished Japanese converts. There is one Japanese member of the College of Cardinals. But there are relatively few Japanese priests and ministers. Many churches are still served by foreigners. Even the starkly beautiful metal and concrete Sekiguchi cathedral in Tokyo, although built by Japan's great architect, Tange, was largely financed by contributions from the German Catholics of Cologne.

For Christianity in modern Japan has proved to be that most frustrating experience of all for the missionary and preacher, a social success but a theological failure. The Christian remains on good terms with Japanese life, but apart from it. It is like the advertising agency vice-president who visits a potential client's house for the weekend, is greatly admired for his wit, intelligence, and his skill at tennis and bridge, but fails completely to get the account. Every reasonably well-educated Japanese knows a bit about the Bible, including a few quotable excerpts. He admires its fine ethical precepts, in a rather selective way. Hymns are very popular at Christmastime and movies like "The Bible," "Ben Hur," and the "Ten Commandments" do well at the box office. Some of the Christian "mission schools" have grown into respected universities, like Doshisha in Kyoto (Protestant) and Sophia in Tokyo (Catholic). They have a certain international cachet, which for many modern Japanese outweighs the disadvantage of being thought different. Many Catholic secondary schools are much admired for the discipline of their teaching, intellectually and socially. It has long been fashionable in Tokyo high society to send a daughter for her education to the high school and college run by the Madams of the Sacred Heart. Each summer, romantic Japanese couples ask the priest at the pretty, rustic chapel in Karuizawa, the fashionable summer resort north of Tokyo, if he will marry them in his church. That they are not Catholic, or even Christians, and have no intention of so becoming does not seem at all illogical to them.

The failure of Christianity to establish itself among the Japanese is a good starting point from which to examine the Japanese religion, a body of eclectic beliefs, quite simple in themselves, but difficult for foreigners to understand. Contrary to some superficial analyses, the Japanese do have a sense of religion and morality, as well as an ethic. They have had their philosophers as well as their teachers of ethics. There are words in the Japanese language which can express the ideals of good and evil, truth and beauty. But they are social ideals, not absolutes. And the premises they proceed from are radically different from our own.

From the tribal dieties of the Greeks and Norsemen to the God of Moses and Augustine and Thomas, not to mention Calvin, Western man has envisaged a more or less helpless and imperfect world looked down on from heaven. The God in Goethe's *Prologue to Faust*, exchanging bons mots with a courtly Satan is about as representative a popular depiction as any. Our system of morality, rooted in the ideals of Plato and the Greek philosophers, has historically consisted of either divine or logically perfect absolute values, toward which humankind can strive, but of which only heaven can judge.

American society at present may have consciously thrown all the old absolutes into the disposal. Most of our present moral judgments are highly relativistic and subject to constant shifts (e.g., as recently as fifteen years ago, stealing government documents could have been thought immoral; the "Last Tango in Paris" would have been heavily censored; and male and female college students found living together expelled). To determine what is moral and what is not, our society seems content to wait for the latest court decision. Yet the tradition of centuries has infiltrated our language and our thought patterns beyond recall. We keep striving for the "good," calling for "rigorous moral standards," or seeking "justice," in spite of changing perspectives. Everyone at a country-club party may be an atheist, or a lapsed tree-worshiper, but it can still draw a thin laugh when someone yells, "God sees you, Mary," as she and Fred noisily slip away.

In Japanese such a joke would not be made and few Japanese would understand what was meant, even on the rare chance that they were really pious Buddhists. The Buddha is not thought of as a personal moral supervisor. To the Japanese, traditionally, it has not been a matter of God seeing you, but of your society and, most especially, your particular group, seeing you. The impulse to goodness is the same among Japanese as others, but it is founded on the sanction and the approval of the group. While this is probably also true in the West today in practice, it is not our ideal. A painted God still watches us from the roof of the Sistine Chapel and the mythological figure of justice still blindly brandishes her weights in front of over-used American county courthouses.

In Japan, however, the ideal and the practice were and are the same. Just as, to the Japanese, the only reality is what he sees around him, the only real justice is the considered judgment of his society. This is not to say that the Japanese rejects the supernatural altogether. He believes in magic and chance, somewhat, and ghosts too. Horoscope watching is at least as popular as it is in the United States. Japan has thousands of practicing astrologers, men and women in studios or crouched at little candle-lit tables in the back lanes of cities; a surprising number of people base business decisions, marriages, and other important activities on their judgments.[2]

The Japanese does not, however, believe much in miracles, and never has. He loves nature and the world around him with a passion, but preferably in small, selective doses. The man who cherishes the dwarf tree in the tiny garden and writes poetry about the sweep of a bird's wing (one bird only) or the literal sound of a cuckoo is a man rooted in the world of immediate sense and sound. He is hard put to imagine any other world, whatever the Buddhist scriptures may say about heaven and hell. Whether in the twentieth century, the sixteenth, or the ninth, this attitude has not changed much.

Thus, Christianity, with its active insistence on an other world, affronted the Japanese idea of a completely finite nature—despite its appeal to the downtrodden of Japan's late Medieval era. The sixteenth-century Jesuits' insistence on every man's submission to the spiritual order set the teeth of Japan's rulers on edge. This was to their feudal society, where vassal depended only on lord, a most threatening sort of anarchism. Christianity's intolerance of other faiths, then as now, has also disturbed the Japanese who still have in their homes two altars, the Buddhist *buttsudan* and the Shinto *saidan*, which are most often tactfully set facing in different directions.

2. The so-called lucky days of the month *Taian no hi*, are carefully plotted in advance on the horoscopes. On these days hotel ballrooms are always heavily booked for weddings and other festive occasions.

The central idea of a suffering Christ, a God who allowed himself to be publicly disgraced, then miraculously rose into heaven is inconceivable to them. I once asked a friend of mine, a Japanese Catholic with degrees from Notre Dame and Georgetown, why the Church had never succeeded in Japan. "Christ hanging on the Cross bothers us," he said, "he looks too much like a loser for a God. It was so embarrassing for Him to do that. It is embarrassing even to try explaining it. Maybe our blood is just too thin in Japan. We don't feel comfortable with a God who sheds blood like that."

What bothered this man was not the idea of God-become-man. The emperors have had their brush with divinity over the years. A society which remembers the prewar days when no one was permitted to look down on the imperial presence, even from the window of a house or an office building, is not unfamiliar with at least semidivine reverence. And there is much of the supernatural implicit in the Buddhist belief in the transmigration of souls, which preceded Japanese Christianity by almost a thousand years. Indeed, the Buddha is the classic example of man-become-God; and Greater Vehicle (*Mahayana*) Buddhism in Japan has its pantheon to rival the Christian communion of saints. The particularly disturbing thing about Christianity, however, was the idea of the man-God as victim who, when resurrected, becomes a completely supernatural God-as-judge. There is little room for the Japanese to compromise with a belief like that.

The Japanese sixteenth-century shoguns, when they persecuted Christianity, focused on the symbol of the cross. Christians could escape death and torture by a simple process of apostasizing. They had merely to step on the so-called *fumi-e* (copper tablets with a crucifix on their face) as a sign that they were renouncing their faith. Before Christianity, crucifixion as a punishment had been unknown in Japan. It was adopted, with conscious irony, as a method for torturing and killing Christians.

That Christianity, with its incomprehensible cross and resurrection, succeeded at all in Japan was due mainly to the courage and moral strength of its first priests. Incorrigible humanists, the Japanese were impressed by the missionaries as men. It was the fortitude of Xavier's priests that made converts.

More than two centuries later, in Kyushu, where the missionaries had landed first, a thin, but strong, tradition of native Christianity persisted, a tribute to the example of the Jesuits. Especially in the islands off the southern coast of Kyushu, scores of scattered congregations, the hidden Christians, had kept up the tradition and what they remembered of the observances of sixteenth-century Catholicism. Nagasaki was the heart of this tradition and the Catholic missionaries who returned there in the nineteenth century were well received. It was ironic that in 1945 the second American A-bomb was dropped almost directly on the Cathedral of the Assumption at Urakami, in that city. All the worshippers were killed.

Buddhism has suited Japan's island climate better. There is no embarrassing God-victim in Buddhism. There are no incarnations or resurrections to puzzle and affront the intelligence. Far from seeking to dramatize suffering, the Buddhist strives to eliminate suffering from his consciousness. Following the example of the Buddha, man can become Godlike by his own efforts. While the Christian thinks in terms of grace and sanctity, the Buddhist thinks in terms of awareness.

The Christian has historically opposed reality and tried to reconcile it with his particular vision of the universals. Reality, after all, is merely the view of people with imperfect eyesight and a restricted vantage point, like that of Plato's famous prisoners in the cave. The constant conflict between apparent reality and partly hidden truth is endemic to Christian civilization.

The Japanese have always resisted the dualism between the ideal and the apparent reality. It will, for instance, be extremely difficult for the translator of this book to render the phrase "apparent reality" into Japanese. For most Japanese "reality" is not a vague idea set off by quotation marks. It is what can be seen and felt and that is all. There is nothing else. That is one reason why, although the Christian faith had its moment of glory in Japan, Buddhism was so much more congenial and, hence, permanent. The Buddhist believes that ultimate truth and awareness must come only out of man's insight into the reality around him. The Christian is shocked by blasphemy or despair. The Buddhist is shocked by selfishness.

In its beginnings as an Indian religion, Buddhism, too, had its strong dualities. The classic Indian mind, at the extreme of Western thought and the beginnings of Eastern, sees appearance as something not related to reality, but hopelessly at war with it. The Indian thinkers felt that enlightenment could only be reached by destroying the senses, as it were, or transcending them. This was the basic ideal of Buddhism. The Japanese, even more than the Chinese and Koreans who first instructed them, brought Indian Buddhism squarely down to earth. More tolerant than Christianity, Buddhism by and large accepted inclusion into the Japanese system of religion, and let itself be grounded, so to speak, like an airplane used exclusively for road travel.

When Buddhism came to Japan from China in the sixth century, Chinese teachers had already modified much of its otherworldly furnishings. The Japanese changed the religion even more, to fit their view of life as basically what one can see and feel and immediately cope with. They easily adjusted their own Shinto world of *kami* (local and ancestral spirits) to the *Mahayana* Buddhist ideas of heaven and hell. Of course, Japanese sects and teachers varied widely in what they preached. But heaven and hell remained very shadowy. They were storybook concepts, tall tales for old men and children. The categories of divine retribution and reward, so carefully spelled out in the religions of the West, would never have attracted the Japanese. Of all classic Western literature, the book I can least imagine a Japanese writing is Dante's *Divina Commedia*.

Japan's ultimate answer to the original Indian Buddhist conflict between reality and appearance was Zen. The Zen Buddhists, too, had developed their thought in China. They were the Buddhist sect most influenced by Chinese naturalism and Confucianism. The Zen Buddhist saw no supernatural world outside his own. On the contrary, he felt that every bit of reality is actually the truth. The great thirteenth-century Zen master Dogen said this eloquently for generations after him, "You should fix your heart in this exercise only today in this moment, without losing the light of time."

Between the twelfth and fifteenth centuries, Zen took hold of Japanese thinking. The ideas of ascetic meditation and self-control in Zen impressed the warrior nobility of the Middle Ages as forcibly as the lofty expressions of Chinese Buddhism had impressed the nobility of an earlier day. The Zen masters set great store on emptying the mind of extraneous thoughts and distracting external things in the search for awareness. Nothingness became in this sense almost a positive quality. One sudden insight, they taught, was worth volumes of dialectical reasoning or disputation. Since the concept of pure nothingness is rather difficult for a layman to imagine, it was only natural that most followers of Zen found their consolation in the contemplation of the simple world of nature around them, a nature divorced of wars, cities, and people.

The Zen approach to religion merged easily with the native Japanese love of nature as the only possible link with anything divine.

Thus Dogen and his successors stimulated the desire to express crystal simplicity in painting and poetry which has characterized Japanese art ever since. The great black and white *sumi-e* ink paintings were rooted in Zen. So was, originally, the tea ceremony which later centuries of Japanese so widely made a sacrament, "a religion of solitude," as one of the seventeenth-century Jesuit missionaries had it. The art of spareness which has characterized the best of Japanese culture developed with Zen, as did the concentrated pursuit of excellence in so many things. The "way" of archery, the "way" of fencing, the art of flower-arrangement have deep Zen origins.

One of the great monuments of Zen culture is the rock garden in the temple at Ryoanji in Kyoto. Designed by Soami in the early sixteenth century, it is a rectangle of gravel about fifty by one hundred feet with fifteen carefully placed stones within it. It represents the Zen ideal of nothingness, of purging the mind of nonessentials. The stones and the gravel are the same, but constantly different. In broad daylight there is one vision; by moonlight another. It is left to each visitor's particular insight to see what there is to see.

Japanese tourists and foreigners go through Ryoanji constantly and religiously spend the five minutes suggested, silently viewing the stones. The purity of the insight offered is particularly difficult to attain, even for Japanese, on Saturdays and Sundays, when group by group the guides explain to their charges how the stones are to be viewed. For foreigners set in their Western ways it is sometimes quite impossible. Although to one eye Ryoanji's garden is the greatest piece of abstract art ever fashioned, to another it may be merely a collection of stones. I visited Ryoanji once with one of Britain's most distinguished journalists who had expressed a wish to see this phenomenon during a brief visit to Japan. He fidgeted for about three of the requisite five minutes, then looked at his watch and asked, "Can we go on?"

The Englishman was expressing the same feeling of perplexity with Japanese thinking that the early Jesuit missionaries felt in the sixteenth century when they discovered that the Japanese Buddhist clergy, unlike Christians, were very little concerned with problems such as the afterlife and the soul, which the Jesuits regarded as central. As St. Francis Xavier wrote, after many discussions with the Zen abbot Ninshitsu, "I found him doubtful and uncertain as to whether our soul is immortal and dies with the body. Frequently he would say, 'Yes,' but again he would say, 'no.' I fear that other learned monks are like him. But it is a marvel how good a friend this man is to me." [3]

Buddhism has remained in Japanese life as a vast cultural—religious backdrop, which sets the scene, but rarely intrudes upon it, like the stage design in a period play. Zen has its great temples and its monasteries. So do the other Buddhist sects. The other Buddhist sects, in fact, far outnumber the small handful who live by strict Zen precepts, who surely qualify as the Hassidim of Japanese Buddhism.

One large religious society, the Sokagakkai ("Society for the Creation of Values") has grown up based on the tenets of Nichiren, the one Buddhist teacher who preached a militant religion and aimed at the extirpation of all other faiths, including rival brands of Buddhism. Claiming ten million believers in Japan and many outside the country, Sokagakkai has satisfied a great need for belonging among the uprooted lower middle classes in Japan's new urban populations with a lavish program of festivals, study groups, athletic and social events, and a self-help system based on cells of believers. But Sokagakkai is as much a social, even a political phenomenon as it is religious (it has its own political party), and it is an exception. For the most part, Buddhism in Japan is more passive than active. It rarely admonishes, almost never evangelizes. But it is always comfortingly there.

Every religion changes as it engrafts itself on people of different nationalities. In Europe, first the word of the popes and later, through Martin Luther, the Writ of St. Paul's Epistle to the Romans, were used to justify people's submission to one corrupt prince after another. The European Jesuits in the sixteenth-century mission to Japan, in fact, materially bolstered their position by assuming the titles of the higher Buddhist clergy. But in the matter of hierarchy, as with the tie between Zen and Japanese naturalism, Japan did more to Buddhism than Buddhism to Japan.

Buddhism, over the centuries, was modified to fit the Japanese ideas of proper rank and station. This was no mean feat, since the Buddha's original purpose was to obliterate such social distinctions. But the cooperative vertical society of Japan could adopt Buddhism where the people of Gautama's own India, hobbled by their uncompromising system of horizontally structured castes, rejected it. As Professor Nakamura writes,

3. As quoted in *A History of Zen Buddhism*, by Heinrich Dumoulin, S.J., a fascinating introduction to this subject.

The most significant thing about the Japanese assimilation of Buddhism . . . was the fact that original Buddhist concepts and sutras tended to be altered, in the process of translation into simple Japanese for infiltration among the common folk, in order to satisfy the native fondness for the rank system. The Japanese word *akirameru* (to resign oneself to, to give up) was derived, so it is explained, from the form *akirakani miru* (to see clearly) under the influence of Buddhist thought. The word, however, is used when one gives up a desire that happens to run counter to the wishes of his superior. The Buddhist expression *inga wo fukameru* (to elucidate the cause and effect of a thing) is used when one advises another to give up his desire and aspiration for the sake of his superior. What the phrase actually means is "account for the wishes of the superior." Causal relations came to be explained in terms of the rank system. This rank system, however, the Japanese accepted as a "Divine Gift," and they were aware that their society greatly differed even from the society of ancient China.

The divinely gifted rank system remains the hallmark of Japanese society today. The bureaucracy could hardly function without it and the modern corporation seems made to order for it. Yet it is not a caste system. People far down on the ladder have their own claims to deference and attention—plus the chance to work their way to the top. (An always viable alternative to edging one's way up in the group in which one begins is to break away completely and, say, start a new company.)

The adhesive that binds this society is virtually a social religion: the unwritten, but highly structured, complex of favors and obligations which the Japanese lump under the heading of *giri-ninjo*. (This concept is at the heart of what the author of the *Japanese and the Jews* calls Japanism). The writers of Japanese-English dictionaries struggle endlessly to explain the meanings of both words, which are, indeed, often used in combination to indicate a sense of mercy, justice, and humanity, as well as a sense of responsibility. *Giri* is most frequently translated as "duty," "a sense of obligation," or "justice." But it is equally widely used to justify acts of honorable revenge and atonement. *Ninjo* is basic "humanity" or "human feelings." As the Japanese understand it, this means a strong feeling for the mutual rights and duties of people, as they are bound by ties of rank and group.

In contrast to American society, where people tend to follow group standards but do not admit it, conduct in Japan is governed explicitly and implicitly by the standards and norms of the group. The obligation, for example, to help a subordinate find a house, serve a superior first at dinner, retaliate for a slur cast on one's group, or fill out the team for the company baseball game, are moral as well as social imperatives. Where obligations exchanged inside one's own group are expected, those which one incurs outside the group can be troublesome and are best kept at a distance. The classic phrase *enryo shimasu*, used in situations where Americans would say, "No thanks, really," "It's too much," or "Honestly, I couldn't put you to that trouble," means literally "I want to keep my distance"—and avoid incurring any additional obligations.

Ruth Benedict's psychological Baedeker *The Chrysanthemum and the Sword*, remains the best study of this Japanese web of contracts and commitments ever produced by a foreigner—despite the fact that it was written during World War II as a kind of enlightened know-your-enemy study and that the author had never been to Japan. She outlines the group consciousness of the Japanese, with her memorable distinction between the Western idea of guilt and the feeling for social shame that is even stronger in Japan. She paints a frightening picture of a world where to receive an *on* ("favor") is to incur a debt of gratitude which must be paid. (In fact, the word *on* means both "favor" and "debt.")

It is not wise to read *The Chrysanthemum and the Sword* completely uncritically, for it is a blueprint rather than a photograph. In some ways the book shows its age. It underrates the degree of open competition in modern Japanese society, for example. More generally, an uncritical reading would lead one to believe that the average Japanese is as predictable in his favor-and-obligation reactions as a sake vending machine: You put in the *on*, and out pours the *giri*.

Modern Japanese society has blurred this system of debts and responsibilities. The young, postwar generation does not feel it so strongly as their elders, quite naturally. Yet it does continue to be the key to Japan's social behavior patterns. Almost thirty years after Ruth Benedict wrote her book, a Japanese woman produced a more accurate and contemporary description of this world. The social anthropologist, Nakane Chie, in works like *Japanese Society* and *Kinship and Economic Organization in Rural Japan*, was able to demonstrate in her own exposition of the "vertical society," how remarkably Japan's old idea of village relationships has cushioned the effects of technology and modern urban change and adjusted them to itself.

The centuries-old Japanese tradition of concentrating on the here and now has done more than its share to keep the vertical society a work-happy people—"workolics," as some commentators call them. The Japanese are famous for their long working hours and reluctance to take extended vacations. Work is the tangible, the immediately achievable. Leisure is good, but only as a transient relief. Few Japanese executives, not to mention workers, take anything like a two- or even a one-week vacation. At that, the Japanese traveler, packing hundreds of miles and hours of waiting, queueing, and on-the-double sightseeing into one-and-a-half- or two-day weekends, is exhausting even to observe. At play, he represents an appalling spectacle of man-at-work.

Japanese religion is geared to give the work ethic ultimate sanction. Just as the Japanese assimilated Buddhist thought into their own system of a this-worldly society, in which social sanctions of the *giri-ninjo* system are the ultimate norm, they managed, too, to turn the passive caste-consciousness of south Asian Buddhism into a religion for boosters and producers. Japan was honoring its hundreds of Shinto harvest gods centuries before anyone ever heard of the Protestant work ethic. In a country where winters can be severe, natural disasters frequent, and hard work is a necessity, it was quite natural to mobilize the Buddhist gods, as well as Shinto, behind the ethics of production. "Do your work," the Japanese unwritten scriptures run. "Better yourself and your family as much as possible within your own social work group, abiding by its rules. Leave your young provided for and your old age secure. That is enough of a heaven for anyone to have."

Americans who are aware of it look on this kind of concrete-things-only world view with some envy. But we have too many demons inside of us to adopt it readily. Assured by theologians, psychologists, and gurus that the God who made him and twenty centuries of his institutions is dead, the American continues in a desperate search to "be yourself." Again he looks outward, if not upward, to some ideal. Freedom is still exalted in our increasingly collectivized society, although we are paradoxically assured that you cannot really be yourself and feel real these days unless you get with a group. Perhaps we are still subliminally scared of hell.

The Japanese handle the problems of modern society less gingerly, because their goals have always been more modest than ours. They do not need to search frantically for God-surrogates and substitutes, because they never much believed in heaven to begin with. They do not need, each one, to be yourself at all costs. In fact, that would mean to be *kawatte iru* ("to be different"), to risk setting oneself outside one's group.

Suicide, or the deliberate seeking of death, is the extreme unction, the last sacrament of the religion of *giri-ninjo*. It is an old and respected custom, from the days when samurai disemboweled themselves after a lost battle or a betrayal of the clan's trust (or vice versa) to the hopelessly suicidal *gyokusai* ("glorious death") tactics used by Japanese soldiers and tens of thousands of luckless civilians of World War II. The Japanese language is full of phrases about this which have centuries of gloomy usage behind them: "To vindicate one's innocence in death" (*mi no akashi wo tateru*); "to make amends by dying" (*shinde wabiru*); or more significantly, "to make one's apologies to the world by dying" (*shinde seken sama ni wabi suru*). There is a special word (*shinju*) for the double suicide of two lovers. Chikamatsu's arresting drama *Shinju ten no Amijima* (*Double Suicide at Amijima*), which was recently made into a brilliant film, is one of hundreds on this theme. Students frequently commit suicide over failure in exams, neglect by comrades, or, on occasion, to follow the suicide of a classmate.

Even when they commit suicide,[4] Japanese generally do so for reasons of honor or social concern. This is what Emile Durkheim, the prime student of this gloomy subject, called "altruistic" suicide rather than the egotistic or anomic motives evident in most modern societies.

On November 25, 1970, Japan's flamboyant, restless, and brilliant novelist Mishima Yukio committed suicide, by the ancient Japanese ritual of *seppuku*, first disemboweling himself with a short sword, then having a follower cut off his head. He did so in the office of the commanding general of Eastern Corps Headquarters of the Japanese Self-Defense Forces, after a patently unsuccessful attempt by him and his small, uniformed private army the Tate-no-Kai, (Society of the Shield) to persuade the largely apathetic officers and men of the SDF Headquarters at Ichigaya to join him in a confused kind of military coup. He had certainly no hope of arousing the SDF soldiers. His speech was largely a gesture.

Mishima was then forty-five. He had just finished a novel and perhaps did not expect to kill himself so quickly. But he had been

4. Japanese suicides in recent years have declined to some 15 per 100,000, a high figure, although less than that in France, Germany, and Sweden. But the figure for suicides among the young fifteen to twenty-four is disproportionately high.

depressed about the materialism of modern business Japan and the political and spiritual apathy of its people. Where most of his fellow intellectuals were leftists, who thought the country should turn to some kind of socialism, Mishima was, partly in reaction to this, a rightist. He felt that a return to the traditional values of old Japan, including fidelity to the emperor, represented the only possibility, not merely of giving his fellow citizens some kind of international social self-respect, but of getting back some explicit, living consciousness of Japan's old religious roots.

As a political leader, he was a great novelist. Some called him the Japanese d'Annunzio. His suicide, as a protest, had an electrifying, if brief impact on his fellow countrymen. Confused, erratic protestor that he was, he made his suicide a protest to his society in the traditional Japanese way. If his countrymen did not agree with him, they respected him for it.

"Why did he do it?" Americans asked. "He had everything going for him. Famous, rich, he wasn't alienated." True enough. Even Mishima's quirks—his body-cult, the esoteric prose, the private army of attractive young men—did not weigh against him in a society which gives considerable extra license of conduct to a literary man (*bunkajin*). He did not kill himself out of lonely individual despair, whatever his personal emotional state. He was trying, on the contrary, to make a public statement.

In a world which, for fifteen centuries, has been highly skeptical about the possibility of recourse to divine justice, (few Japanese ever walked off-stage calling down the judgment of heaven on their persecutors) suicide is honor's ultimate defense. The world of *giri-ninjo* is still a tribal world. In the tribal world, which acknowledges neither masses, nor psychiatrists, nor legal absolutions, nor universals, one's honor among one's fellows is of ultimate concern.

In the summer of 1972, during one of the periodic work-to-rule slowdowns for which the Japanese railway unions have become distinguished, the crew of a train carrying the crown prince and princess on one of their periodic tours was censured by union officials for refusing to obey strike orders and delaying the imperial timetable in the process. The censure was tabled almost as soon as it was given. No one in Japan is ready to make much of an issue over the imperial system. Even the Communist party, in its new drive for bourgeois identification, contents itself with perfunctory gestures in the direction of anti-imperialism, like questioning the emperor's trips or the practice of his addressing the Diet when it convenes as too "political."

In the past twenty years, there have been few incidents that call to mind the prewar days, when to look down on the imperial presence from a higher altitude was forbidden, when school principals endangered their lives to drag the emperor's picture from burning classrooms, and a train crew that missed the emperor's schedule by even a few minutes was an object of national criticism. The throne is almost professionally anonymous. Even the widely heralded trip to Europe in 1972, with all its headlines and hours of expensive NHK color television coverage, had an impact on most Japanese city people little more lasting than that of the spring festival stage show at the Kokusai Theater in Asakusa.

Indeed, everything is done to keep the emperor away from the limelight. His devoted corps of palace chamberlains, in the Imperial Household Office, do their best to see that he avoids any action or even appearance that might have the slightest political connotation, or in any way offend some of the electorate. At this work they are most skillful.

Each year, on the emperor's birthday, the palace grounds are thrown open for His Majesty's faithful subjects to enter, bow and write their names in the visitors' register. In days gone by, the crowds stretched as far as the eye could reach. Even after the World War II defeat, during the height of the occupation, the faithful came by the thousands, walking over the neat gravel, past the moated gates where generations of older Japanese had come to venerate, to expiate, sometimes to die. At the moment the crowds have settled down to an average 50,000 per birthday. They are mostly old people from the country, busfuls of the *nokyo-san*, the dowdy, noisy, shuffling groups from the farmers' cooperatives. There are also busloads of children, of whom a certain number are bound to end up their historical tour of Tokyo viewing the palace on the great day. The numbers are not very large, nor are the people particularly reverent. Their presence is sympathetically noted in the newspapers, with the same courteous optimum crowd estimates that are made for the organizers of union struggle demonstrations.

Is the modern emperor a kind of unnecessary appendix to the Japanese body politic or does he remain, whoever he may be, an unprominent, but oddly vital organ, without which the national metabolism would be fearfully unbalanced? Although many people on the left in Japan believe the former and great numbers of the citizenry are passive in their feelings toward the imperial system, no one is anxious to tamper with it. No one is quite sure what the effects of an excision might be. In a country whose current political workings are singularly barren of both personal charisma and rooted political legitimacy, the emperor may indeed be a very, if not a Vitally Important Person.

In April 1973 NHK television produced an interesting, hourlong show on Japan's favorite version of the "whither-are-we-drifting?" theme: "What Are the Japanese?" An NHK poll rated four things in which most Japanese took pride. They were, in this order, (1) economic progress and expansion; (2) native industry; (3) the emperor system; (4) the beauties of nature in Japan. The ratings were 33, 25, 21, and 20, respectively. In a poll taken five years previously, the same four were at the top—then rated 35, 24, 20, and 30. Obviously, pollution had taken its toll of the beauties of nature, in many people's estimation. The emperor's rating, so to speak, had changed only slightly, and that for the better.

Later that year the emperor did break into political debate, in several connections. Nakasone Yasuhiro, then minister of international trade and industry, had said publicly in an interview that Japan has "a monarchical form of government." This brought protests from the parliamentary left, as a possible infringement on democracy and the postwar constitution, which it is their public posture to protect. Most people, indeed, felt that it was in the best traditions of Japanese understatement not to say whether the country was a monarchy or not.

At the same time there were some even on the far left, who looked on the old prewar days when Japan was really an empire, with something oddly akin to nostalgia. In a dialogue published not long before, an anonymous senior high-school radical wrote wistfully, "Postwar democracy in Japan is a fake. During the war soldiers could die with a cry 'Long live the emperor.' But today we don't have a state spiritually united behind the emperor as they did. Our present society is just a set of explosive relations." Which may serve to prove the late Mishima Yukio's point that the far right and the far left in Japanese politics have few differences between them.

It is all a far cry from the 1930s, an era when one of Japan's most respected legal scholars Minobe Tatsukichi (the father of the present governor of Tokyo) could be drummed out of his chair at Tokyo University for merely expressing a widely approved theory that the emperor constituted the "supreme organ of the state." The reason was interesting. Rightists held that to call the emperor the highest organ of a political body somehow interfered with his manifest divinity.

At the end of the war, Hirohito renounced his divinity and assumed the role merely of "the symbol of the state and the unity of the people." He presides at ceremonial functions as the head of state, receives ambassadors, along with every other foreign dignitary, real or fancied, who can persuade the Foreign Office and the imperial household that he is worth a few minutes of the imperial presence. He continues to handle ancient state-religious rites like the imperial rice planting and, on occasion, visits the grand shrines, personally or through family representatives. The Crown Prince Akihito has begun to take some of these burdens over. And other imperial princes, Hirohito's brothers notably, are available almost weekly to inaugurate museum openings, appear at selected company anniversaries or cut tapes at cultural festivals in department stores.

Akihito is an extremely conscientious, tactful man whose principal diversion is a maddeningly consistent game of tennis. The

Crown Princess Michiko, a beautiful commoner from a rich business family, is quite popular among the Japanese. Their children are regarded with something akin to fondness. Yet the emperor of the future is destined to be a man sitting in a half-empty shrine, without either the occasional popularity of European constitutional monarchs or their official orbs and scepters. (The three traditional religious symbols of the Japanese emperor and nation—the mirror, the sword, and the necklace—are still safely in the care of their priests, however.) No more parades or ceremonies, such as those through which the world first knew Hirohito, an uncomfortable spectacled man in a well-tailored general's uniform, walking his white horse before the newsreel cameras. His personality is to be communicated to the people cautiously, with the press ritually running pictures of birthdays, new palaces, or other anniversary celebrations. It is all very underplayed.

The memory of the general's uniform is, of course, one reason for the Emperor Hirohito's low posture after the war. Granted that he exerted himself on behalf of peace, at least toward the end of World War II, and that his intervention was critical, he previously served as a symbol for militarism for two generations. The high-school activists of the seventies may envy the 1.5 million Japanese soldiers who died "for the emperor." But there are many families inside Japan, families who were left fatherless and brotherless, who continue to wonder whether the emperor's writ was worth what they lost.

As a Japanese friend once put it, a man who had, incidentally survived three years in New Guinea in the emperor's service, "He made a bad mistake to go with those generals before the war." The imperial presence at all those military reviews reflected his close contacts with Japan's military leaders. He was something more than a passive by-stander.

Yet the shrine, in a sense, has always been sparely furnished. It is a peculiarity of this ancient dynasty that assertion ill becomes it. Old clothes and a lack of panoply have historically been its best insurance for survival. The single imperial dynasty, although woefully inbred, has never been replaced, displaced, or succeeded since, according to the legend of the Kojiki, the sun goddess Amaterasu-Omikami first sent her grandson down from heaven to rule the land of the reed plains. By his origins and judging from the activity of most predecessors, the emperor is more of a pope than a Caesar. Quite early in Japanese history the emperors were supplanted in actual rule by regents, then Shoguns, and relegated to ceremonial religious roles. Although their political influence was potentially great, it was traditionally best exercised behind the scenes, and by someone acting, variously, as the emperor's regent, agent, or surrogate.

Significantly, the last premodern emperor who really tried to rule as well as reign, Go-Daigo II, caused the nearest thing to a supercession of the dynasty. For almost two generations, rival emperors ruled in the north and the south of Japan, until the schism was finally ended with the Ashikaga Shoguns in firm control. That was in the fourteenth century, oddly enough only a few decades before the Great Schism in Europe, when the papacy was divided for a time between Avignon and Rome.

When the Emperor Meiji came into his patrimony with the Meiji Restoration, he briefly restored Go-Daigo's tradition. As it happened, Japan was well-prepared for Meiji's essay at direct imperial rule. Paradoxically, it was during the peace and enforced isolation of the Tokugawa Shogunate that Japanese scholars and religious thinkers, with much encouragement from the feudal magnates, rethought the vague doctrines of Shinto and turned their ancient ancestral beliefs into an explicit religious-nationalist cult of the divine country with the divine representative at its head. Partly the product of Japan's isolation, the Shinto revival was also in part a delayed nativist reaction against the long primacy of Buddhism.

The Tokugawa Shoguns themselves were responsible for emphasizing Confucianism and Confucian studies in Japan. They very consciously wished to foster a religious set of values that stressed obedience to a hierarchical system more efficiently than Buddhism. Chinese Confucianism led back to native Japanese studies, however. And, ironically, as the Japanese scholars read more, they began to see the Tokugawas as the weakest link in the whole system.

Almost inexorably, the *Kokugakusha* ("scholars of the national learning") went back past the Chinese moral and ethical scriptures, to the study of their own ancient religious texts. Reading them, they developed their theories that the cornerstone of the Japanese polity was imperial rule. It was thus under the banner of a return to primitive religious Japanese nationalism—Arnold Toynbee chose to call it "a resuscitated pre-Buddhist paganism"—that the Meiji modernization received its sanction.

The development of these ideas from the seventeenth to the

nineteenth centuries is still critical to any study of Japan. For, if modern Japan was symbolized by Meiji, it was nurtured under Tokugawa. It was Tokugawa society whose structured ideals and hierarchical value system put such a strong stamp on the Japanese character, a stamp which successive modernizations of the twentieth century have yet to efface. The vertical society may have existed before Tokugawa, but it received its unique conformities during that era. Not the least reason for the revival of the emperor's active primacy was the growing realization among scholars that the Tokugawa system of ordered values lacked real legitimacy. There was weakness at the top.

It is interesting to make comparisons between the pre-Meiji period and what went on in Europe and the United States at that time. Japan's revival of the imperial system made modern Japanese nationalism more cohesive and aggressive than the European variety. But there was more. The European ideas of the divine right of kings began to collapse during the Reformation, when the premises of European Christianity were smashed along with its unities. As the Bourbons and the Stuarts were expelled and the Habsburgs seriously weakened, the divine right collapsed of its own weight. It was succeeded by the rights of man. The Industrial Revolution was advanced on a wave of aroused consciousness of democracy and individual freedoms, which were at least partly divorced from their religious underpinnings. It found its culmination in the United States, the first modern polity to be established with democracy as its base. The very idea of kings was destroyed for Europe during World War I.

In Japan, in a sense, it worked the other way. Industry and the beginnings of popular awakening were both ushered in under the emperor's banner. It was not in the name of revolution, but on the crest of a wave of reinvigorated absolutism that Japan sailed out to meet the world.

In *The Study of History*, Toynbee singled out Japan as a classic case of the conflict between the "Herodians" and "Zealots," which he found to exist in various civilizations at various times. The former, named after Herod, the first-century despot of Galilee, wanted to take over foreign learning. The Zealots, called after the early Jewish Zealots, were the champions of native culture and were uncompromisingly against any foreign importations. Toynbee evidently had second thoughts about the fitness of this Israelite analogy to

Japan, as he himself recognized when he called the Meiji Restoration "a pursuit of the Zealot end by Herodian means." [1] In Meiji Japan for the most part, the Herodians and the Zealots were all part of the emperor's party.

With the death of Meiji, the idea of direct imperial rule receded. Yet throughout their later struggles in the 1920s and 1930s with proponents of democracy, Japanese rightists kept trying to propel the emperor back to center stage. The young officer rebels of 1936 advocated what they called the Showa Restoration, the substitution of direct imperial rule for the corruption of the political parties. World War II, with defeat, finally brought repudiation of this rather Polynesian counterpart of the European divine right of kings.

Now the emperor is alone. The exaggerated reverence and the aura of semidivine inspiration have gone; but in a nation of professed unbelievers, a subliminal papacy remains. No one is anxious to disturb it. The imperial line is not merely the link with the past. It still, in a sense, represents the continuity of Japan's past, present, and future. There is no other obvious principle of legitimacy left in a country whose majority obviously shrinks from any effort to substitute ideology or secular dictatorship or for that matter Jeffersonian democracy as a replacement.

The emperor has now reigned for more than a half-century. He celebrated his 50th Jubilee in 1976, but he is back in his old clothes of long-past centuries. They may be severe and ill-cut, but they harmonize with the bare sanctuary from which Meiji, wisely or unwisely, so boldly stepped forth. The emperor is, after all, something special. He represents something that no other people have. He may indeed reign, for "Ages Eternal," just like the old books said, as long as his posture remains comfortably low and he tries no new clothes on again.

---

1. He added, "The difficulties inherent in the application of our alternative terms, which at first seemed to present so simple a dichotomy, became apparent wherever we turn." Then he went on to other examples.

Chapter VI

Dependence Begins at Home

> Into hell may I fall; punishment by the gods may be upon me.
> I pray nothing but to serve my lord, with utmost loyalty.
>
> —Hagakure

Underlying the admirably structured landscape of Japanese society, like a geological fault which is continually felt but is rarely mentioned by the real estate agents, is a major national failing quite as widespread as the sturdy work ethic, the extraordinary capability for consensus, and the loyalties of *giri-ninjo:* It is a corollary craving for dependence. The urge to depend shows itself in many and sometimes wondrous ways—from the indulgence which the Japanese give to badly behaved small children to the unnerving reliance of big businessmen on their bankers or the righteous indignation of trade unionists when management weakly attempts to discipline them for half-wrecking the plant during a boisterous demonstration. It also appears, from time to time, in Japan's conduct of international relations: in ancient days with China and Korea, in recent times with the United States.

The Japanese have their own word for extreme dependency, which has a simple equivalent in no other language. The word *amaeru* is related to the word *amai* ("sweet"). It means literally, "to presume on the affections of someone close to you." When the Japanese say someone is *amaete iru*, or in extreme cases, an *amaembo*, they mean that the person in question has an excessive need to be catered to, protected, or indulged, but not by just anybody. The person you depend on, the object of your passive *amae* is invariably your senior. He may be your father or your older brother or sister (the case of the dependent member of the family, who sponges on his relatives from cradle to grave, is familiar enough in any society, but more so here). But he may just as well be your section head at the office, the leader of your local political faction, or simply a fellow struggler down life's byways who happened to be one or two years ahead of you at school or the university.

The *amae* syndrome is pervasive in Japan. It is a mark of this collectivistic society that most of its members expect to be in some way taken care of. The minority who do not expect this have a stern and relatively unprotected role in Japanese life, although to put it mildly, an essential one.

This kind of built-in feeling of dependency, the inheritance of centuries of tight, group living is not, however, based on the conscious notion of giving one's due in return. There is no tradition of independent give and take in Japan. The dependent individual, it is true, has a built-in obligation to work hard and do his job as he sees it. But he looks to the man above him to assure his own stability. It is the job of the person over him in the society to sense his needs and attend to them, almost irrespective of the service he performs. At least in an emotional sense, Japan's society has realized quite naturally the Marxian slogan so vainly preached to the Russians and other assorted peoples. From each according to his abilities, to each according to his *amae* needs.

In past generations, whether during the feudal Tokugawa period or the thirties, *amae* represented some hope, at least, for the little man whose options were limited. In the less structured society of today, however, the question of who imparts the security is growing complicated. How does one judge the young Japanese who is vocal about his discontents and insists that he is an individualist, but almost unconsciously expects to receive the comforts of *amae* too? To non-Japanese, in any case, the undercurrent of *amae* among Japanese is an almost constant source of confusion.

If an American department head in a company wants to push the promotion of a subordinate, he is consciencebound, if not dutybound, to justify it on grounds of efficiency. Even when the manager of the sales department wants nothing more than to secure a soft berth in purchasing for a college friend, he has to argue on the basis of efficiency, however thin the argument may be. A Japanese executive does not feel bound by such appeals to efficiency (unless he has been considerably influenced by American-type management courses). Quite apart from efficiency, he will simply point out the proposed promotee's seniority and good attendance record, stressing the precious intangibles of loyalty and sincerity. He will quite possibly also tell his senior that his own preferences and peace of mind are at stake and should be taken into account, not to mention his loyalty. In the best traditions of *amae*, he expects the senior to rate these last-named factors very highly.

Distinguished Japanese businessmen will walk into the president's office and argue passionately for the promotion of a grossly inefficient subordinate on very simple grounds: Mr. Ikeda has worked a long time directly under Mr. Sakai. If Ikeda is not promoted, it is a reflection on Mr. Sakai. Another division head in the same firm contends with equal passion that a department must be put under his jurisdiction, primarily because this arrangement will gain him peace of mind! Provided the company decides his way, the man argues, he would be free of doubt and confusion caused by worry over the pending switch of departments. He could then go on to handle the problems piling up on his desk with redoubled vigor.

An American response to such a request would be quite uncomplimentary—unless that is, you happen to be an American in business in Japan, in which case you ignore this kind of reasoning at your peril. Such requests, based on an awareness of *amae*, have tremendous feeling behind them. ("After Mr. Yamamoto has worked so hard and so sincerely in preparing this plan, how can the president disregard it, even if it is almost unworkable?") A rejection creates great resentment. Indeed, the only parallel is the requests that a small child makes of a parent. "Please make this decision, as I have requested, so my mind will be clear" is a phrase which I have heard more than once in requests of this sort. They are made equally often by old ex-bureaucrats and by young ex-revolutionaries.

Japanese executives take such demands quite seriously. In fact, most good Japanese managers will anticipate them with such skill that they rarely need to be made. Japanese businessmen would have no trouble explaining why they cater to an employee's sense of *amae*. The highest efficiency, they argue, is achieved when the group is working with a single mind. Therefore, even when a course urged on one may be difficult or inefficient, in the long run its acceptance will help the company. Its advocates will be pleased, hence more constructive members of the corporate community, once their request is granted. Their minds presumably cleared of doubt, they will return cheerfully to their voluntary round of ten-hour days.

When such demands for indulgence are denied, the lower decks in Japan can turn surly and become quite dangerous. This is true not merely in business, but in almost any walk of life. For *company* substitute *neighborhood, ministry*, or *faculty*. In recent history this was shown in the most extreme way by the agitation of the young military officers that reached a climax in the revolts of 1932 and 1936. Apart from their role in setting Japan's course toward World War II, the apparent influence of these captains and lieutenants on their superiors was quite extraordinary. They constantly provoked incidents and arranged confrontations in China and Manchuria, as well as in Japan itself, turning quite cross when the generals were reluctant to send the troops marching with the same trusting stamp of their seal normally given to plans for peacetime maneuvers. If the generals did not accede to the young captains' and majors' requests, why should they not be punished? *Amae* has its own logic.

When a modern Japanese business or political leader initials a decision which he feels unwise, largely to please subordinates, then goes off to the golf course, does he do so partly from an uneasy sense of history? Sitting at his desk in one of Marunouchi's gilt-glass skyscrapers, does the company president ever visualize himself behind General Nagata Tetsuzan's desk in the military affairs section of the War Ministry, on an August day in 1935, about to be literally chopped down by a lieutenant colonel from the provinces who felt that he and other lower officers were dishonored because their wishes were not indulged in a dispute, ultimately, about the direction of the army's future expansion? [1]

As Japan went to war in the late thirties and the forties, the defects of the *amae* system could be seen with awesome clarity. From the time of the provocation of the Chinese at the Marco Polo bridge in Peking and Lieutenant Colonel Hashimoto Kingoro's free-lance artillery shoot at the *U.S.S. Panay*, relatively low-ranking officers increasingly took the law into their own hands. Their elders, in both the military and the civil governments, indulged them in a frightening way. To the end, the Japanese officers, even in relatively high positions, expected superiors to cover for them and safeguard their sensibilities, at almost any cost.

During the disastrous Imphal campaign in Burma in 1944, Lieu-

---

1. In the matter in question, it is interesting that the killer, Lieutenant Colonel Aizawa Saburo, expressed considerable remorse for the slaying of his superior in one respect only: He had failed to dispatch the general with a single stroke of his sword. Friends and well-wishers sympathized with his problem.

tenant General Mutaguchi Renya, commanding the 15th Army, was visited by the Area Army Commander General Kawabe, just as his failure to capture Imphal was becoming disastrously evident. With two of his divisions shattered, it was finally obvious to Mutaguchi that the offensive he had so long pushed for should be called off. But he found it impossible to tell his superior that he had lost. As related in Kojima Jo's book on the Pacific War, "General Kawabe would surely have been of a mind to call off the operation, if Mutaguchi had asked him to. And Mutaguchi was many times on the point of asking. But he could not screw up his courage to make the admission of defeat and he deferred to his superior, hoping that Kawabe would take the fateful decision out of his hands.

"Both generals later remembered the tense situation. According to Mutaguchi: 'I had hoped to let General Kawabe understand my true feelings about continuing the operation and felt that he could adequately catch my mood, but I couldn't really express what was in my mind. I really hoped to leave it to him to sense my feelings from my expression.' " [2]

Japanese writers themselves, discussing their prewar and wartime history, have been quick to point out the failings of leaders overindulging their subordinates. Professor Maruyama Masao, one of Japan's most distinguished historians, notes in an essay on the wartime leadership, "the men who held supreme power in Japan were in fact mere robots manipulated by their subordinates, who in turn were being manipulated by officers serving overseas and by the right-wing *ronin* and ruffians associated with the military. In fact, the nominal leaders were always panting along in a desperate effort to keep up with the *faits accomplis* created by anonymous, extra-legal forces. . . . the shots fired at the Liu T'iao River and Marco Polo Bridge reverberated louder and louder until they engulfed the entire continent in war; and the plots of the outlaws were successively ratified by the top officials—as *faits accomplis*, so that they came to represent the supreme policy of the nation.

"This brings us to the phenomenon known as *the rule of the higher by the lower*, which paradoxically became more and more pronounced as the anti-democratic, authoritarian ideology, centered on the military, began to make headway on all fronts." [3]

Some twenty years after writing this prescient comment, Professor Maruyama's own study, along with many others, was rifled

2. *Taiheiyo Senso*, vol. 2. 3. *Thought and Behavior in Modern Japanese Politics*.

by student demonstrators at Tokyo University, in the course of prolonged rioting, strikes, and other disturbances. In the case of Tokyo, the trouble lasted for seventeen months in 1968–69. This same wave of violence closed down or seriously disturbed more than fifty major Japanese universities during this period. The forerunner of similar university disturbances in the United States, Mexico, and Europe, the Japanese outbreaks virtually put higher education out of business for a year.

The most significant aspects of the university crisis were the apparent inability of the university faculties to control their students, as well as their near-paralysis when faced with the choice between either calling in the police (whom many of the professors had been telling their students, for some years, represented the forces of nascent fascist reaction) or letting their classrooms and libraries be blockaded, vandalized, or both. In fascinating succession, many presidents, deans, and professors submitted to kangaroo courts by rude mobs of students. In most cases they confessed to real or fancied failings, after long hours or even days at bay, with many abject promises of self-reflection. Generally they neither resisted nor called for help, with a few outstanding exceptions such as Professor Hayashi Kentaro (who later became Tokyo University's president).[4]

Most of the professors, especially those in political science, could hardly have been considered politically antipathetic to the students' denunciation of American imperialism, Japanese ruling circles, and the like, even though they were responsible for the depersonalized bureaucracy which the students were also, with justice, complaining about. Like the old Japanese Army generals who had sown the seeds of the China incident among the young militarists with their firebrand lectures at the Military Academy, the professors themselves had done much to lay the ideological groundwork for the burning and the barricades on their campuses; and they could, similarly, hardly bring themselves to check the excesses. Like the generals before them, they too were experiencing the bitter fruits of national dependency on the bottom-up ethic, *Gekokujo* ("the rule of the higher by the lower"). It was a classic case of the *amae* system in action.

Most of the students expected, as a matter of right, that the professors, if not the whole society, should be *amai* ("soft") in hand-

4. Professor Hayashi, then dean of liberal arts, was imprisoned in his office for a week, by a mob of left-wing student demonstrators in No-vember 1968; he refused either to apologize or to indulge them in any way, despite constant attempts at intimidation.

ling them. They set out to burn down Tokyo University and barricade whole streets in downtown Tokyo with the same sense of security from punishment enjoyed by the small Japanese male child who cheerfully smashes the crockery in a restaurant under the eyes of his doting elders. Their professors, by and large, did indulge this feeling, far more than did American college officials under similar circumstances. In Japan's compartmentalized society, the feeling that the university should be a world in itself, free from government and police interference, had remained strong. So had the instinctive Japanese tendency to indulge the dependent. Unfortunately, the ability of the university to solve its own problems had concurrently weakened.

Examples of this craving for dependence could be cited here and there in any country. A visiting Japanese sociologist might justly have called the behavior of American students in roughly the same period "*amaete iru*." But only in Japan is this need to be loved and cared for so deeply rooted and so widespread. Well-to-do gentlemen in Europe and America, for example, may have mistresses set up in separate establishments, with varying degrees of formality and acceptance. In Japan the auxiliary wife occupies a far stronger position. Modern Japanese use the term *ni-go-san* ("the number two mrs.") to describe this kind of permanent mistress. Many *ni-go-san* bear and raise one or more children as an auxiliary family, faithfully supported by the father. Without a strong, native sense of being *amaete iru* among the women, and a corresponding feeling of *amae* among the men, this peculiar social infrastructure would hardly be as strong as it is.[5]

Some of the most egregious *amaembo* (literally, "people who crave dependence") are found among Japanese trade unionists. Unlike unionists in the West, however, they rarely resort to open violence. That would bring prompt police intervention. But they regularly take advantage of the seasonal "joint-struggle" [6] periods

5. Shortly after his election, Prime Minister Tanaka Kakuei was taken to task by an impolite opposition Diet member (Communist) for having a *ni-go-san*. It did not hurt him; most of the voters, like the Americans who had championed Grover Cleveland in a similar situation, supported his defense that he was a good provider for the lady and her household. The newspapers, in the best establishment tradition, hardly mentioned it.

6. Every spring, and occasionally at other times during the year, Japanese unions, spearheaded by the national union executives, set aside special periods of mass-meetings, work stoppages, and sometimes strikes, in an effort to make management accede to new proposals for increased wages, better working conditions, etc.

to hold mass meetings during office hours, disrupt work by impromptu rallies, and blockade executives in their offices. Sessions in the so-called *dantai kosho* ("group negotiations") often turn into improvised kangaroo courts. One or a handful of company representatives are surrounded by the entire membership of the union, or groups of unionists working in shifts, and held for hours on end, in the hope that they will finally break down, exhausted, and sign a favorable agreement. Even when one allows for Japan's prounion postoccupation labor laws, the incidence of company representatives calling the police to break up such mob intimidation is remarkably small. Like the reluctance of Japanese professors to ask for police protection against rioting students, this is, to the foreign mind, almost inexplicable. So, too, is the attitude of the union struggle committee members, who feel that such tactics are their right and expect, after the kangaroo courts are over, that everybody will end up smiling.

That, in fact, is what usually happens. The pattern of indulgence toward unruly behavior in Japanese life runs directly from the child breaking crockery unpunished to the struggle committee chairman's violent tactics being ignored, if not condoned, by those in authority. They are *amai*. In justice, it should also be said that most *amai* Japanese authority figures, whether parents, premiers, or company directors, tend to look with contempt on the views of children, citizens, or company employees; and they disregard them when making final decisions.

Japan has very few psychiatrists, just as it has relatively few lawyers. The standards of modern psychiatry grew up in the West. It is equally difficult to analyze the Japanese character by the norms derived from Sigmund Freud's Vienna as it is to work out a Japanese legal agreement by norms derived from Justice Holmes's Boston. The number of psychiatrists in Japan is increasing, however, and they prosper. As they set out to analyze their own society, the phenomenon of *amae* has understandably engaged them.

Dr. Takeo Doi, professor of mental health at the University of Tokyo, as well as one of Japan's foremost practicing psychiatrists, published, in 1971, a fascinating book called *Amae no Kozo* (*The Structure of Dependence*) in which he analyzes the national dependency feeling in rather stern terms.[7] He explains the element of

7. This book recently appeared in English, under the title *The Anatomy* of *Dependence*, in an excellent translation by John Bester.

self-gratification in *amae* with its concomitant reluctance to accept responsibility; and he brilliantly develops his reasons for regarding the concept as "vitally important in understanding the Japanese mentality." "This dependence," Dr. Doi writes, "the *amae* of a child toward its parent, of a student toward his teacher, of a company employee toward his superiors, of a junior toward his senior [in school or university terms] is considered utterly natural in Japanese society. *Amae*, a Japanese would say, is surely something essentially innocent, something indispensible in cementing human relationships."

As Dr. Doi sees it, however, this cementive force is far from innocent. With its roots in the parent-child relationship, the "passive love" of *amae* generates a continual feeling of dependence. A person expects, as his right, to be emotionally helped and supported. As the doctor points out, the person who is *amaeteru* exemplifies the opposite of the Western proverb, "The Lord helps those who help themselves."

The *amae* feeling is central to an understanding of Japanese society. It is closely connected with the Japanese emphasis on shame, more than on individual guilt, in a society where the apology is still all-important and its acceptance almost mandatory. The person with *amae* is constantly, if not consciously, concerned with "the desire not to lose the other's good will, to be permitted the same degree of self-indulgence indefinitely."

The national sense of *amae* displays itself, also, in a basic Japanese sympathy for the underdog, the weakling, the *yowai mono*. This is probably natural in a society where the members of each group are generally assumed to move at a common pace. With certain obvious exceptions (like prewar Japanese Army sergeants booting stragglers on routine training marches), most groups will try to help their weak members, even if helping them reduces the general efficiency. Where the American, if he is in a tight situation, tends to come on strong—"bulling it through" used to be a favorite phrase—the Japanese often tries to appear far humbler and weaker than he really is. "When you get a lawyer," a Japanese legal expert once advised me, "either get a very good one or a very bad one. The very good one will doubtless win your case on sheer ability. Conversely, if your lawyer is quite inept, the judge is apt to help him out, so you may be just as well off in the end that way."

In a more general sense, the very mention of the word *giseisha* ("victim") predetermines a Japanese audience to genuine sympathy. The word has a magic ring in Japan. The victim is, after all, a person whose sense of *amae* has been betrayed and abused. The man who trusted too much deserves protection, just as the man who has overly prospered on people's trust should be put down. Japanese society has its own stern sense of ecological balance. An American, on the other hand, tends to think of a victim more in terms of pity, which is not necessarily a compliment. Where the American likes to call his protest groups "defense" leagues, the Japanese tends to think in terms of victims. The number of victim leagues, unions, and associations in Japan is huge. They range from completely legitimate and tragic victims, like the sufferers from the so-called Minamata disease in Kyushu, who successfully sued the Chisso Chemical Company for polluting their food supply with mercury, to hastily fabricated "victims" groups organized for political purposes, to prevent a building project or to attack a company to which they are opposed.

In normal times, through most of Japan's history, the national sense of *amae* has tended to be inconspicuous. When things run smoothly, the dependency feelings of the average Japanese are quite sated. He is doing his job. The organization of which he is a member prospers. He is respected and, in a sense, loved—or at least he feels that way. His sense of well-being, whether he is a truck driver, a cabinet minister, a coal miner, or an advertising man, plays its own infinitesimal part in solidifying his society. He is emotionally smug, almost.

When trouble comes, man's normal reaction is to want protection. This the Japanese seeks from those above him, whereupon *amae* comes out in the open. Whether he has earned the protection rarely enters his calculations. He feels he is owed it. Desserts, merits, rewards, or failings are considered far less than they are in most Western relationships.

A striking modern instance of *amae* looking for an object to depend on was the behavior of so many Japanese prisoners during World War II. As part of their indoctrination, Japanese soldiers had been taught to fight to the bitter end. Civilians as well as military men were systematically led to believe by the government at that time that capture in wartime was literally a betrayal of one's duty to the emperor. This indoctrination was reinforced by officially fostered stories of American atrocities against captive soldiers and civilians alike. With the whole weight of the establishment mobilized to tell him that surrender was the ultimate shame, the Japanese who was taken prisoner, a few intellectuals excepted, felt that his links with home were forever severed. Indeed, had Japan won the war, that assumption would have been proved correct.

"*Hazukashii desu*" ("I am full of shame"), was almost a universal first statement made to American interrogators. Far from wishing their names to be sent home through the Red Cross, the majority used false names, generally grabbing at the most obvious name they could think of. One American prison-camp officer, taking a roll call of prisoners after the Marianas campaign in 1944, was startled to find that there were no less than twenty people named Hasegawa Kazuo in his camp—until he recalled that this was the name of one of Japan's best-known actors.

Once the shock of capture wore off, the Japanese prisoners not only adjusted to their new situation, they showed a strong tendency to attach themselves, in terms of personal loyalty, to one or more of their principal captors. As I noted in an earlier book, *Five Gentlemen of Japan*, "The ties of respect and subordination which a man had felt for his old company commander, Captain Yamamoto, he might easily transfer, or try to transfer to the prison camp commander, Captain Smith. Members of a clan society, they hoped to join another society by adoption." The Captain Smiths involved were at first puzzled. For, with the transfer of loyalty, went a subtle assumption of a man much in need of *amae* that his needs and wants would be taken care of by this new authority.

This kind of dependence later seemed to be projected, on a national scale, in Japan's postwar relationship with the United States. When the American occupation troops landed in Japan in 1945, they had a lot more going for them than they realized. The complete failure of Japan's first total war had, in a sense, uprooted some eighty million dependent psyches from their ultimate source of security. The emperor's role was publicly reduced and the national structure of Japan (*kokutai*) as children had been taught it in the thirties, shattered. The behavior of the American occupation was peculiarly well fitted to take advantage of this void in the Japanese cosmos by offering a substitute father image, as the Americans, contrary to Japan's fears, set out to rebuild the country instead of permanently enslaving it.

The grand gesturing of General Douglas MacArthur never rang true to most of the troops under him, still less to the American electorate. But it was readily accepted by a people looking for someone to depend upon. Long after MacArthur had gone, a certain national feeling of *amae* toward America continued to grow. Far more than any other country before, America, the people-to-learn-from of the Meiji days, came to fit into the postwar scheme of things in Japan as a logical and worthy object of dependence. The policy of directive-plus-incentive-plus-aid, which MacArthur originally symbolized for Japan, restored and redirected a feeling of *amae* that was all the more difficult to explain because it became so deeply embedded.

How else does one explain the behavior of journalists and academics who consistently take an anti-American position in politics, but at the same time appeal to the American sense of democracy and fair play in their polemics (while awaiting their next invitation to attend a seminar at Stanford or Princeton)? How else does one explain the attitude of Japanese businessmen at the beginning of the seventies, when the Nixon administration declared economic war on Japan? Not only did they refuse to stop flooding American markets with their goods, which is understandable for capitalists making good profits. They also refused for a long time to believe that the Americans would react against them for this, despite conspicuously crude warning signals. How else does one explain the apparent head-in-the-sand attitude of the Japanese government, which waited until disastrously late in the seventies before trying to head off growing, obvious anti-Japanese feeling among the American people, stimulated by the success of Japanese economic competition? The Japanese long found it hard to understand why the United States, pressed by real economic difficulties, insisted on at least equal treatment with Japanese business, (e.g., if a Japanese firm could do unrestricted business in the United States, an American firm should get the same clearance to do business in Japan). For this is not the way an *amae* relationship works. It is the elder, the parent, the teacher, the rich uncle who must give. It is the child, the pupil who must be indulged. Similarly, American political spokesmen, through the sixties and well into the seventies, have constantly asked, even demanded, that Japan make a larger contribution, preferably in the form of military forces, as a *quid pro quo* for the security offered by the American nuclear umbrella over Japan. But the average Japanese, who grew up under an avowedly benevolent occupation, are apt to think of American protection as not a bargaining point, but something that is just there, and justly theirs.

In return for thus being *amaeteru*, it is true, the child is loyal to the object of his dependence. The basic strength of pro-American sympathy in the Japanese people, it must not be forgotten, is more deeply rooted than almost anywhere else. When the United States does something unfriendly or falls into difficulties, the Japanese tend to take the affront or the disaster almost personally, just as Japanese have in the past rejoiced in American successes. When Neil Armstrong landed on the moon, the popular joy in Tokyo was as great as in New York. Watergate evoked a corresponding distress. Yet the elder, by the rules of the game, must be prepared to indulge the dependent's feelings, just as does the professor who welcomes yesterday's student rioter back to the university as if nothing had happened. *Amae vincit omnia.*

Within Japanese society today, the phenomenon of *amae* is more and more obvious, probably because it is less supported (and less disguised) by the direction of strong authority. Both the army rebels of the 'thirties and the first students bold enough to throw rocks at an occupation car early in the fifties had workable and, whatever their failings, highly visible authority systems to revolt against (and depend on). Whatever their faults, both the prewar emperor and General MacArthur could be seen on any clear day.

This is not true any more. Barring the premierships of Yoshida Shigeru and Ikeda Hayato, the Japanese postoccupation governments have grown weaker, not stronger, in their ability to form new policies and lead the nation in the way an American or British administration is prepared to lead its country, which includes making unpopular long-range decisions. Yet the average Japanese, as a legacy of his background and history, tends to expect a strong administration near the top or, at least, a malleable, but believable, supreme authority at the very summit.

Even Japanese feudal society used to rely heavily on an ethical consensus. As Sir George Sansom pointed out, "It is indeed a special feature of the [Japanese] system that rights grew out from below rather than descended from above." A direct line runs from the villagers, with their feelings for the traditional rights to collect wood from a certain district and to bring their offerings to a certain shrine, to the modern trade unionists, with their strong feelings about the sanctity of their meeting halls in the company offices. The Japanese have always maintained a feeling of people's rights, which higher authority should respect.

The Japanese still willingly queue up for a bus or a taxi and would not think of crossing a street against the light. After a while, even veteran New Yorkers transplanted to Tokyo become obedient and wait at crossings. (It took me three years.) But, in present-day Japan, the man with a sense of *amae* is growing increasingly irritable, as it becomes apparent that the basic problems of his society are not being taken care of.

A concomitant loosening of family discipline seems to be taking place as well. One may not wholly agree with Dr. Doi when he writes of Japan, "Paternal authority today has receded entirely into the background." In comparison with American and other Western societies, the authority of the Japanese household still appears strong; the Japanese sense of family, social, and national solidarity seems enviable. Yet father-son arguments are now commonplace, and they sometimes end violently.

It was a sign of the times that, early in 1973, the Japanese courts threw out the hallowed law that parricide, the slaying of a parent, was to be penalized as an especially heinous crime. In the same year, crimes of infanticide and the killing of small children by their parents had increased in Japan to a previously unheard of degree. On the average, thirty cases of infanticide or abandonment are reported in Japan each month, where twenty years ago, despite the hardship of the occupation days, there were hardly any.

During the Tokugawa era, the practice of infanticide, called *mabiki* (literally, "thinning"), was widely practiced by the peasantry. It was an index of the farmer's poverty that he could not feed an extra mouth. Now, however, a senseless kind of *mabiki* happens in the affluent cities. Babies' bodies are found daily in railroad station lockers, occasionally with a scribbled note saying, "I am sorry." Small children are starved or beaten by enraged parents with appalling regularity in the country that has traditionally been regarded as a children's paradise. As Doi notes, the basic reason for this is the number of "childish parents," young people with a strong need to be dependent themselves, who cannot face the concurrent responsibility of providing a stable foundation of care and authority for their own young.

There is little question that the youth of Japan, brought up in a society whose affluence is without precedent in Asian history, expect to be indulged. But they are increasingly less disposed to accept the counterpart of indulgence, the old Japanese notion of responsibility toward one's work, one's group, and one's superiors —not to mention one's own dependents. Older Japanese gentlemen shake their heads at the apparent aimlessness of the young and their vocal discontents. The *Musekinin Jidai* (the "Age of Irresponsibility") is what they call their times. But they offer little hope or inspiration. Worse yet, there are fewer relatives or conscientious neighbors or superiors to offer help. The young for their part, may answer with some justice, "Whom do we have to look up to? Businessmen making money?"

One can readily ask, "What happens now? What happens to a structured society, determinedly hierarchical, where high have traditionally been accustomed to indulging low, when everybody wants to be dependent? Is there a sociological domino theory?"

Sick, I'm telling you, this whole country's sick. They don't pay any attention to women. They ignore you. It's just literal oppression. Why do they take it? I wouldn't take it—would I, Henry? Not for a minute I wouldn't take it.

—Chicago matron visiting Tokyo

Just before the witching hour of 6:00 P.M. hairdos glistening from noontime appointments and stomachs somewhat tight after a late afternoon gossip at the coffee and pastry salons, the ladies begin to file into the cramped dressing rooms in the rear of the gilded bars, cabarets, and the world's most public members-only clubs. It is suiting-up time on the Ginza, time to begin another evening of high-priced illusion for thousands of would-be adolescent boys aged thirty-five to sixty. In place of last night's kimono, which must be laboriously unstitched and cleaned to get rid of six ounces of Scotch and champagne, which a fun-loving, fifty-five-year-old managing director splashed down her back, Hanako has decided to wear a new, long Mori Hanae print. It cost ¥140,000, about what an assistant section head is paid per month in a first-rank Japanese company; but since she has taken home more than that in a good week, and has a rather elastic tax reporting system, she did not feel the price exorbitant.

Hanako's establishment is small, but first-class. It is not so exclusive as Rundell's, where an imperial prince looks in occasionally, but it is less crowded than Club Hana or Le Rat Mort. It is worlds apart from the big, bawdy Crown, the Queen Bee, or the lower-class emporia in Shinjuku and Ikebukuro where a girl would need chain mail to keep her dignity intact after eleven thirty P.M.

Hanako is a hostess. There are now in Japan about five hundred thousand others—full-time and part—like her, staffing the countless places where Japanese men come to drink with their friends and business associates. She is in no sense a call-girl or prostitute, no more than the dwindling number of geisha who still ornament the traditional tea houses in Akasaka or Shimbashi (and who are even higher-priced). But her job is to please men.

From shortly after 6:00 P.M. to a little before 12:30 A.M. she will greet customers, sit with them, take their drink orders, and encourage them to order more. She may drink a bit herself. A little dancing to the piano or, in big places, the large floor-show band; a little friendly banter, as she fends off half-serious indecent proposals with feigned shock; hand-holding, occasional bear hugs, and fitful knee-rubbing [1] are all part of her job. But mostly she listens and condoles, applauds, laughs, and empathizes.

Hanako need not follow the example of the proper schoolmaster's wife in *Tea and Sympathy* all the way and make the ultimate sacrifice, although many girls do, either at one of the hundreds of nearby *abeku* hotels (Japanese-French for *avec* someone, hence a romantic rendezvous) or, more grandly, on a hot-springs resort weekend. Hanako is twenty-three and most attractive. She is extremely careful about how far she extends her charm. (Most of the Ginza bar customers would be shocked themselves if their joking indecent proposals were taken up.) If Hanako or any of the first-rank Ginza hostesses sleeps with a man, it is apt to be the beginning of a long, well-selected, and, for him, expensive relationship, the ultimate cost of which can only recall J. P. Morgan's famous comment about the man who wanted to buy a yacht.

In the end, she hopes to marry or, at the least, emerge as the *ni-go-san* of a rich and very *amai* patron. For the present, however, she dutifully plays the field. At the club, she dodges the too-cozy corner, but cheerfully accepts the random embrace, laughing at her boys' well-heard jokes and congratulating them on their latest achievements, business or social, real or fancied. She is kind and understanding as any boy's mother—and fifty years younger.

Suzuko is a more conventional working girl. In her late twenties, she is a copywriter in an advertising agency, where she has worked

---

1. Sometimes the attentions of the kindest hostess are misdirected. This happened once in my presence, at Tokyo's *Getsu-sekai* ("Moon world") cabaret. An enthusiastic hostess, after a half-hour's friendly nuzzling and exchanging of toasts, elicited an embarrassed comment from the visiting American guest of honor, a distinguished scholar who had been seriously wounded in World War II. "Would you mind telling her somehow," he whispered, "that she's spent the last fifteen minutes pawing my artificial leg."

for eight years. She is good at her job and, on the whole, likes it. There are ample opportunities to meet people and she has money enough (¥160,000 monthly and bonuses, with no family responsibilities) to travel a bit. She is quite attractive and has enjoyed an active social life, which includes several new boyfriends, one spectacularly broken engagement, and two persistent old suitors ("spare-tires" in Japanese girls' slang, always useful in an emergency). She worries, sometimes, about drinking too much.

Like copywriters in New York, Paris, or Sydney, her ingenuity is taxed finding ever-new superlatives for the archetypal housewife to use in the agency's commercials, lauding the perfect detergent, headache remedy, or seasoning sauce. In Japan, too, advertising is usually couched in a family context. It is not enough, for example, for the housewife to cure herself. Generally she finds the new digestive powder and smilingly gives it to her husband on his way to work or watches the children laughing in appreciation at the new whiteness of their blouses or enjoys the convivial slurping of father and children as they try the new instant noddles she has bought.

Grinding out the continuing episodes of Japan's new print and film folklore is not easy; but Suzuko and her colleagues readily share ideas and problems. Her work is appreciated.

This is not to say that she is one of the boys. Although her wage scale is no different from that of men doing the same work (Japanese law has been very strict on this point, since 1946), she knows she has missed becoming an account executive several times. "Able," her bosses will say, "but you know, an account executive has to go out entertaining and drinking with the clients and the media people and you can't expose a woman to that sort of thing. It's too embarrassing for everybody." (They mean the men.) When ideas are called for, she speaks last and increases the polite, honorific content of her language in direct proportion to the originality of her proposal. At meetings, she is always the one who orders the refreshments, or in the absence of a secretary, serves the tea. She half-resents this, although she would feel even more awkward if one of the junior men were to perform this function.

She likes her apartment and is jealous of her independence. When she goes to school or college reunions, she bores quickly at her friends' competitive preoccupation with their husbands' respective promotions and their children's school examination successes. Her best friends are OLs ("office ladies") like herself, who have their own careers and are either spinsters or divorced. They frequently go out on the town together, to a *yakitori* ("skewered chicken") restaurant, or their regular bar, feeling generally closer and more relaxed than when they are with men. But when she chooses, there are plenty of men available to escort her, on state occasions, to expensive go-go establishments like Mugen in Akasaka, or, more frequently, to dinner and a movie and perhaps a tour of the more modest bars in Shibuya.

For about four months she lived with one of her boyfriends, but she broke it off because the relationship seemed pointless. She would not mind marrying, but she has grown very particular. Her mother, whom she sees frequently, tells her she has not much time left to pick and choose. She agrees, but half-heartedly. A girl should be married; but life at the agency is still interesting, even if it is a man's world.

Kazuko, in her early thirties, is slightly older than Hanako and Suzuko, but much prettier. She is also quite bright. At middle school and high school she generally ranked ahead of the boys in her class, and easily passed the entrance examination to Keio University, where she majored in European history with considerable distinction. She was one of the relatively small number of girls (13 per cent) at Keio, a first-line Japanese university, and at one time thought seriously of taking the tests for the Foreign Office. Instead, she married at twenty-four. She lives with her husband, an equally bright economics graduate of Kyoto University, whose prospects are excellent, but whose current salary in a large Tokyo trading company (¥190,00 monthly) is barely enough to rent the half-comfortable, four-room apartment in a "mansion" in Omori, about an hour's commute from his office, and keep up payments on their car. They have two children, small boys.

Tonight, as most evenings, Kazuko's husband will not be home. Just breaking into the lower-middle-management level which does the heavy work in Japanese business, government, and everywhere else, he works well after closing time, checking figures and writing reports. Once or twice a week he goes out to modest bars (you do not visit Hanako's place on ¥190,000 a month) with his contemporaries or immediate superiors in the company, for those long, gossipy, generally frank discussions which ease the in-hours formality of the Japanese company and forge its peculiar social unities. Kazuko does not expect him. He is doing his job, she knows it is hard and there is, after all, the weekend. She concentrates on her two small boys.

Ken-chan, four, is starting nursery school and Kin-chan, the baby, is not yet two. Happy, alert boys, by American standards they are hopelessly spoiled. Since they were born, their mother has been with them constantly, with only two or three brief vacations alone with her husband; then her mother or his came to watch the children for the two- or three-day weekends they were absent. They rarely cry for long. When a baby cries, a Japanese mother almost compulsively picks it up. She does not put her child to bed, she sings to him and rocks him patiently, until his eyes slowly close.

Her manner with them is authoritative, but polite. They are, after all, boys. "Please come here." "Please stop that." "No, no, let's do it this way." Shopping for clothes, reading to them, washing them, bathing with them, planning for their schools, playing, constantly playing with them, she tends to shrink her whole life to the dimensions of the child's world, which thus becomes her own. The social life of couples as Americans know it—getting together for cocktails, bridge, bowling, or dinner—is still rather rare in Japan. Even in the case of intellectual moderns like Kazuko and her husband, it is still rarer after they become families. The Japanese tend to go out with members of their own sex. On occasion, Kazuko gets together with college classmates or women friends from the office where she once briefly worked, while her husband is out with the boys. She spends considerable time cleaning, reading, or going over the family finances, which, by common consent, she controls. But always, it is the children who come first.

Kazuko is a superb mother. Her children are growing up with a sense of security which most of their American contemporaries will only read about in the psychology books. They will never forget the cotton-wool closeness of their early life. At school and college they will miss it; long after they have gone out into the world, it will be a wistfully remembered vision of total protectedness. They will have acquired a permanent feeling of dependence—a sense of *amae*—which they will always try to satisfy. The ear to whisper to, the shoulder to cry on, the hands to caress. It is a large job to handle all of these needs, which explains, among other things, why girls like Suzuko have to bring the tea at the office, and girls like Hanako can earn ¥200,000 in a good week.

Hanako, Suzuko, and Kazuko are three faces of the Japanese Eve. There are thousands upon thousands like them. It is true that most entertainers live on a humbler scale than Hanako. Most of the 19.8 million Japanese women who work for a living have far less interesting jobs than Suzuko—from conventional factory and farm labor to part-time child care and caddying on golf courses. Most of the wives' past educations and future prospects are much lower than Kazuko's. Yet the stereotypes can fit.

Faces and forms vary widely. Japan's uniracial society neverthe-less contains within itself a whole spectrum of physical human types and dimensions. Kazuko has a round, cheerful face with a snub nose and wide, inquiring eyes. She is small and recalls the image of the tiny, doll-like charm lovingly handed down to pos-terity by Pierre Loti's *Madame Chrysanthème* and, in general, every French male visitor who could get his hands on some writing paper. Hanako is rather tall, by contrast, with long legs and the full-breasted figure which is far more common among modern Japanese girls than it was with their ancestors. Suzuko has the nar-row oval face and the slightly roman nose of a classic Heian Era beauty. There are sculptures which look like her in tenth-century temples. Although foreigners might call it samurai beauty, she is not from a samurai family. Her grandfather was a fisherman. Ha-nako does come from a samurai family, albeit one in reduced cir-cumstances. If her mother had not died when she was a child, she would have been safely married off by now, and probably never would have seen the Ginza clubs except in the movies. Kazuko is the best-educated of the three, and is now socially the highest. Her family were not samurai, however, but farmers.

A certain social mobility is a strong point of modern Japanese society. Yet the golden mean is the norm. In most cases, marriage partners come from the same slice of society, whether office fam-ily, neighborhood family, or university family. A good 40 per cent of Japanese marriages are still arranged, most by parents, close friends, and go-betweens, as in the old days. For a really modern type of arranged marriage, the Japanese have invented public com-puter-mating sessions, like the recently popular television show "Gateway to Marriage" ("*Kekkon no tobira*") which, weekly, brought forth scores of volunteer would-be fiancés and fianceés, matched up their cards by computer (educational level, tastes, fam-ily, etc.), and splashed interviews of putative couples on the screen, while parents watched proudly in the background. Watching this program (which was taken from an American prototype, at that) I often wondered what would be the reaction of the earnest democ-ratizers during the United States occupation. Feudal practices with a computer. Is this what their efforts have come to?

The purpose of such arrangements is to avoid mistakes—by see-ing at least that the people's backgrounds and circumstances are reasonably well matched. Although divorce is far more prevalent than it used to be, it retains a stigma of failure in Japan far stronger than in America. Many Japanese young people, like the hopefuls on the computer TV show, prefer an arranged marriage as safer. And if anything goes wrong, they are not to blame.

Women can now, however, obtain a divorce as easily as men. It is a far cry from the prewar days, when a mere accusation of, say, infidelity, was enough for a man to divorce a wife by serving her with a short bill of divorcement—the infamous "three-and-half liner" (*mikudari-han*). (For women, a divorce was far more diffi-cult, if not virtually impossible to obtain.) Alimony and settlements are now as commonplace as they are in the United States, although the amounts are smaller and the American bias against the husband, in children's custody cases, here works the other way around. It is indicative of the Japanese approach that barely 2 per cent of di-vorces ever find their way into the courts. Ninety per cent of them are by mutual consent, the remainder are settled by arbitration.

Putting legal protection and safeguards aside, however, the posi-tion of women in Japan would still seem to be a subordinate one, and far from enviable. But that is by American standards. The Hanakos and Suzukos of Japan—and certainly the Kazukos—are not noticeably protesting their lot. They are said, many of them, to enjoy it. Their position in Japanese society is, historically and actually, far stronger than meets the eye. In a sense, they run it.

In the beginning Japan was a matriarchy. The fact that the leg-endary founder of the country was not a god, but the sun goddess, Amaterasu-Omikami accurately symbolizes the historic depen-dence of the Japanese upon women. Until the Heian period, in the ninth century, it was the customary thing for men to move to their wives' houses when they married. The sophisticated society which the brilliant woman novelist Murasaki Shikibu wrote about in those days was a world in which women not only moved as relative equals to men, but even retained a certain superiority.

As the unity of imperial Kyoto was cracked, then shattered and Japan plunged into a series of civil wars, so was the memory of the matriarchy. By the fifteenth century, Japan was totally ruled by the principle of might-makes-right.

Woman retreated to a subordinate position within the home; but she built up the home, ultimately, as her castle. The familiar pattern of the young wife coming to live with her husband's family, then being tyrannized by her mother-in-law, has been justifiably a favorite of Japanese novelists. Nor is it wholly dead today, in a society where 20 per cent of Japanese families live with either the wife's or the husband's parents.

As the woman drew back into the home, she strengthened her hold on it; and men recognized it as her domain. More than most Westerners can imagine, the parent and child relationship, at least in the early years of life, became a mother and child relationship. Men are indulged, as they have been since the warrior code began to govern society. But, as they are indulged, flattered, and waited upon, they develop a dependence upon woman that is the deeper for its being implicit. Allowing for the normal pulls and tugs of biological urges, the Japanese woman, today as in past history, is vastly more self-sufficient than the Japanese man.

In modern times, women's liberation has been literally a struggle, with its own heroines and standard-bearers. The inevitable growth of freedom that came with industrialization and the spread of Western ideas about the Meiji Restoration, was consciously checked by both the Meiji bureaucrats [2] and the militarists of the thirties. It was not until 1948 that the United States occupation granted equal civil rights to women and impelled the repeal of the codified tradition which kept a woman subordinate to any male member of her family—whether father, brother, husband, or son, it had to be a male who had the primacy in the household.

The number of women going on to high school education in Japan has long been the same as men, if not slightly higher; and by now the percentages of women and men going on to college are equal.[3] Thanks partly to increased educational opportunities and

2. When the newspaper *Hochi Shimbun* daringly hired a few women writers in the early nineteen hundreds, the editors kept quiet about it. "If it were known that women were the authors of the paragraphs read by the general readers," it was explained, "silly prejudice would destroy the effect of the writing."

3. This must be qualified, however. More than three-quarters of the students attending four-year universities are men. For the two-year junior colleges, the reverse percentage is true; women are in the great majority. Many of the junior colleges, in fact, are regarded as finishing schools, where girls go before they are married.

partly to the labor shortage in Japan, women have become doctors, bureaucrats, professors, and businesswomen in numbers their mothers, not to mention their grandmothers, would have thought fantastic. And even more than in the United States, the Japanese woman is the target of the consumer marketing man. The reason is simple: She controls most of the money.

Today in almost 90 per cent of Japanese households, the husband brings home his monthly pay envelope and turns it over to his wife. She doles his pocket money out of that. She decides about the family purchases and does most of the purchasing. The modern Japanese family stays together more than it used to. A revealing sign of the times is the fact that the Ginza bars now are busiest on Friday, not Saturday, so that more men can spend weekends with the family.

Nonetheless, the family is still something of a closed corporation as far as the husband is concerned. So is the rite of birth and upbringing. The Japanese mother still thinks of the baby as something uniquely hers. Courses on teaching fathers how to participate in the rites of childbirth, so popular in the United States, would horrify Japanese, both male and female. The father is not wanted in the delivery room. It is not his job.

Unquestionably the relations between the sexes in Japan have become far freer than they were. But this is not to say that they are becoming Americanized. Men sometimes do the dishes, it is said. And it is commonplace to see couples walking along the streets of Tokyo holding hands, where fifteen years ago this was a rarity. But on company outings the men stick together and so do the women, by choice. Sexual permissiveness is newly allowed to women, by Japanese standards; it has always been allowed to men. But the amount of sexual activity in Japanese schools and universities is still far less than in modern America.

In the same era when a new genre of women's magazines in the United States have brought phallic worship back to its pre-Christian popularity, the most popular subjects in Japanese ladies' magazines still remain fashions and homemaking, mild scandal—in prose, not pictures—the inside stories of movie stars and popular singers, and, most importantly, sewing and knitting. Although dirty ladies' novels from the United States and Europe are appreciatively translated, the hard porno books and movies with which Japan abounds, are left to men. Some beauty parlor waiting rooms, however, feature some rather explicit comic books, and at least one popular Japanese ladies' magazine does include a periodic special section on relevant matters, such as "Chinese sex techniques and power," ostensibly for the readers' husbands. In their novels or on television, Japanese women still prefer idealized, romantic episodes about family life, in which the suffering wife, daughter, or sweetheart saves the men from making utter fools of themselves through a combination of skill, hidden stamina, and weepy patience. For many of the girls, indeed, it is a true story.

The relative segregation of the sexes in Japanese society would certainly be regarded as unhealthy by aware social commentators in the United States. In America, the institution of marriage is constantly called into question. The best aim advanced for wedlock, one hears, is the chance for two people to fulfil themselves and become, as the manuals say, "friends, companions, and lovers." With the Japanese that is secondary. The basic aim of the family is to create descendants and insure their survival. The woman is the custodian of the home and of this mission. She will tolerate a lot from an erring husband to keep her mission going.

If she lives in a society outwardly dominated by male assertiveness, it is a society which she herself has helped to forge. And at its core, the need for unanimity, the tendency to compromise, the imperative to homogenize opposing views, the delicate sense of beauty and harmony in things: all this surely shows a woman's hand. Amaterasu's perhaps. And her faithful sons'.

Chapter VII

A Matter of Language

> Commodore Perry was reminded that Japanese did not act with the same rapidity as Americans did, which was thus illustrated: Should several Japanese meet together, desiring to visit the American ships, one would say: "It is a beautiful morning." To which another would add, "How pleasant it is!" Then a third would remark, "There is not then a wave to be seen upon the water." At length a fourth would suggest, "Come let us go and see the ships."
>
> —*New York Daily News*, June 13, 1854

The Japanese and the Americans share honors as the world's worst linguists. Both nationalities stand out as such, even among neighbors with similar linguistic backgrounds and problems. The classic British traveler, with his stubbornly Anglican French or German, will be able to make his wishes clear in a remote Alpine inn, while the American is still politely fumbling in his Berlitz phrase book. The visiting Korean or Chinese student at Harvard or Columbia will be talking to American friends at parties in workable English, while his Japanese colleague is sitting up late with his dictionary, trying to figure out what the professor said in a lecture he heard two days before.

Yet the Japanese and the American both worry about language problems a great deal. They start out from premises (or prejudices) which one might think would make communication between them easy. The Japanese, as the inheritor of one of the world's most confusing languages, tends to write off any foreigner's efforts to learn Japanese as basically impossible educationally—and, indeed, something of a danger socially. Just as most Japanese families oppose marriages with foreigners, few really enjoy having foreigners master their language. The better they speak it, the more intrusive they seem to the Japanese subconscious. This is partly because Japanese, with its many grades of polite and familiar expression, is such a personal, indeed, intimate language. It is also partly because every Japanese knows his language is very difficult. He is aware from folklore if not from experience how easily foreigners can make mistakes. He is resigned, therefore, to the necessity of using a foreign language himself when communicating with non-Japanese. Few Japanese would argue with the statement of the Yale historian, John W. Hall, that "Japan is the only truly world power which does not have a 'world language.' . . . In many ways Japan remains the most culturally distinct and intellectually inaccessible of the great powers and this fact impedes greatly the exchange of ideas and sentiments with the rest of the world."

The reasons why Japanese and Americans so often are inept in foreign languages are completely different. The American's failure is due largely to lack of incentive. Because so many people in the world, for better or worse, speak English, he feels a minimum need to learn a foreign language. When he does, he can do well. Americans have the advantage of some of the world's most scientifically planned language courses and a tradition of leadership in linguistic scholarship dating from modern pioneers in linguistics like Edward Sapir and Leonard Bloomfield. Yet for all of this, the number of Americans who really learn foreign languages remains small. When we do, out of instinct, background, and availability, we generally turn to French, Spanish, German, or another of the European tongues. The American, however, remains conscious of the problem. He is apt to enunciate loudly and clearly, when speaking English to foreigners, for instance.

The Japanese, on the contrary, have every need and incentive to learn a foreign language. And, for the great majority, English is it. They study it relentlessly. They have been studying it ever since the dawn of the Meiji days, when English replaced Dutch as Japan's basic means of communication with the outside world—much to the consternation of Japan's diligent Dutch scholars.[1] By

1. Dutch was used because of the small, but secure monopoly which the Dutch trading post in Nagasaki held on Japan's official contacts with the outside world, during two centuries of Tokugawa seclusion. Through their tiny window on the West, at Deshima in Nagasaki Harbor, a few Dutch books, notably dictionaries and technical works, found their way into Japan. Until the mid-nineteenth century, a Japanese almost had to learn Dutch if he wanted to read a book on modern engineering or history or philosophy. More than a century before the Meiji Restoration in 1868, the *Rangakusha*, scholars who learned Dutch, had begun to introduce Western learning into Japan. They took their name from the Japanese word for Dutch (*Oranda*).

1859, Fukuzawa Yukichi could write in his autobiography, "as certain as day, English was to be the universal language of the future." The Japanese militarists banned the study of English during World War II. But after the postwar United States occupation began, it was resumed with ever increasing intensity.

American rather than British English is and will remain Japan's basic channel of communication with foreigners. British and Australian accents are generally next to incomprehensible to Japanese speakers of English. In 1974, the private English-language schools, institutes, courses, and tutoring services could be numbered in the thousands, not to mention the normal heavy dosages of English given in schools and universities. English was a required language in middle school, the Japanese equivalent of junior high. The English-language teaching industry in Japan, in fact, did a business with estimated sales of half a billion dollars annually: schools, institutes, tutors, books, records, home cassette study guides, and all.

Yet the results of all this energy and zeal are not impressive. Ironically, in a country where the English-language sign over a shop or on an advertisement has the kind of cachet that French glosses like *Bon Ton Cleaners*, *La Maisonette*, or *Chez Louise* used to have in the United States, the reading knowledge of English is relatively poor. And Japanese do badly, on the whole, at English conversation. They are characteristically hypersensitive about making mistakes in public, calling undue attention to themselves, or committing themselves prematurely to a position which may be wrong. All three of these national characteristics are deadly handicaps to anyone trying to learn a foreign language.

The problem of communicating is made really awesome by the differences between English and Japanese. As a matter of basic linguistic structure, Japanese is related to no modern language, except Korean, which it resembles strikingly. But even with Korean, the resemblances date very far back. There is almost no common ground to begin with as there is, say, for an American studying French or German.

Borrowed words are another matter. The Japanese borrowed words extensively from China and adapted the Chinese writing system to their own needs. There the similarity stops. Grammar, word order, and likely as not, the meanings of words and phrases taken from Chinese originally are totally different, not to mention the Japanese alphabet-type kana syllabary, which the Chinese do not possess. As far as a true understanding of the language goes, a Chinese trying to figure out Japanese has about the same relative advantage that a late Empire Roman would have in learning German or Polish.

English and Japanese are in many ways polar opposites. The Japanese do not care about the same things in language that Westerners are taught to care about. The structure of the phrase does not count so much as the manner in which the words are directed. Japanese is an "I and Thou" language. Its purpose is to establish a feeling of communication between two or more people, and in so doing it employs marvelous subleties to express their relationship. For example, although pronouns are sparingly used in Japanese, there are ten different ways to say "I" or "we," depending on the degree of politeness one wishes or is compelled to use. Verb forms are subtly varied for the same reason: to denote the speaker's rank and relationship to the person spoken to.

Yet underlying all the niceties of phrase and the tens of thousands of sophisticated Chinese-borrowed compounds, the basic structure of the language is astonishingly primitive. The distinctions between verb and noun, for example, are not strong. Tenses are not clearly defined. Adjectives can be inflected like verbs and relative clauses can be vague to the point (for a Westerner) of absurdity. There are no relative pronouns in Japanese; the action is simply pressed adjectivally in front of the subject.

English grammar is rich in classifications and distinctions; it can be precisely analyzed. Not so Japanese, although since the eighteenth century Japanese grammarians have tried hard to do so. Words are thrown into the conversation in a seemingly heedless confusion of nominatives, predicates, and modifiers. Japanese gives far less consideration than most developed languages to syntactical or logical distinctions, as the Western mind understands them. There are neither definite or indefinite articles, for example. A noun can be either singular or plural. The distinction between present and future tense is often elusive—not to mention past and past imperfect. The possessive can be irritatingly vague; for instance, *kashu no shiriai* is translated either as "an acquaintance of the singer," or "an acquaintance who is a singer." It is up to the listener to sort it all out, depending on his relationship to the speaker and the context and intensity of their conversation.

Japanese is not confined in tight-fitting, Aristotelian categories, as English is. As a language, it is rooted in the here and now. It is interested in moods, rather than judgments. It is concerned more with sensibilities than sense. When you use English, you are using a language that constantly makes logical value judgments and invites value judgments to be made in turn. In Japanese, on the contrary, you have a language that is shy of making logical, legal, or philosophical judgments. Japanese did not possess an adequate verb "to be," for example. Concepts like being and reality had to be artificially made up of Chinese-borrowed compounds at the time of the Meiji Restoration. Even the word for *concept* (*gainen*) had to be artificially formulated.[2]

On the other hand, Japanese enjoys a wealth of distinctions in purely concrete matters. Take the question of numerical classifiers. Weak though it may be in singulars and plurals, Japanese has a multitude of words to express enumerations of different kinds of concrete objectives. *Ni* or *Futatsu* means "two;" but two animals are *nihiki*, two birds are *niwa;* two rifles, *nicho;* two drinks, *nihai;* two stones, *niko;* two boxes, *futahako;* two cigars, *nihon*, and so on.

Social judgments and inferences are always important. When Mr. Nakamura says to Mr. Shirai, "*Shirai-kun, sono ken ni tsuite Aoki-san ni sodan shite ano mondai o umaku shori shite kure.*" ("Shirai, talk with Mr. Aoki about that matter and take care of the problem nicely, please."), the Japanese may not be too specific about who is to handle the problem and what problem or problems are to be handled. The language is extremely precise, however, about the authority relationship between the speaker and the other two persons involved.

Of course, sentences like "the sky is blue" or "the man walked across the street" can be translated quite literally from one language to the other, even though basic things like word order are opposite; for instance, "*ano otoko* (the man) *wa* (denotes subject) *michi* (street) *o* (denotes object) *koemashita* (crossed). After that, however, things begin to get difficult. Should you translate "*Matsumoto san ga sono teian o yoku kangaete itadakitai desu*" literally as "I'd like Mr. Matsumoto to reflect on his proposal"? Or, given the situation and the word usage, should you translate it more accurately into American thought patterns as, "I think Matsumoto's dead wrong and he'd better come up with another idea." On the other side of the ledger, how does Mr. Matsumoto himself trans-

2. In his book on prose writing, *Ronbun no Kakikata*, the Japanese sociologist Shimizu Ikutaro makes this and similar interesting observations about the difficulty of adjusting Japanese to modern usage.

late simply a legalistic English sentence like "I'd like to know the precise degree of liability which each of the accused was aware that he had incurred." Japanese lawyers do not have an easy life.

As one of the foremost interpreters of Japanese and English, the scholar and TV-radio commentator, Kunihiro Masao put it, "English is a language intended strictly for communication. Japanese is primarily interested in feeling out the other person's mood, in order to work out one's own course of action based on one's impression." Frequently the Japanese use language itself to convince someone of the ease or difficulty of an action or a problem. Whether the language is Japanese or English matters little in this case. For example, not so long ago, a Japanese businessman working for a foreign company in Tokyo produced a series of incomprehensible English-language memoranda about a particular problem. His foreign superior read them and said, "I'm sorry, but I don't understand anything in these memos." The Japanese smiled with relief, "I am glad you don't understand them. I don't know how this problem can be solved, and I hoped to convey this feeling to you by writing this kind of memo."

Where English is a good language for categorizing and decision making, Japanese does better with protracted discussion. A classic case of the contrast was an ambiguous translation during the 1970 discussions between Premier Sato Eisaku of Japan and President Richard Nixon, at San Clemente, California, in their notably unsuccessful attempt to settle the issue of large Japanese textile exports to the United States. When Nixon explained his problems with the textile imports (which had been offending many of his political supporters in the American South), Sato answered, "*zensho shimasu*," a phrase literally translated as "I'll handle it as well as I can." To Nixon this meant, "I'll take care of it," that is, Sato would settle the problem and find some way to curtail the exports. To Sato, however, it was merely a polite way of ending the conversation. Of course, he would not handle the problem badly. But it was simple politeness to use a nice, forward-looking phrase like *zensho*. To Premier Sato, or any Japanese, this meant only that he would think about the situation and try to cope with it in as graceful a way as possible, without necessarily making any decision.

Few other nations have suffered so many serious diplomatic problems over the translation of words. Much has been written about the confusion caused during the last-minute negotiations between the State Department and the Japanese envoys in Washington just prior to Pearl Harbor, because Ambassador Nomura Kichisaburo probably misinterpreted the diplomatic politeness of Cordell Hull (the American version of *zensho*, one might say) as a genuine offer to negotiate further. A war later, at the end of July 1945, what slight chance remained of averting the bombing of Hiroshima and Nagasaki, and the subsequent Soviet entry into the war was ruined when the Japanese news agency Domei mistranslated a key word *mokusatsu* in Premier Suzuki Kantaro's reply to the Potsdam Declaration as "reject" instead of "ignore."

The slightest shift in a Chinese character can change meanings radically. The original Japanese translation for the United States containment policy toward the Soviet Union in 1947, for example, was *sekitome* ("checking, stemming"), which correctly interpreted what at least started as a defense policy of reaction against Soviet aggrandizing in Europe. Later, Japanese newspaper editors, pushed by their "progressive" reporters, quietly changed the word to *fujikomi*, which has more the meaning of "trapping" or "confining." In the same way, the word used by *Asahi Shimbun* editorialists for the United States military intervention in Cambodia, *shinnyu*, meaning "invasion" had a far harsher meaning than a similar word (also pronounced *shinnyu*, but meaning "advance into") when used to describe the Soviet invasion of Czechoslovakia.

English has a clear distinction between affirmatives and negatives. English speakers tend to express themselves in declarative, flat statements which, according to the roles of the English language, should be either accepted or contradicted. Japanese, on the contrary, shuns flat commitments and dotes on multiple negatives. A Japanese response to a question rarely is either 100 per cent acceptance or 100 per cent contradiction. One prefers to examine the proposition discussed and suggest various approaches that could be made to it.

"It isn't that we can't do it this way," one Japanese will say.

"Of course," replies his companion, "we couldn't deny that it would be impossible to say that it couldn't be done."

"But unless we can say that it can't be done," his friend adds, "it would be impossible not to admit that we couldn't avoid doing it."

Such a conversation, oddly enough, can be reproduced in Japanese with ease and naturalness. It may even result in a positive course of action.

Japanese habitually understates. English habitually overstates.

The comparison of a small Japanese thirty-one-syllable poem with one of Shelley's or Walt Whitman's or, for that matter, one of Robert Lowell's or Allen Ginsberg's is informative. In a more prosaic situation, a Japanese lady will ask politely, "Do you think the decorator's taste really fits this house?" She actually means, "What awful taste. Can you really stand having him on the job one more instant?" A Japanese referring to himself, when he invites someone to dinner says, "Please forgive the poor quality. It's really nothing at all." He does not mean that. It is just by way of putting oneself in a humble position from which the Japanese expects to be rescued by his conversational partner. "No, no, what a magnificent meal!"

American English, of course, has its own kinds of eccentricity and overstatement. I remember a genial old Russian gentleman who worked for an American news magazine and spent his spare time collecting American speech foibles. When I left once for a trip, I said, "I'll be seeing you." He stopped me with a mischievous smile and said, "When an American says, 'I'll be seeing you,' he really means 'I'll not be seeing you,' doesn't he?" The thought had not previously occurred to me.

The American use of polite hyperbole is lavished on any situation, but always rather casually. ("What a present. You shouldn't have done it."). Japanese hyperbole is ritualistically designed to make the other party feel good and to diminish one's own part in the proceedings, even though the real intention may be far different. In fact, *ingin burei* ("polite rudeness"), a Japanese word in considerable currency, denotes the extensive habit of using very honorific language to put someone down or to insult him. The honorific complexities of the Japanese language are such as to discourage the Japanese themselves, who are condemned to live from cradle to grave within a framework of verbal pretense and affectation. Thanks to the long-established codes of etiquette, there are fixed forms of polite circumlocution, for even the simplest of functions.[3]

3. It was inevitable that the Japanese take the various American circumlocutions for toilet, for example, and run them down the drain. The basic words for toilet, like the Chinese *benjo*, and the Japanese *go-fujo*, (literally "the unclean place") are now frowned on in polite society, as their counterparts are in the United States. Most Japanese now use the English adaptation *toire*, or the native *otearai* ("washroom"). Sophisticated places like Tokyo's Hotel Okura now label the men's rooms "*Shinshi no kesho shitsu*" (literally "gentlemen's powder room").

Letter writing remains Victorian in its conventions. A reminder to pay a bill or an announcement that a new man has joined a bank's board of directors are equal occasions for the obligatory opening remarks about the rude necessity of imposing on one's time or the change of seasons. "Now that spring promises to come," the letter begins, "the first buds of the cherry blossoms can be descried climbing the far-off mountain. Although we hesitate to break in on your very busy schedule, troubled as you are with many worthy pursuits, it is time that necessity compels us to remind you that your firm's indebtedness . . ."

Many small companies still consciously employ one man as a letter writer. In one firm I knew of, the letter writer was an aging gentleman, nominally in charge of the personnel section, whose practical use to the organization in this function had long since been evaluated as counterproductive. Yet he was the only person out of some two hundred who had the necessary background to write one of those polite letters in which, given the vagaries of Japanese society, the slightest slip of grammar could lose the company several prestige points. Mr. Imamura was also a master at handling complaints and apologizing. Most Japanese groups contain within them an apologizer, useful for occasions when a soft, well-spoken answer can turn away a lot of wrath. Since Imamura's company sold household appliances, through the agency of some one thousand fast-talking salesmen, the old man often had some very angry customers to soothe. He rarely failed. Bathed in the polite polysyllables of Imamura's gentle phrases, relaxed by the utterly ambiguous wanderings of his honorifics, refreshed by the flattering balm of his concern for the other party's health, living arrangements, and future familial prosperity, the most irate complainant would generally go away happy, his order uncanceled.

For all of its complications, Japanese is one of the world's richest, fastest changing, and intellectually stimulating languages. Starting out as a primitive island tongue with some still unproved suggestions of Malay-Polynesian antecedents, it is classed with Korean as a member of the Altaic family. First through Korea and then directly, the Japanese borrowed heavily from China over a period of at least five centuries, much as the raw European tribes of the early Christian era borrowed and then swallowed up the vocabulary, along with scraps of the stored knowledge of ancient Greece and Rome. Just as Latin came into the European languages almost inseparably from the Christian religion and Greek philosophy, the Chinese language came to Japan along with Buddhist theology and the Confucian way of life.

Yet it is too facile to equate the Japanese use of Chinese with the English reliance on Latin. Even the parallels are far from exact. Albeit removed on its island, English grew up surrounded by diverse European influences. The Saxons imposed their language on the Romanized Celts and in turn, took admixtures of Danish, German, and, after the conquest, heavy dosages of Norman French, which already owed much to Latin itself. The directly borrowed words of Latin and Greek origin had several layers of precedent to build on.

No one imposed anything from without on Japan. There were no parallels to either Anglo-Saxon invasions or Norman conquest, not to mention anything like a Roman occupation. The first Japanese scholars realized themselves that their rough, if complicated, language was inadequate to express the subtleties of the newly imported Buddhism or to describe the cultural niceties of the visiting Chinese and Korean artisans. Above all, they had no native system of writing. So they adopted the letters and the learning of a highly sophisticated Chinese language, dialects, characters, and all, which was structurally poles apart from their own.

To accommodate Chinese to their own language structure, the Japanese developed two parallel kana syllabaries which now express some fifty-one sounds and which serve as a phonetic alphabet. (The exact number varies, typically, depending on whether some old forms are used or not.) For the English analogy to be valid, one would have to imagine the native Celts, still unable to express themselves in Runic script, suddenly embracing an importation of Latin and Greek lexicons in manuscript, hundreds of Church Latin psalters, and the works of the Athenian dramatists in Attic Greek. Nor was there any Norman army to enforce conformance to a new language.

Such borrowings from the Chinese took place at different times in Japanese literary history. The first waves of major borrowing had been incorporated into the structure of the language by the time the great tenth-century collection of poetry, the Manyoshu was written; its imagery reflects the subtle cultivation of T'ang Dynasty Chinese civilization. In the early eleventh-century novels of Murasaki Shikibu and Sei Shonagon it is used to the full. The last great borrowing from the Chinese was made, purely for convenience's sake, at the time of the Meiji Restoration, when words for modern devices like the telephone (*denwa*) were manufactured—*den* ("lightning" or "electricity") plus *wa* ("talk")—from Chinese character roots in the same way that Europeans devised their modern scientific terms (like telephone) from manufactured Latin and Greek.

Chinese is not, however, the only source which the Japanese language has drawn on. Portuguese and Latin words arrived with the sixteenth-century missionaries. Some of them are still current, like *bateren* from *padre* for old-time priest and *tempura*, the now famous batter-fried fish and vegetable dish, from *tempuro* ("seasoning"). The Dutch left behind words like *biiru* ("beer"), *koppu* from the Dutch *kop* ("glass") and *buriki* from *blik*, meaning a sheet tin roof. Student and medical vocabularies still show marks of nineteenth-century German. *Arubeito* from *Arbeit* is used everywhere for a part-time student worker and *Karute* from *Karte* for a patient's record in a hospital. But from the pre-Meiji days to the present, it is English that predominates; since the occupation, American English exclusively. *Restoran* ("restaurant"); *infure* ("inflation"); *torankii* ("tranquilizer") *besu appu* ("base-up," "salary raise") *midoru* ("middle management"), *doraibu in* ("drive-in") *hisuterii* ("violently neurotic")—the borrowings from English wend their way through every section of Japanese life. And their number increases.

Chinese, of course, has a more organic relationship. The sophisticated Chinese compounds and the native Japanese idiom have lived for almost fifteen centuries within the same house. Combined and distilled, as by the poets or in the great novels of more recent times, they can express subtleties of thought and meaning which have no parallel in any Western language since ancient Greek, in which the use of grammatical particles, not dissimilar to those in Japanese, could convey extraordinary shadings of expression. Reading the works of Mishima, for example, who loved difficult characters and richly archaic turns of phrase, is a little like tackling Theocritus in the original.

But listen instead to a couple of Japanese friends describing what went on at a party the night before and the language becomes almost violently primitive, laced with onomatopoeic phrases, defying translation. It is like the conversation of U.S. Marine Corp sergeants with its supportive use of pungent Anglo-Saxon words. This reversion to a more primitive expression is partly a necessary supplement to a language which relies so much on the visual symbol for its meaning. As Sir George Sansom wrote in his epic *Historical Grammar of Japanese*, "The ideograph itself is so tersely expressive that its users are apt to rely on the visual appeal of symbols rather than the aural appeal of words."

There is basically something very different about seeing a language in terms of pictures and seeing a language in terms of lines. The European languages are linear and intended essentially to get between one idea and another as quickly as possible. They are simply used as vehicles for conveying ideas which exist independent of them. Asian ideograph or picture languages are not like this. The picture itself means a great deal. When you are trained to think in terms of pictures, when you have grown up to speak in terms of drawing rather than conceptualizing, you are less interested in exactitudes and more interested in flourishes. You are less interested, so to speak, in how quickly a line goes from point to point; but more interested in how well it is drawn.

In the spring of 1946, swimming high on the tide of optimism that characterized the early days of the United States occupation, an education mission, twenty-seven strong, arrived in Tokyo from the United States. The American educators in the mission were genuinely shocked at the vast amount of time given over to the study and practice of Chinese characters in the Japanese schools. One of their first recommendations in the school democratization program was to substitute the Roman alphabet for the characters: "It is recommended that some form of Romaji be brought into common use . . . bringing about a more democratic form of the spoken language. . . . The need for a simple and efficient means of written communication is well recognized and the time for taking this momentous step is perhaps more favorable now than it will be for many years to come."

Needless to say, this well-meaning effort never got off the ground. The Japanese language is so rooted in homophones, it is hard to see how any way other than a system of codified picture writing could maintain the distinctions. Consider, for example, the following meanings of the word *senko*, as found in *Kenkyusha's Japanese English Dictionary*, which I offer without further comment:

senko 千古 n. (E) remote antiquity; all ages. 2 eternity. . . .

senkō 先考 n. (E) one's deceased (=late or lamented). . . .

senkō 染工 n. a dyer; a dye-works hand.

senkō 船 工 n. a shipwright; a ship-carpenter.

senkō 線 香 n. a joss stick; an incense rod. . . .

senkō 閃 光 n. a flash; a glint; . . .

senkō 戦 功 n. military merit (=exploits). . . .

senkō 専 攻 n. special (=exclusive) study research; . . .

senkō 潜 行 n. (L) 1 [traveling in disguise. . . .

senkō 潜 航 n. submarine voyage; navigation under. . . .

senkō 選 考 n. selection; evaluation. . . .

senkō 詮 衡 n. choice; selection; screening. . . .

senkō 旋 光 n. polarization; optical rotation. . . .

senkō 跣 行 n. (L) -suru v. go barefoot. . . .

senkō 先 攻 i. batting first. . . .

senkō 選 鉱 n. concentration (=dressing, . . .

senkō 洗 鉱 n. ore washing. . . .

senkō 鮮 紅 n. scarlet.

senkō 穿 孔 n. perforation; boring; punching; rupture. . . .

senkō 銑 鋼 n. pig iron. . . .

senkō 先 行 n. preceding; going first; walking ahead of. . . .

The only way to keep one's sanity in this kind of situation is to visualize in terms of ideographs. The process is undeniably cumbersome, which explains why so many sophisticated Japanese, in the
course of daily conversation, are constantly drawing characters in
the palm of their hands, to explain to fellow Japanese what they
mean.

Although some Japanese dictionaries of Chinese characters have
listed more than 40,000, in actual practice the number of characters
known to a Japanese college graduate is about 5,000 (and growing
fewer every day, too, thanks to relaxed standards.) Even encyclo-
pedias think in terms of no more than 10,000 characters. In 1946,
the Education Ministry prepared a list of 1,850 basic Chinese characters which should be thought essential. (Most proper names and
place names were not included, however.) Of these 1,850, exactly
881 are to be learned in the first six years of schooling. This was

probably as close to a reform of the cumbersome written language as the Japanese will ever come.

The younger generation is predictably not so strong in its knowledge of characters as its elders. Recent tests showed however, that even complicated characters, as long as they were shown regularly on television, had a wide degree of comprehensibility. Another result of television, not wholly injurious, has been to standardize yet further the pronunciation of the national language. This standardization process has been going on since the educational reforms of the Meiji Restoration. Despite the complications of its written language, spoken Japanese is probably the most dialect-free major language in the world. Although the older people, especially in the north and in the extreme south, still speak with local *namari* ("dialects"), the standard Tokyo pronunciation is stamped on the country far more extensively than is the standard English in either Britain or the United States. The insular commitment of most Japanese to this principle is reinforced by their stern sense of the fitness of things.

Foreigners are not expected to speak or understand this complicated language. Even the best Japanese, if spoken by a foreigner, produces something akin to shock in its hearers. Traveling in the country, I have often stopped a farmer or a workman and asked directions like, "If we take the next right turn and then go left, can we get to Matsumoto?" The farmer will look at me uncomprehendingly, say instinctively in Japanese, "No, you have to take the second left, after you pass the city hall," then add with a hand waving across his face to denote total bewilderment, "Engurishi no speeku."

The phenomenon is not limited to the countryside. Not long ago, I went to buy some tickets to a James Bond rerun at one of Tokyo's downtown movie palaces. I noticed a sign in Japanese saying that there were unreserved seats only in the first few rows, so I asked if any of them were still available. The girl at the window froze and said nothing. When my wife, who was with me, repeated the same question in better Japanese, the girl remained mute. I just was not supposed to have been able to read the sign. No problem. I have had the same experience hundreds of times.

Admittedly, the fumbling efforts of foreigners to speak Japanese (including my own) are generally difficult to understand. Perhaps the caution of most Japanese in dealing with foreigners' Japanese is inevitable. But there is a certain amount of morbidity in the way they handle these attempts at communicating with them. Away from the small group of people-who-know-him, the Japanese-speaking foreigner is apt to feel disembodied, not quite a piece of ectoplasm, but certainly not entitled to flesh-and-blood communication. Perhaps some special insignia, like a yellow star, should be given to those foreigners who can be certified as speaking intelligible Japanese. It could be carried on their suit jackets, as a warning.

Most foreigners, of course, hardly try. A general apathy about learning the language of this host country is prevalent among foreigners resident in Japan. Americans are particular offenders, although the Europeans do not do much better.

In the early occupation days, when I first came to Japan, there existed, thanks to the accidents of war, a rather sizable group of Americans who knew something of the Japanese language and were not bashful about practicing it. One might have predicted that the several thousand Americans trained in Japanese by the army and navy during World War II would have continued to develop this language and impart some curiosity about what they had learned to those at home. Develop it they did, and thereby produced a rather remarkable generation of Japanese scholars, with some diplomats and businessmen as well, but few others followed them. I have yet to notice any conspicuous upsurge of younger Japanese scholars to take the place of our wartime generation, although of late there has been a slight increase. As of 1973, barely 5 per cent of American universities have Japan or Japanese represented in their curricula in any significant way. There are about five hundred American scholars working fulltime in the field of Japanese studies at American universities—a number far larger than that in other countries; for instance, the Federal Republic of Germany, fifty-nine; the USSR, twenty; Australia, thirty-three—but far, far less than the American connection with Japan would seem to warrant. There are probably less than one thousand Americans who can handle the Japanese language, written and spoken, as a working tool. There are less than one hundred Americans who could write passable Japanese prose.

I lived with the American community in Tokyo first in 1945 and 1946 as a navy officer, later in 1949–50 as a journalist. Then I went away. When I came again to live in 1966, I was surprised to find that the proportion of Americans who are proficient or even vaguely aware of Japanese seemed to have shrunk.

The average American who comes to Japan to live is at first disposed to give the language a try. He begins either by going to one of the crash courses for a month or two ("after eight weeks of excruciating toil, you, too, can speak . . .") or by hiring a tutor who will come to the office at specified times. It sounds simple and appealing: Just forty-five minutes three or four times a week and you will soon know what all those people around you are talking about. By about the third or fourth month, generally, discouragement sets in. It is fueled by the comradely consensus of one's peer group at the American Club, the Press Club, the Keyaki Grill at the Hilton, or whatever extraterritorial oasis one frequents, that "the language is just too damn hard."

"I'll tell you, Harry, I have this teacher and I have been trying these phrases, but I can't make head or tail of it. It's all 'mooshie mooshie' to me."

"Fred, when you try to use it, they don't understand you anyway. Besides, my boy Fuji, in the office, speaks damn good English. He tells me it's really bad manners to try using the language, because it doesn't look good for the boss to make mistakes."

"You know, George, that's a point. I hadn't thought of that."

"What the hell, fella, as long as we know enough for the bathhouse?"

And so it goes. Similar exchanges occur between resident American women after the Thursday morning ladies' doubles at the tennis club, at the meeting to plan the annual Print Show of the College Women's Association, or *sotto voce* at the various social clubs which keep the foreigners' wives out of trouble, cutting up paper and watching flower arrangements.

After the four-month drop-out, the language is rarely tried again. Perhaps when the businessman is reassigned after his first two-year term, he starts to think he might better give it a second try. But then he shrugs and joins what has come to be known as consensus number two: "Well, of course, it's too late. If I'd known I was going to be here for five years at first, I would have worked on it, but after all, Mr. Nakamura speaks English better than I do."

The extent to which foreigners in Japan are thus the prisoners of Japanese interpreters can hardly be understated. On occasion, the interpreter is skilled enough in both languages, as well as in the study of human nature to do his job most effectively. A Canadian sales manager of my acquaintance was wont to tongue-lash his Japanese employees when they failed to deliver their order quotas. Face purple and arms flailing, he would begin, "You are stoopid.

Stoopid, I tell you. How you people call yourselves salesmen I dunno. You don't know the first thing about sales. You've paid no attention to anything I've been saying. I'm surprised that you can all stand there and dare to look me in the face with your lousy records. Now what do you plan to do about it?"

As rendered into Japanese by his resourceful interpreter, who was a salesman himself, this exhortation came out somewhat as follows: "While I have the highest regard for your basic abilities, I would not be completely honest if I did not permit myself to express my disappointment at your recent performance. Despite your fine efforts, the final results were far less than we had hoped for. You should now reflect on your performance."

The effect was the same. The salesmen were quite aware that they had done badly. The apoplectic countenance of their foreign leader was its own explanation. Indeed, the translation was quite accurate, given the differences in style between the two civilizations. For the interpreter to have translated literally would have involved a real loss of face for all concerned—the sales manager for losing control of himself, the salesmen for having been thus publicly insulted, as well as the interpreter for being the vehicle of such intemperate communication.

There are, however, relatively few interpreters possessed of such gifts. Most do a fairly good job of translating the gist of what is being said, with special attention given, in business conversations, to numbers and figures. But most nonprofessional Japanese speakers of English, often with long residence abroad and years of study behind them, cannot even do that. The outstanding exceptions to this rule, in business and politics as well as among the academics, are few and overworked.

Even when they use English a lot, most Japanese assume a special kind of frozen personality when they speak it. They tend to stick to certain areas with which they are familiar, like the businessman who can talk fluently only about his golf handicap and the length of time his visitors plan to stay in Japan or the hostess who greets foreign businessmen at *Le Rat Mort*, the expensive Ginza cabaret, with "How are you? You have trouble with dollar shock?" It is most dangerous to assume a general knowledge of English by a Japanese, unless he or she has been thoroughly tested. A man fluent in banking terms can no more discuss literature than the English-speaking hotel waiter can discuss banking. An engineer may have all the vocabulary needed to talk about turbines, but will not understand an English-language film or news broadcast.

Context and connotation pose fearful problems when going from one language to the other. Picking up a recently published dictionary of famous sayings, I noticed the entry under Wellington. The translator had picked up the Duke's memorable phrase, "The Battle of Waterloo was won on the playing fields of Eton." To an American or European this conveys a world of meaning: Wellington meant that the British aristocracy, which officered his army, learned courage and leadership through the English class-segregated public school system. The phrase was often used to justify the class system and the whole British ideal of the well-born generalist, always in control of the situation. The Japanese translator, however, simply gave a literal rendition: "The cause of victory at Waterloo was the education given by Eton Middle School." He added an explanation that Eton was famous because of its students' learning and their enthusiasm for sports. Of the real meaning of the Duke's saying, nothing.

This mistake was obviously an honest one, but all too typical. Spoken English is especially hard for the Japanese, because of its directness. The typical Japanese conversation goes around in circles, widening or narrowing depending on the interest of the participants. The central topics to be discussed are repeated over and over again, like a fugue. Not so English, where all too much depends on the single sentence, the sharp question, or the unitary paragraph. One misunderstood clause can result in total incomprehension.

In his recent book, *Understanding and Misunderstanding*, Nishiyama Sen, a former United States embassy official who has probably interpreted more important conferences, business and diplomatic, than almost anyone in Japan, paints an unforgettable picture of the English-speaking Japanese who nods his head in assent at what the other party is saying, expresses his own statement in excellent English, but misunderstands the principal point of the conversation. This happens more often than anyone cares to admit. The resultant requests for confirmation from the interpreter ("Didn't he want some changes in the draft?" "No, he really didn't." "Hadn't we better ask him again and confirm it?") not only sidetrack the conversation, but generally give the foreign party an unwarranted impression that his Japanese opposite numbers are dumb, sly, or stalling. ("Hell, Pete, Shimizu speaks English as well as I do. He *must* have understood what we were saying.")

To avoid such misunderstandings, it has become the fashion

lately to run some of the innumerable Japanese-American symposiums or seminars with simultaneous interpretation. This has the advantage of glamour and neatness. Everybody feels comfortable with his adjustable earphone tucked next to the water pitcher and often develops a real illusion that an exchange of views is taking place, just like in the United Nations. While not denigrating the abilities of the Japanese simultaneous interpreters, some of whom display something akin to genius at their art, I have myself rarely found this to be successful. For the translator of Japanese, as Nishiyama puts it, must not merely transmit information, but must literally "re-express" the thought. Translating from English to French, for example, one has merely a problem of retailoring the thought to suit the other language, like altering a suit of clothes. Going from English to Japanese, or vice versa, you need to unravel the material and start weaving all over again. This can take time. Unless the interpreter is very skilled indeed, the principals in Japanese-American dialogues end up in the position of two people looking at each other around a corner, back in the mirror room at the fun house.

Although I have an ample vocabulary, my own Japanese has drastic limitations. At its best, I speak a Japanese version of broken Shakespearean English (thanks partly to my having swallowed Sir George Sansom's *Historical Grammar of Japanese* whole, at an early stage of my Japanese education). At its worst, my Japanese sounds like the fluent, but incomprehensible, English of a Hungarian movie director who has spent twenty years in Hollywood without taking an English lesson. Despite these handicaps, I have always preferred to do business in Japan in Japanese. While a good interpreter can tell me almost 85 per cent of what is said, his very presence diminishes my chance to find out how the Japanese I am talking to feels and what he might be thinking. Given the nature of the Japanese language, this means losing contact with him at the very beginning.

Just a little bit of language helps. I used to try pointing this out to American friends, urging them to keep up their studies in Japanese, particularly the written characters, without which the language is a jumble of words, all sounding alike. Over the years such efforts have been generally unavailing. "Sure, I tried studying it," says the fellow shaking his dice cup in the club's stag bar, "but it's impossible. We're only going to be here for a year more—and where can you use Japanese after you leave."

The Men with the Flags:
Few Leaders for Good Followers

The Japanese like flags. Flags waving, flags on buildings, flags on automobiles. Corporations have their flags, just like universities; newspapers have their house flag fluttering on the fenders of the company cars, serving as a combination press pass and prestige symbol. In World War II, soldiers would write their names and their comrades' on a small flag, and carry it into battle, again as a combination souvenir and talisman. Even in this age of low-calorie patriotism, the national flag flies everywhere on holidays and almost any festive occasion—far more than in most countries. House flags exist in profusion. Trade unions, sports teams, and summer resorts show their flags and lavishly distribute small sample banners to members, rooters, and visitors.

Foreigners visiting Japan, or watching Japanese tourists visit their own countries are quick to notice that virtually every group of Japanese sightseers moves along briskly through whatever environment—railroad station, temple, national park, fen, crag, torrent—following a man or a woman with a flag. The flag bearer may be merely a guide, a bus attendant, or a fellow company employee who knows how to get to the company outing. But the party, as a group, is temporarily in the flag bearer's charge. They follow his flag faithfully. It would be regarded as poor form either to deviate from his instructions or to leave his care. Where an American might think of this group flag-following as irksome, the Japanese happily depend on it.

The groups following their flag bearers make a tempting symbol for Japanese society as a whole. Granted the deep dependency feelings in Japan's group-conscious society, it is natural to see the whole people in terms of innumerable groups, each marching securely but docilely behind the man-with-a-flag at their head. Every village has its reliable head man; every company its strong-minded president; every union its assertive chairman. Every government raises up its party strong men, every ministerial bureaucracy has its senior vice-minister. Some notable Tokyo University presidents like Nambara Shigeru have seemed like laws unto themselves in the educational community.

There are, of course, the great business leaders: Matsushita Konosuke, the creator and long-time chairman of the huge Matsushita Electric Industries; or Yoshida Hideo, the visionary man who turned Dentsu into a world power in advertising; Nagano Shigeo, who built Japan Steel into a world corporation; Imamichi Junzo, whose name has become almost synonymous with the rapid growth of the civilian, that is, nongovernment, broadcasting industry in Japan; Honda Soichiro, whose motorbike put a good deal of the world's population on two wheels. The apparent *Führerprinzip* is fully as active on the left. The untimely assassination of Asanuma Inejiro in 1960 took the life out of Japan's Socialist party, possibly forever. The Japanese Communist party counts its landmarks not in terms of theoretical shifts, but in the names of strong leaders such as Nozaka Sanzo, Tokuda Kyuichi, Shiga Yoshio, and more recently, Miyamoto Kenji.

The leader's responsibility for attending to the needs and wants of those under him is indeed great. In return for the *amae* he satisfies and indulges, he exacts strong loyalty. He gets a big press, in a society which prefers people to principles more than most. He is constantly deferred to. One goes back again to the prototype of the *oyabun* ("the boss," "the parent") and the *kobun* ("the child," "the follower"). Such a feudal association is inescapable in a country which is still organized vertically, its kindred groups tied together in interlocking pyramids of descending importance.

So the analogy of the man-with-a-flag is tempting. Japanese make good followers. Their society would seem to be made for strongman leadership. But is this really so? Or is the leadership role generally more like that of the real man-with-a-flag in the tourist trade, someone who is temporarily empowered to guide, but whose rights to authority extend merely over one well-defined route? For the matter of authority in Japan is not all that simple. Despite the appearance in their history of strong men and dynamic leaders, the Japanese have displayed an extraordinary ability to surround authority with checks and balances, to dilute it, to contain it, as well as to keep ultimate formal authority and actual day-to-day authority rather far apart. Japan is not the sort of country where dictators thrive. It is the Chinese and the Koreans (especially the latter), the Japanese say, who need strong men to govern them. When a potential Japanese Führer springs up, on occasion, it is surprising to see how quickly he is headed off, diverted, or otherwise brought down to normal measure. The exceptions are a few men who are obviously outstanding and a few others who are clever enough to run a one-man show, while respecting all the appearances of the group-decision ritual.

The closest thing to a dictator in ancient Japanese history, aside from the early, legendary emperor and empress figures, was Prince Shotoku. A nephew of the Empress Suiko, Shotoku was prince regent through the first quarter of the seventh century. Like other princes, Shotoku waged war and won battles, but he was far more of a cultural and a religious than a military leader. A unique figure in world history, this exemplary Japanese Buddhist somehow combined the detached wisdom of a Marcus Aurelius with the legislating propensities of a Justinian and the missionary energy of a Constantine. His like was not seen again.

Although later leaders came to power, they generally ruled as part of a family or a clan hegemony, rather than as individuals. Even Japan's three sixteenth-century military despots, Nobunaga, Hideyoshi, and Ieyasu first came to power in tandem, as members of a successful coalition. In a country where ostentatious humility is practiced to the point of exasperation, "high posture" was always frowned upon, even on the part of successful generals.

Through the worst of their internecine wars, the Japanese displayed a constant preference for group counsel and consorted action, instinctively true to Article 1 of Shotoku's Constitution: "Matters should not be initiated by a single ruler." Even the strongest of military dictators was never quite given the absolute power which the feudal military leaders of Europe enjoyed in similar circumstances. Oda Nobunaga had to listen to his council; he would disregard their advice at considerable risk. (For one thing, the rejected advisors might take their men over to the other side.) Tokugawa succeeded in making his Shogunate permanent only because he was wise enough to build up his own bureaucracy and institutionalize it. Ultimately, his descendants become so surrounded with ingrown privileges and protocol, that it was a rare Tokugawa Shogun who could duplicate the high-handed behavior of his dynasty's founder.

The men who led the Meiji Restoration of 1868, like the leaders who rose from the post-Mediaeval civil wars, were mostly from lesser samurai families. They were revolutionary only in their desire for foreign technology—aside from that, they were mostly cautious men. What is so interesting is how well they worked together in harness: Fukuzawa Yukichi and Okuma Shigenobu the liberals; Ito Hirobumi the impatient constitution-maker; Saigo Takamori and after him, Yamagata Aritomo, the military champions; and, through it all, Okubo Toshimichi, the man who kept Meiji policy in the middle of the road with some of the deftest political footwork in modern history. Any one of these men, and others like them, could alone have led a large country. What is so extraordinary about the Meiji period is not their disputes, which were many, but their capacity, at least through a critical time, to rule and work together.

The men of Meiji, led by their emperor, constituted a working oligarchy. So did the militarists of the 1930s, who resolved in the worst possible way the contradictions between democracy and absolutism that had been left unsolved in Prince Ito's Meiji Constitution. It is significant that, at a time when Europe was being mobilized under one-man dictators, Japan's dictatorship was an oligarchy, whose military members—fine, medal-heavy gentlemen though they were in their day—are remembered by history principally in the names of their factions. For instance, one need know only that a general belonged to the *Toseiha*, the ultimately victorious "control" faction which wanted to attack the Soviet Union. When General Tojo Hideki became premier, he was little more than one successful factional leader who served as a symbol for others.

What happens to the man-with-the-flag in this world of oligarchies and factions? He is present, but he sticks to his defined route. He is the regimental commander of prewar days or the department head in Japan's current business society. He may, indeed, given enough authority and force of character, be the cabinet minister or the president of the corporation. Yet he has to work hard if he wants to make his own decisions. In every Japanese organization, the network of councils, advisors, and petty, but firmly centered, local authorities make it no small feat for a company president or a government minister to escape being something like the king on a Western chessboard, valuable for his symbolic leadership, but unable to move by himself more than one square in any direction.

It is in times of emergencies that the checks and balances tend to fall away and national leaders appear, to pull the country along with them. This sort of leader appeared in fairly recent times in the defeat and postwar reconstruction decade after 1945, which was potentially as dangerous to Japan's future as Meiji or the fifteenth-century civil war period. Despite the benevolence of the occupation (articles used to appear in quite serious Japanese journals in the early postwar days, proposing that Japan be made the forty-ninth state), it was quite clear that the country could only survive if it tried to pull itself up by its own bootstraps, or put better perhaps, take advantage of the boost given it by the Americans.

At this pass, Yoshida Shigeru appeared as prime minister. He proved to be anything but a rubber stamp. It was Yoshida who first asserted the independence of the Japanese government under occupation, then successfully lobbied for the peace treaty of 1951. In the negotiations preceding that treaty, he resisted pressure for a rearmed Japan. He equally resisted internal pressure toward socialism.

Like others who grew into their situations, Yoshida became an almost professionally cantankerous man. People were afraid of him. He deliberately ignored conventions on which others set great store. He played outside the rules, a license Japanese society gives only to the *oyaji*, an old man with power. The classic instance of his shouting "*Bakayaro*," the Japanese equivalent of "You stupid son of a bitch," at an opponent during a Diet debate became famous in a country where prime ministers now spend most of their public Diet time (Japanese Diet debates are televised) either explaining or apologizing.

Yet he became, in time, a national father figure who remained powerful even after his retirement. He had succeeded in giving his beaten country back its spiritual self-respect. In so doing, he paved the way for the mechanical prosperity that was to come. At a critical time, he carried the flag for everybody.

Whether others will recreate this kind of role is a question. Charisma, whether political or otherwise, has not been much in evidence since Yoshida left office. There are, it is true, local politicians who have ample appeal in their own areas (for instance, the Socialist governors Minobe Ryokichi of Tokyo and Ninagawa Torazo, whose more than a quarter century in office in Kyoto amounted to a dynastic record). Edging closer to the modern image of a charismatic political figure is Oda Makoto, the magnetic leader of the anti-Vietnam War movement (*Beiheiren*); or the writer-turned politician, Ishihara Shintaro, whose dynamic platform appearances in his white blazer and white gloves brought out the younger generation in droves ("Kennedy, Kennedy" they called him) to make him the Liberal Democratic party's biggest new vote-getter. Even closer is the leader of the Sokagakkai, Ikeda Daisaku. Ikeda has welded the millions in his Buddhist society into a powerful, assertive, and vastly rich political-social-business institution. To visit with Ikeda is an impressive experience. For better or for worse, he possesses to a rare degree the mix of intensity, intelligence, and organizing ability that national leaders are made from.

Such men are exceptions, however—along with the few visible tycoons in Japanese business. The present eminence of Japan was largely achieved through group effort, the work of a nation with many thousands of able majors and colonels, but few generals. "Japan's progress," the late essayist Oya Soichi has written, "Its success in creating the only nonwhite modern nation-state, has depended less on individual charismatic leadership than on the charisma of the Japanese people themselves." It is a proud statement. But as Oya himself conceded, group charisma is a rather fragile commodity.

Chapter VIII

How to Succeed in Business by Trying
The Truth about Japan, Inc.

> People are the castle
> people are the walls
> people are the moat . . .
> —Takeda Shingen,
> sixteenth-century military leader

In 1850 a rough country boy named Furukawa Ichibei walked to the imperial city of Kyoto from his village of Okazaki, to get a start with his uncle in the textile business. Ichibei was then eighteen. He had received virtually no schooling, having worked most of his life as an apprentice to a merchant selling bean curd to the local people; but, since the age of eleven, he had harbored the ambition to make it as a businessman. He was a quick learner and almost frighteningly hard-working. People noticed him. At the age of twenty-seven, a rich relative made Ichibei his adopted son (a common practice in Japan, then and now). This added social push was all he needed. At thirty-one he became Kyoto purchasing agent for the Ono Gumi, then with the House of Mitsui one of the two leading merchant houses in Japan. He handled all of Ono's raw silk market dealings for the area and was, by all standards of the day, a success.

The Meiji Restoration of 1868 brought a rush of opportunities for foreign trade as well as domestic expansion. Japan's businessmen, whose energies had been increasingly frustrated in the late years of the Tokugawa Shogunate's seclusion policy, scrambled to get in on the ground floor of the new building. Furukawa was by then chief clerk, that is, general manager, of the House of Ono. He had Ono build Japan's second modern silk-reeling mill at Tsukiji, in Tokyo, and he prepared to jump into the competition to produce a finer-reeled quality of export silk. Furukawa's silk mill became a model for the developing Japanese industry, but this and other expenditures proved to be bigger than Ono had bargained for. The firm became overextended, its notes were called, and it went into bankruptcy. (The Japanese government had not yet developed its pneumatic systems of credit and control to forestall such major catastrophes.)

Furukawa paid off as much as he could in company assets to the newly founded Dai-Ichi Bank, which had extended the credit. He added all his personal wealth, even his household goods, because he felt personal responsibility for the loss. At forty-four he was penniless and, by all the accepted standards of capitalism, a ruined man. Not in Japan, however. Shibusawa Eiichi, the founder of the Dai-Ichi Bank was Ono's creditor. One of the most able of the early Meiji entrepreneurs,[1] Shibusawa was an enlightened moralist who believed that sound character and honesty were just as important to a business as technical ability, if not more so. He was so impressed by Ichibei's sense of responsibility that he helped him get another start, this time on his own. In 1877, backed by Shibusawa, Furukawa received a government grant to develop the copper mines at Ashio. Other good things followed. Within ten years he was known as the Copper King, at a time when copper was still Japan's number two export item.

Under his slogan of "new business above all" (*Shingyo Senitsu*) Furukawa parlayed his original mines into a network of related companies. In contrast to modern conglomerates, they grew organically, out of necessity and favorable circumstance. Copper was used in cables, so the mining company was followed by the establishment of a cable company in Osaka. The reorganized Furukawa Electric Company made tractors and earth-moving equipment and used conveyor belts heavily. This led to the founding of the Yokohama Rubber Company in 1917. Since Furukawa wanted to expand its new electric company's wire output to include machinery

---

1. He was probably one of the few bank presidents who got his start by raising a force of five hundred infantrymen. He recruited them for the fading Tokugawa cause in Kyoto in 1867. After the Restoration, Shibusawa was in turn recruited by the winning side into the new Finance Ministry, which he finally left, however, because of his opposition to heavy military expenditures. As a financier and businessman, he developed Japan's cotton textile industry and set the course for nationalization of the railroads. He also introduced the modern corporation to Japan in the form of the joint-stock company.

an agreement was signed with Siemens, the German electric manu-facturer in 1923, which resulted in the Fuji Electric Company. So it went, all in the name of the founder, who died in 1905, leaving to his descendants in the company the modest explanation for his success as "luck, dullness, and perseverance" (*Un-don-kon*).

As of 1973, there were thirty-two member companies in the Fu-rukawa group, including an aluminum company, a light metals company, a plastic manufacturer, a chemical products company, a paint company, as well as makers of batteries, foil, diesel engines, electromagnets, and computers. In 1972 the largest of these, Furu-kawa Electric had sales of $750 million. Add to the member com-panies of the Group 238 smaller related companies—subsidiaries, old suppliers, or various semipermanent subcontractors, in the Jap-anese tradition, and one gets an impressive total. There are in all 225,678 employees. Annual sales of the Furukawa group come to about $4.5 billion, which is not a bad memorial for a boy who started out selling soy bean snacks.

On the third Wednesday of every month, the group's executive committee, fittingly called the Third Wednesday Club, has its meeting.[2] Its members are the presidents of the ten largest Furu-kawa companies. They do not constitute a formal board of direc-tors, but in a loose way, they try to keep the group's activities pointed in the same direction through discussion and consensus. This they generally succeed in doing, thanks largely to their com-mon corporate backgrounds and objectives, in a way that could never work in an American company or any other European set-ting (with the possible exception of the Mafia). There are almost no heavy stockholders among them. Like the directors of most Japanese companies, they are basically salarymen, however ample their perquisites.

Nonetheless, old-timers at Furukawa feel that they suffer from a lack of central direction since they lost their own group's bank and trading company. (Both were dissolved in the 1920s.) In other groups, like Mitsubishi or Sumitomo, the members tend to orbit around these central agencies, the one controlling expansion funds and the other handling general marketing responsibilities, in Japan as well as overseas, for everything from small consumer items to heavy industrial plant. Both are institutions peculiar to Japanese

2. Japanese companies are fond of such terminology. The Mitsubishi Group's governing board is called the   Friday Club, Mitsui Bussan has the Monday Club, and so on.

business. Their presence made it relatively easy for the old *zaibatsu* companies to regroup, even after the United States occupation trust busters had destroyed the formal ties among them.

Although Furukawa misses the bank and the trading company, group members use the closely connected facilities of the Dai-Ichi Kangyo Bank, Furukawa Ichibei's original creditor. Larger than even Viscount Shibusawa could have forseen, Dai-Ichi Kangyo is now number one among non-United States banks in the world and fourth in size after the three largest banks in the United States. For 1976 its total deposits were $29 billion, total assets more than $43 billion, representing a sizeable increase over the year before.

Another compensation for the lack of a group bank is the Furukawa Group's claim to one of the country's newest national assets, the Fujitsu Company, which manufactures Japan's only completely home-grown computers. IBM still holds onto 40 per cent of the Japanese computer market, with most Japanese-made computers, like NEC and Toshiba, manufactured under American license. Fujitsu, with 10 per cent of the market, is the only Japanese producer not bound by any licensing agreements to restrict exports. Ironically, the reason for its independence was the old Furukawa connection with the German Siemens company. Although Siemens was lagging behind in computer and electronic development, Furukawa management was reluctant to break an old commitment and deal with its newer American competition. The ultimate result was that Fujitsu had to learn the business by itself. Its computers, after a long period of travail and trial, are now highly competitive with American models, and a great deal more compact. In the eyes of the protection-minded Ministry for International Trade and Industry, for one, it is a very special company. By supporting its development, the government is continuing a long tradition, dating from the days when Mutsu Munemitsu, Prince Ito's foreign minister, introduced Furukawa Ichibei to the better politicians of the time and Hara Kei (who was assassinated while prime minister in 1921) held directorships in the Furukawa Mining Company, while running one of Japan's two leading political parties.

Like Furukawa, the first generation of the giant Japanese business groups (including the other family *zaibatsu*) had their origins in the industrialization of the Meiji Era. Other great Japanese companies grew up long after Meiji and often in circumstances that the men of Meiji could hardly have foreseen. There was no foreign minister assisting Matsushita Konosuke when he began the Pana-sonic empire with his first electric goods shop in Osaka in 1918. The post-World-War-II Nissan automobile was an unexpected by-product of the heavy industries development company founded by Aikawa in the thirties to assist the colonization of Manchuria. The Meiji planners might even have been distressed by the reduced circumstances in which a small group of businessmen and former technical officers in the army and navy, led by a military contractor named Ibuka Masaharu and ex-navy Engineering Lieutenant Morita Akio, started Sony's spectacular scientific and business development from the wreckage of World War II. Yet neither the shape, the thrust, nor the success of these and other new Japanese companies would have surprised them very much. For the Meiji planners, in their half-premeditated, half-improvising way, had cast the mold for Japan's economic miracle of the 1960s and 1970s almost a century before.

Those young samurai who did not go into government in the 1870s and 1880s found a ready outlet for their energies in trade. They were encouraged to do so by their friends and, more especially, their rivals in the government. Quick learners, they naturally made common cause with the already existing merchant aristocracy of Japan. The House of Mitsui, after all, had been founded in 1637, long before anyone in Britain or France had ever heard of the House of Rothschild. Mitsui money financed the Emperor Meiji's forces in the decisive battles against the defenders of the Tokugawa Shogunate.

Iwasaki Yataro, the young Tosa samurai who founded the Mitsubishi group, got his real start with thirteen ships the government gave him to ferry troops for the Formosan expedition in 1874.[3] By 1890, the government had leased and sold the Nagasaki shipyards to Mitsubishi. In the same year, Mitsubishi agreed to buy some rather unattractive land beyond the imperial palace moat. Today this comprises the heart of Tokyo's Marunouchi business district. By the turn of the century, Mitsubishi was branching out into other fields. At present, the forty-five companies of the Mitsubishi Group include manufacturers of airplanes, beer, automobiles, petrochemicals, synthetic fibers, and office equipment, as well as an insurance and a credit card company. Combined annual sales of the group

3. After a series of mergers and further government subsidies, the original thirteen ultimately multiplied into the Nihon Yusen Kaisha, which remains today Japan's premier shipping line and a perennial anchor for bullish rallies on the Tokyo stock exchange.

account for 3.6 per cent of Japan's yearly industrial total; its employees make up 1.5 per cent of the country's total labor force; its assets come to about 8 per cent of the total. Broadly speaking, Mitsubishi's business activities now cover about 10 per cent of the total business activity in Japan.

When the Meiji industrial expansion began, government officials and businessmen were all, in effect, members of the same club. They have kept things that way for more than a century. The club's membership is never frozen, however. As in the rest of Japanese society, it is always possible for a loner to make his way. The Hondas and the Matsushitas have been world figures for a long time. Others come along to join them. Ten years ago, no one had heard of Yoshida Tadao. Now that his YKK zippers sell $780 million annually (1977) after an amazing international growth story, he is a member of the club in good standing.

Where the American tradition holds government and business to be perennial adversaries, in Japan they have always been closely related. The American outlook on government-business collaboration has, of course, changed with the times; but it is always an uneasy relationship. If not the extreme of determined free enterprisers denouncing government interference, we present the more distressing spectacle of cost-plus contractors cozying up to government agencies or corporations readily intimidated into mailing extravagant party campaign contributions. The Japanese have more open connections. Because their heavy industry has been largely created by the government, it has always seemed natural to accept the administrative guidance of the Ministry of International Trade and Industry, or its prewar predecessors, as a complement to doing business, rather than an impediment. Government bureaucrats continue to accept senior posts in business (largely sinecures) after their early retirements at fifty-five, a practice cynically called "descents from heaven" (*amakudari*) by hostile newspaper commentators (who only rarely receive directorships when *they* retire).

There are obvious strains and abuses in such a system and they are growing worse. Yet it is impossible to overestimate how this collaboration between businessmen and bureaucrats has smoothed Japan's road to industrial progress. Most of their contact, even today, is informal, conducted in thousands upon thousands of meetings, casual conversations, and semiofficial "guidances." Their sum effect is to give Japan the benefits of a planned economy without either the inefficiencies or oppressiveness of a totalitarian bureaucracy.

All this cooperation would be of little use, without funds to invest, buy, and develop. This problem was taken care of by Meiji's remarkable financial expert, Matsukata Masayoshi, when he founded the Bank of Japan and, indeed, the whole national banking network, in 1882. Matsukata rejected the idea of having local national banks, which at first had been uncritically borrowed from the United States. He substituted a strong central financing system, butressed by specific development banks for funding different sectors of industry. As it started, so the system has continued. There is a bank at the center of every Japanese industrial complex, or close by. In most cases, the bank calls the tune. It provides the money for expansion and, when needed, takes steps to contract or even (but rarely) to liquidate. Gaining the confidence of a Japanese bank takes a long time, rather like taming a tiger, and you can never be sure when the animal may revert to its primordial habits and snap at you. But the effort is a necessary one. Once trust is established, the bank becomes a virtually inexhaustible source of capital.

Japanese bank branch managers, who have considerable autonomy in their operations, wield power which most American bank presidents only dream about. The bank's people know their enterprises on a day-to-day basis, but rarely interfere as long as the general course is set and appears to be in the right direction. Although the rate of borrowings to equity in most Japanese companies—well over three to one—shocks most American businessmen, it bothers few Japanese. Like the bureaucrats and the businessmen, the bankers and the businessmen know each other. And, thanks to the national savings habit, the banks generally have money available.

The government continues to sit at the controls of the whole operation, although it is, of course, constantly talked to and influenced by the huge trade associations and other business interests. When it is time for subsidies, they are forthcoming. When an industry needs protection, it is quietly helped and protected. Working through the Ministry of International Trade and Industry (the virtually all-powerful MITI, as it is familiarly known), the government assigns informal quotas, "suggests" useful mergers, and extends needed financial support by tax benefits, helping with raw material imports, plant expansion credits, technical help, and other ways too numerous (and often too devious) to mention. This finely calibrated support is one of the many secrets of Japan's successful export trade. When it comes time to deflate the economy and apply the credit squeeze, on the other hand, word goes out from the Bank of Japan and the money starts to dry up like a vanishing waterhole. (Because of the high ratio of bank loans to equity capital, the drying-up process works far more quickly than similar efforts in the United States.) The Bank of Japan's tight money credit squeeze of 1973 and 1974 was basically the same kind of deflationary exercise that Prince Matsukata ordered to halt the inflationary crisis of the 1880s. In 1973, however, Japanese companies had accumulated a great deal of excess liquidity from export trade balances and the task of locking the pump at home was more difficult.

To the young economists of the American occupation in 1945 and 1946, this interlocking structure of government, industry, and finance represented at once an obstacle and a challenge. Anxious to take Japan out of its feudalism and into a new kind of democracy, they tried out programs which even the New Deal had yet not been able to put across at home. In some areas they succeeded. The sweeping land reform, for all of its inequities, destroyed tenant farming and created a whole new class of independent farmer. By now much of the land may have been sold for bowling alleys, golf courses, and suburban villas and its owners' sons moved to Tokyo; but the structure remains.

Similarly, the occupation succeeded (all too well, Japanese businessmen say) in recreating a new, aggressive labor movement in Japan. Unfortunately, the United States directives were unable to grow any Walter Reuthers overnight to handle Japanese labor's new power with imaginative, economically oriented leadership. Most of the labor leadership, ultimately, sank into a conservative marriage of convenience with the Socialist party and has long overemphasized cloudy political goals to the exclusion of long-range union planning. Yet the net effect was probably good, in that it revived a sense of freedom and participation among Japan's workers. Insofar as the movement tended to fit in with strong company unions, it became, in an odd way, a force for solidarity within the Japanese corporation.

In trying to break up the Japanese *zaibatsu*, as well as to decentralize the banking system, the reformers ran into trouble. Even had American policy toward Japan not shifted in 1949 and 1950, the men from SCAP would still have had rough going. As it was, most of the *zaibatsu* groups sprang back, like vegetation reconquering a briefly swept clearing in the Matto Grosso.[4]

By the middle of the fifties, the world of the Meiji planners was restored, almost as if the war and the occupation had never happened. There were, however, two important aftereffects. First, Japanese business, having been trapped by the military into the venturesome disaster of the 1930s, was resolved not to go into the military-industrial complex business again. Secondly, the occupation purges of top management released a great deal of latent energy among the middle and lower ranks. These men in their early forties now began to rebuild from scratch, exactly the way the young samurai of Meiji had gone about realizing their blueprints. As G. C. Allen notes in his economic history, "The release of fresh energies by the post-war political reforms may be compared with the effects of revolutionary political changes at the time of the Restoration. The new freedom kindled many fires."[5]

The fires were fueled from other sources, also. In 1949 Joseph M. Dodge, the Detroit banker, came out to Japan to right the economy with a massive deflation and cost-cutting program. The Dodge Plan was based on rigidly balanced budgets and industrial rationalization, with an idea of getting Japan on its feet as quickly as possible. But by cutting off government aid and subsidies, thousands of small and middle-sized firms were forced into bankruptcy. Even large companies were threatened. Had this economic purge been administered by any but an all-powerful foreign occupying authority, it would have brought considerable civil disturbance—a fact of which the grateful Japanese Finance Ministry bureaucrats who cooperated with Dodge were only too well aware. As it was, the purge was accepted with little protest. Japanese industry was duly rationalized.

In 1950, after the outbreak of the Korean War, the country which only five years before had exhausted its industrial resources trying to defeat America became, in its way, an unexpected arsenal

---

4. The actual prewar *zaibatsu* companies (*zaibatsu* means literally "financial clique") were about twenty family-owned trusts, the foremost being Mitsui, Mitsubishi, Sumitomo, and Yasuda. They were run by holding companies. Only Yasuda failed to survive the war as a group, although the

Fuji Bank, currently Japan's second largest, was originally a Yasuda enterprise. Although holding companies are still illegal in Japan, the so-called groups preserve the spirit, if not the letter of the *zaibatsu* organization.

5. *A Short Economic History of Modern Japan.*

of democracy. The $3 billion boost given to the Japanese economy by United States and United Nations military procurement between 1951 and 1955, had the effect of a propellant. From then on into the seventies, the economy rocketed up and up and up.

Through this period, the Japanese government, consulting and cooperating with business in the best Meiji tradition, laid down a series of plans and guidelines without parallel in a non-Communist country. They differed, also, from most similar totalitarian efforts in that they worked. The first Draft Plan for Economic Recovery, which took in the period between 1949 and 1953, aimed to tighten the economy and begin modest increases in production. This was succeeded by the Five-Year Program for Economic Independence, which the Economic Planning Agency put out in December 1955, to cover the years from 1956 to 1960. Its aims, too, were modest: (1) to stabilize the country's balance of payments without further reliance on the United States military's special procurements and (2) to step up production and increase employment opportunities consonant with normal population growth.

Both succeeded. By 1958, the modest 5 per cent growth rate originally projected had been almost doubled. This encouraged the planners to put through a new Long-Term Economic Program, designed to keep the growth rate at no less than 6.5 per cent annually between 1957 and 1962.

By 1960, this plan was threatening to go off the tracks, as the growth rate hit 10 per cent. In the decade ending with 1961, Japan's industrial production quadrupled. So did its manufactured export goods. Home consumption kept up the pace. From a subsistence society, the Japanese people were turning themselves into what would ultimately become one of the world's great domestic consumer markets.

Whereupon followed Prime Minister Ikeda's famous double-your-income program, a ten-year exercise extending from 1961 to 1970. His announced aim was to keep the growth rate at a steady 7.2 per cent. The annual growth rate during these years actually averaged 11 per cent and the figures rarely stopped climbing. A GNP which stood at ¥19,000 billion in 1961 rose to almost ¥80,000 billion by 1972. The world not only started noticing Japanese radios and cameras, but grew used to ordering Japanese ships and machine tools. At the Toyota plant in Nagoya, engineers had worried for years over Japan's answer to the Volkswagen. By 1965, the new Corona was rolling into the showrooms. And journalists all over the world had begun to crank out the articles on Japan's economic miracle. Most striking was the shift in the export pattern. From textiles to transistors to steel to televisions to machine tools to chemicals, computers, and calculators, the Japanese shifted the weight of their exports over the past twenty years to wherever the action was, by a combination of government planning, private initiative, directive, subsidy, and judicious bank financing.

There was nothing in the law to say that Japanese business had to do things the way the government's planners wanted. But since the first days of Meiji they had been doing it that way. Like the veteran crew of a racing yawl who reef the mainsail or prepare the spinnaker without being told, the businessmen and the bureaucrats set and shifted their courses, with their extraordinarily single-minded press and communications establishment making sure that the public was well-informed about the significance of what its establishment leaders were doing. The public, in the best Japanese tradition, did what was expected of it. It is hardly surprising that angry foreign competitors began to use the phrase Japan, Inc., in describing the national phenomenon.

Mere solidarity and planning, however, were hardly enough to create the Japanese economic accomplishment. Two other factors were key: productivity and development.

Since Japanese industry had been almost wiped out during World War II, the postwar technicians and planners had nowhere to go but up. They put an extraordinary amount of their national product into capital investment and started ahead of the rest of the world. The steel industry's modernization plan, for example, began in 1951. Japan now owns the world's largest LD converters and the world's second largest network of strip mills and tandem cold reduction mills. By 1965 the coke ratio in Japanese steel plants had been reduced to 512 kilograms; the world's average fluctuates between 600 and 800 kilograms. The pace continued. Since 1960 Japanese capital investment has regularly stayed at more than 30 per cent of the GNP—as contrasted to well under 20 per cent in the United States. More than 70 per cent of Japan's machine tools are less than ten years old, as against barely 35 per cent of American machine tools. In some industries, like shipbuilding, the Japanese have put themselves almost beyond overtaking. An American engineer who recently visited several of the major Japanese yards put their productivity as about four or five times that of comparable

American plants. From the huge six hundred-ton moving cranes for assembly to superior safety standards and locker and desk space for the workers, he found the Japanese far beyond the United States competition. "They could pay their yard workers $20 to $30 an hour," he said, "and still be more competitive than we are."

Between 1951 and 1970 Japanese business investment in new plant and equipment increased fifteenfold, a building program probably without precedent in economic history. Thanks to their new plant, new methods, and an originally undemanding labor force, Japanese corporations were able to operate until very recently at a high rate of productivity. Between 1966 and 1970 productivity rose between 10 and 15 per cent a year. Increases in output per man ascended in scale: Given a 1960 base of 100, the output per man in Japanese manufacturing now hovers somewhere around the 400 mark, while that in the United States is about 150. Between 1966 and 1971 unit labor costs decreased, in contrast to conspicuous increases in the United States and Europe. The great productivity surges enabled the Japanese to keep their wholesale price level extraordinarily even for twenty years, at least until economic troubles began in 1973. In the five year period 1967–71 Japan's wholesale prices increased by 5.9 per cent, while those in the United States increased by 13.9 per cent and in Britain and France considerably more (Bank of Japan figures). This stability explained a lot about the meteoric rise of Japanese exports.

Wages soon began to rise as sharply as productivity, however. By the late 1960s, as the consciousness of good times spread, labor began demanding more, and a growth-conscious management rarely refused. By 1968 a pattern of escalation had set in. Wage increases, running at more than 15 per cent annually, overtook the rising productivity curve. In 1973 and 1974, with labor shortages becoming more obvious, wage increases bumped against the 20 per cent mark, and in some cases hit 30 per cent—the high water mark.

Nevertheless, productivity rose; the labor productivity index stood at 139.3 in 1973, based on 100 in 1970. It was extraordinary that Japan managed to keep up productivity at such a consistently high level. But it is doubtful that the spectacular productivity gains of the sixties can be repeated. Nonetheless, in the face of almost impossible inflationary pressures, the Japanese showed that they could keep producing more for less over a long period. They are almost unique in the world in so doing.

Japanese industry's second biggest asset lies in its almost obsessive pursuit of new techniques and discoveries. A century after the Meiji reformers scoured the world for blueprints and experts, plans and inventions, their descendants, the Japanese companies and government ministries, are doing exactly the same thing. The Japanese dote on *jōhō*, an all-embracing word which means "intelligence" and "information" in the widest possible sense. We are living, they say, in the *jōhō shakai* ("the information-intelligence society"). We are developing, in its later stages, a *jōhō sangyo* ("information-intelligence industry"). Among themselves almost professional communicators, with a fanatic respect for something new, the Japanese have continued to move their industry in new directions.

The start was easy. Since 1945 more than $6 billion worth of patents, rights, and inventive royalties have been acquired from the United States, which is regarded far more highly than Europe in such matters. Government and industry combined, the Japanese spent $1.3 billion on research and development in 1965, $3.8 billion in 1970, and it is estimated, an annual $12 billion expenditure by 1975. Japanese business is innovation-minded, helped by its relative disregard for this year's profits in the interests of long-term growth. The revolutionary Wankel rotary engine, for example, was up for grabs in Germany, where it was originated, for many years. American car makers virtually ignored it. It was left to a Japanese company, Toyo Kogyo, to develop it and put it under the hood of the moderate-priced Mazda, which set a new standard in pollution-control.

Americans have widely regarded Japan as a nation of borrowers. It is, in a sense. Borrowing has been a big factor in the postwar economic growth of Japan. In the world's technology trade, Japan continues to be a heavy importer. By 1971 Japanese industry was paying out $.5 billion a year for foreign patents, rights, and techniques. More than borrowers, however, the Japanese are adapters. Their skills lie in taking existing discoveries and fitting them very selectively to their own uses.

As time goes on, the Japanese will make more discoveries on their own. Already their soaring average of patent applications has passed that of the United States. They will have to innovate more. Yet their talent has a special advantage in a world where so many of the big discoveries, the major breakthroughs in science and technology are already made. What counts in the space age is how quickly one can assimilate, absorb, and communicate the information.[6] That happens to be the *jōhō shakai*'s specialty.

Much has been written about the Japanese worker who supports and, indeed, provides this formidable industrial superstructure. Phrases like "disciplined work force," "the age-old work-ethic," "feudal survival," "life-time employment," "fanatical devotion," or the familiar "workolics" are used continually, often by the Japanese themselves, to describe the Japanese worker. Sometimes Japan's work environment begins to seem like a kind of transistor-operated anthill. But it is extraordinary only to us, looking in from the outside.

To understand the Japanese worker, we have to scrap the Western view of work as we know it. This includes both the Marxist idea of a society divided into giant horizontal levels, in which a recognizable class called the proletariat is struggling against the domination of another class called the bourgeoisie, and the free-enterprise version, in which work is seen as a struggle by the individual, sanctified or not by the Protestant work ethic, to better himself and, if possible, to become a capitalist in turn. In Japan the job is the society. The society is the job. Every man who enters a company equally shares in it. It is not so much a question of whether Mr. Sakamoto is a director, a junior department head, or the fellow who drives the company bus. What counts is that he works for Mitsubishi. If he works for Mitsubishi, he is a Mitsubishi man. Most of his friends come from Mitsubishi. He drinks with them, golfs or bowls with them, and shares his troubles with them. He competes with them, surely, but like siblings competing within a family which no one would think of leaving. With the exception of his relatives, and possibly a few school friends, most of his associations—and often those of his family's—go on within the framework of the company.

As a perceptive Japanese diplomat-turned-writer, Kawasaki Ichiro, noted, "In Japan work is a ceremony. . . . To the Western worker, the job is an instrument for the enrichment and satisfaction of the real part of his life, which exists outside the place of

6. It has been estimated that the discoveries made in the course of the American space effort alone, now all reposing in the files of the National Agency for Space and Aeronautics, will take decades even to communicate, let alone be utilized by relevant agencies of government and private industry.

work. For the Japanese worker, life and job are so closely interwoven that it cannot be said where one ends and the other begins." [7]

There is nothing feudal about the Japanese work society, in the Western sense of the term. In fact, within the structure of the company, there is a great deal more working democracy than in most American companies. The rights of those in the lower echelons are very strong. The *amae* dependency syndrome works nowhere so thoroughly and often so disastrously as in a Japanese corporation. Because the company is such an enveloping social institution, policies are explained on a wide level. Intracompany education is essential.

The Personnel Department is all-important. A worker entering a company from school or college is checked over like an applicant for membership in the Union Club. School, background, family, health are all rigorously examined—even in Japan's competitive work market. The screening must be good, for someone is, literally, being taken into the family. Once the first three months' trial period is successfully passed, the worker is essentially in for life, until retirement age.

The best way to understand how the Japanese company is different is to watch its union. Japanese unionists are far noisier and more demonstrative than their opposite numbers in American unions; but they strike far less often. In 1972 the total working days lost from strikes and labor disturbances in Japan was 5,146,668, or 150 days per 1,000 employees. This was a big increase over most past years, partly due to the 92 day seamen's strike; but it was still far below American totals, or the shocking 24 million lost working days in Britain.

Each "joint-struggle" period—in the spring and fall, when the union presents its demands for wage increases and other improvements, is preceded by snake dancing, harangues, cheers, conspiratorial meetings, a profusion of red flags and derisive cartoons of management.[8] Efforts to head off such performances are futile. One thoughtful American company president in Tokyo offered his union double what they asked if they would sign a two-year contract and forego the semiannual joint-struggle ceremonies. His offer was indignantly refused. He never could understand that the

7. *Japan Unmasked.*
8. My own company union's most successful effort was two posters, one of them depicting me as a gorilla holding a cigar-smoking human mask in front of his face ("Take off the mask of culture") and the other as a storm trooper, holding an ax (the firing ax) dripping red with the blood of employees.

struggle was as much a part of the job as the contorted expressions of the performers are a part of professional wrestling.

The union's right to strike and demonstrate is unquestioned. In fact, the basic labor laws which the U.S. occupation pushed through between 1945 and 1947, modeled on the Wagner Act and similar American statutes, reflected the SCAP reformers' zeal to protect the worker, with little or no thought to the rights of employers. No provision is made, for example, for adjudicating "unfair labor practices" by unions—only by employers. Abuses have been considerable.

Japanese management has reluctantly made its peace with the laws, now virtually unrepealable. The executives of a company union, indeed, are watched closely by management for their talent in handling negotiations. The chairman of the struggle committee who drives the company to its knees by hard bargaining will be promoted into management. If he behaves himself thereafter, five years later he may be handling labor negotiations for the company.

The union's demands are negotiated and compromised on, generally with substantial raises resulting that fit the rising pattern of the economy. When a company is making money, the union knows it and wants big bonuses. When a company is losing, that too is taken into account. Few Japanese unions strike a single firm as an example to others. In fact, with the exception of the government workers' unions, which are very political, the average union is primarily a company union. Its ties to the nationals are loose.

Although Japanese workers rarely strike for political reasons or matters of principle, they will and do strike to preserve their status. Shave wages, announce unpopular policies and management may get away with it. But to fire one man or otherwise tamper with full-employee status is dangerous. The Chase Manhattan Bank in Tokyo was tied up for seven years in a strike over firing some employees. In 1973 an agreement was finally signed and the red flags taken down, after Chase paid seven years' back wages to some of the people it had dismissed. In the same year Tokyo University's Earthquake Research Institute failed to settle an angry three-year strike by fifty temporary employees which put the institute out of action. Why strike? They wanted to be permanent.

In more than half of the Japanese companies the seniority system remains the determining factor in salary increases. Merit has very little importance, except down the road, when it comes to making department head. It is true that more Japanese companies are making promotions according to merit, and this is not wholly due to the infiltration of modern management theory. Partly it is because the younger Japanese tends to have more wants and aspirations than his infiltration of modern management theory. Partly it is because the younger Japanese tends to have more wants and aspirations than his father or grandfather and he needs money to satisfy them. Partly it is due to the tremendously competitive labor market.

The Japanese do not want to import foreign labor like the Germans or the French have; they are going to tremendous lengths to utilize what native labor force they have. The labor market is steadily tightening, however. Due not only to the rapid growth of Japan's industry, but also to the steadily rising level of industrialization, skilled workers and supervisors have a seller's market such as few labor forces have ever enjoyed. A good department head in a Japanese company is now highly prized. Faced with rising costs of living and a profusion of consumer luxuries as well as irritations, the Japanese company man is more disposed to go over to a competitor than he ever was before. However, compared with the American worker, he is still rooted in his job and his own company family.

The other side of the Japanese worker is the Japanese consumer. He is a very important person. In the big-time power economics of the late twentieth century, the Japanese consumer plays something of the same role—heroic, if unsung—that the British grenadier played in the political wars of the eighteenth and nineteenth centuries. He is the backbone of an entire system, although he is rarely mentioned in dispatches. The more dashing manufacturers and financiers gallop off with the headlines, but the consumer is the man on whom they all essentially depend. He has almost endless reserves of endurance and curiosity. He buys and buys and buys. He saves and he buys again. There is no one in the world quite like him.

The modern postwar consumer rose from the debris of Japan's wartime defeat quite spectacularly. In 1955 the average consumer had only about $200 of disposable income per year. In 1973 almost 3 million Japanese consumers bought private cars. In 1974 the disposable income per capita in Japan was close to $3,000. In the intervening years the consumer has equipped himself with washing machines, TV sets (120 per cent saturation, including black and white and color), and relatively expensive tastes in food and drink. Where 70 per cent of disposable income in 1955 went for a subsistence level of food, the 30 per cent of income reserved for food in 1973 now buys meat,[9] imported oranges and grapefruits, and milk and dairy products. All of these are, doubtless, factors in

9. Japan imported 175,000 metric tons of beef in 1974.

increasing the size of the average Japanese to the point where th
Education Ministry has had to revise standards for the size of th
desks in the schools and dress store proprietors worry about th
lack of large sizes in younger styles. ("It's all this bread they ea
now, instead of rice," a clerk in an Akasaka boutique muttered a
he desperately flipped through sizes 11 and 13 for some college-gir
customers, sizes that only a few years ago would have been though
more suitable for sagging matrons who were better off with envel
oping kimono.)

Japanese men now buy Pierre Cardin suits as well as ties. Gradu
ally the striped shirt and the gaily patterned tie are cropping ou
among the sober white-shirt, dark-tie business set. Departmen
stores stock so many foreign fashions, it is small wonder that Japa
nese textile companies keep protectively diversifying into suc
fields as resort tours and language laboratories. Between 1971 an
1972 Japanese whisky manufacturers increased sales over 100 pe
cent, not to mention what they ordered from abroad. In 1959 Japa
imported some 90,000 cases of Scotch whisky annually. In 1973 th
number rose to 1.2 million. And, if current sales estimates are o
target, by 1979 almost 3 million cases will be landed to solace th
thirsty executive with Johnny Walker, or lesser brands,[10] after h
comes back from a hard day on the links with his Ben Hogan clubs
his Christian Dior golfing shirt drenched with honest sweat.

This huge urban consumer market—in a country where mor
than 75 per cent of the people live in cities of over 50,000 popula
tion, read the same ads in the newspapers, and have their nativ
cupidity sharpened by the same television commercials—underlie
the strength of Japanese home industries. Although they are des
perately dependent on raw materials from overseas, their basi
market remains at home. Despite the intensity of its export drives
Japan exports only 30.1 per cent of its total basic industry produc
tion, as opposed to 54 per cent in Britain and 36.6 per cent in Ger
many (1970 United Nations figures). There are National or Pana
sonic distributors for televisions, radios, and other electroni
equipment all over the world, not least of all in the United States
But the fact remains that overseas business accounts for no mor

10. In matters of brand selection the
Japanese are as hierarchical as in most
other things. The clear preference,
for example, for Johnny Walker
Black Label over Red Label dates
back to prewar days, when the Im
perial Navy, which for a time too
over imports of Scotch from Britis
trading firms in China, reserved th
Black Label for officers only.

than 21 per cent of the Matsushita Electric Company's $5.7 billion annual net sales (1976 figures).

Given the wherewithal to buy, as Japan's income began to rise in the early sixties, the Japanese consumer responded with the zeal of a starving man let loose in a cafeteria. After almost a generation of privation, from the thirties onward, he felt it was time to make some acquisitions. There were for the first time in modern history no fleets to maintain, no armies to provision and no one to tell the Japanese consumer that he had to hold back for the good of the country.[11] Money started to circulate. Exports increased, thanks to a yen rate of 360 to the dollar that lasted from 1948 to 1971 and grew increasingly favorable to Japan as Japanese productivity rates went higher. This led to more spending and wider aspirations.

The advertising business began to take hold in the late fifties and, predictably, added fuel to the consumer's desires. (Dentsu, Japan's largest ad agency, is now also the largest in the world, with annual billings of $950 million in 1973, most of them domestic.) In the United States, people had spaced out their acquisitions. In the thirties you finally got your washing machine. In the forties there was a bigger radio in the house. By the early fifties everyone wanted a television set. In Japan the washing machine and the television revolutions hit almost simultaneously. The stimulus to the country's manufacturers was immense. Indeed, the huge surge in plant capacity that was fueled by the consumer revolution in Japan made an export drive almost inevitable, even if it had not been subsidized.

By the end of the sixties hundreds of thousands of Japanese were traveling to foreign parts each year, in a tourist outburst that has, among other things, kept Japan Air Lines one of the few really profitable international carriers, and has fearfully taxed the supply of overseas Japanese language interpreters. In 1977 some 2.5 million Japanese tourists visited the United States, Southeast Asia, and Europe—with special attention paid to tour favorites like Guam, Hawaii, and Hong Kong.

Despite the energy crises and warnings of desperate environmentalists, the car factories continued to produce their wares. In

---

11. Japan spends just over 1 per cent of its annual budget on armaments, which gives it quite a competitive advantage over countries like Britain and the United States, which spends quite a bit on the nuclear "umbrella" which is still the basic guarantee of Japan's security.

1973 production of private passenger cars alone ran close to 2.5 million. By 1975 a total of 28 million vehicles were in action on the roads, which are already choked with pollution and inadequate to the traffic already on them. Yet as a demonstration of sheer consumer buying power, at least, Japan's car ownership is an impressive statistic.

In the very act of spending his way past the level of Western European consumption on to that of the American, the Japanese consumer has managed, at the same time, to keep up the world's highest rate of personal savings. The Japanese still put a substantial part of their national income into the bank. Savings have continued to increase with the years. In 1966 the average saving household in Japan had just under ¥800,000 in the bank. In 1972 the figure was ¥1,640,000 ($5,460). It has since increased. The average rate of private saving, as a percentage of disposable income, is more than 20 per cent, as against barely 6 per cent in the United States. The Japanese wage earner still saves almost half of his semiannual bonuses, even though he is also prone to spend heavily at bonus time.[12] Each year money goes into the till against the purchase of a house, the children's education, or old age care, with the result that more investment cash flows into the Japanese banks, who loan it to the companies to put into more plant, so they can sell more to the consumers who made the savings.

Japanese intellectuals, who lean toward theoretical Marxism almost as a status symbol, have long looked askance at this grabbiness of their fellow countrymen. The late Ryu Shintaro, a brilliant essayist who was for many years chief editorial writer on *Asahi Shimbun*, once wrote a book called *The Flower-Viewing Binge Economy* (*Hanamizake Keizai*) in which he likened the whole consumer economy to the genial drunks who set off for Ueno Park with a barrel of sake to sell during flower-viewing time, at ¥100 a glass. Before any customers appeared Yaji got thirsty and asked for a drink, giving his partner a ¥100 coin. A few minutes later, Kita

---

12. Almost all Japanese companies give employees semiannual bonuses, in addition to their regular salaries. The bonuses are normally large and employee unions try their best to make them larger. Two or three months' salary at each bonus period is considered modest. Companies which have had a good year often give as much as five or six months' worth at a time. It is highly possible that a Japanese worker may receive, with bonuses, double the salary for which he is listed. Thus to quote a Japanese worker's monthly salary alone is a most inaccurate way of estimating his total take.

felt a bit parched and asked Yaji if he might have a drink. To keep accounts straight, he gave Yaji ¥100, the same coin he had received. On they went, until the sake was drained. They congratulated themselves on finishing off the barrel so quickly, then sobered up in shock when they realized that there was still only the one ¥100 piece between them.

Ryu died before he could see the worst excesses of the consumer economy he warned against. He was right about the evils of a consumer economy, of course, by Marxist standards; but Keynes might have suggested some mitigating factors. The inflation of the mid-seventies has given more point to the criticism. Tokyo's governor Minobe Ryokichi, who was originally voted in by promising to cut prices for housewives, still darkly talks of the affluence around him as the last gasp of capitalism (although he does not say so in elections). But the consumer economy continues to roar on, through crises and recessions, with the ¥100 pieces still apparently multiplying.

In a sense, the Japanese consumer is a sitting duck. He lives in the most tightly knit society in the world, where density of population is combined with generally accepted levels of taste and aspirations and a high educational standard that tends to produce consumer gullibility, rather than skepticism. Brand names have a national influence. Just as the average Japanese gets his information from one of the three major national newspapers, *Asahi*, *Mainichi*, and *Yomiuri*, he drinks basically one of three beers, *Asahi* (no relation), *Kirin*, and *Sapporo*. For all the weaknesses of the distribution network, which is heavy with middlemen and subcontractors who, to the Western mind, are totally useless, the manufacturer knows his market and his marketing targets very well.

Paradoxically, the Japanese is a frugal person, but one who loves frills. He lives in housing which is far below Western standards. Unlike the American consumer, who spends most of his life trying to move into a better house or apartment, the Japanese will purchase living quarters only once in his life, if he is lucky. He allocates most of his savings toward that end. The urge to buy one's own house is almost overpowering, especially with the specter of fifty-five-year-old retirement at minimal pension rates always hovering in the distance. In 1973 a Bank of Japan survey noted that well over 80 per cent of all Japanese families were saving toward the purchase of land or a house or apartment. Given the crowding of the cities and the lack of significant government help on mortgage loans or any pretense at rent controls, this very demand has pushed up the price of land to almost unreachable levels. Land in a residential area near Tokyo's center has gone well over $200 per square meter and $100 in suburbs 30 kilometers outside of town. The low-cost housing of Tokyo's Housing Supply Corporation drew 26.8 applicants for each rental housing unit, 84.6 for each housing unit for sale.

A recent pamphlet by the Japan External Trade Organization states, "Perhaps the major point of difference between the Western mode of living and that in Japan is the higher level of accumulation of wealth which characterizes life in the West and the lack of this accumulation in many areas of Japanese life." Central heating in homes is still a rarity. The average Japanese family lives its life in a welter of gas heaters, electric space heaters, hot water heaters (separate ones for each sink), electrically heated dining areas (kotatsu), where people can at least keep their legs warm, and such bizarre devices as the popular electrically heated, synthetic fur slippers.

Resigned to this austere environment, he spends money readily on other luxuries. He heads for the department store on Saturdays like a soldier back from the front visiting the PX; and he buys accordingly. The average Japanese house may include a clutter of expensive electric clocks, tape recorders, hair dryers, and more bulky luxuries (almost 20 per cent of Japan's households possess an electric organ), but it may have only the most primitive kind of plumbing. It is an interesting comment on the world's fastest-growing affluent society to note that the neighborhood bathhouse is still a fixture, due to the short supply of adequate bathrooms and washing facilities in so many Japanese homes.

Given the constants noted above—the efficient, well-educated worker who doubles in brass as a formidable consumer market, the combination of sound government-business planning and subtle changes of direction, the massive financial capacity and awesome research and productivity potential—there would be no reason to expect that Japan could not fulfill the extraordinary projections which some students of the future have outlined. The flaws in the picture are obvious, however. Before Japan moves on to overtake American production and consumption in the 1980s, the Japanese planners must not only face the problem of the world's dwindling natural resources, which they are swallowing with rather offensive speed. They must also face gathering problems of direction and concern among their own people, who are now vocally consumer-oriented, intensely worried about pollution and land use, not to mention a rising concern about the virtues of working patiently for Japan, Inc., on the transistorized anthill. As the *Economist* wrote in one of its impressive Japan surveys, "The real question is whether the economy can be reconstructed in the 1970s, whatever the growth rate is, so that it ceases to be a hosepipe squirting a jet of exports onto the world and so that some of the pressure generated by that extraordinary Japanese energy is diverted to cleaning up Japan itself." [13]

The extraordinary export drive of 1976–1977 made the hose-pipe analogy all too real. By 1978 the increasingly angry reaction from Europe and the United States had made itself felt almost everywhere in Japan. The society is finally aware of the problem. Discussions, meetings, and much antenna-waving are now taking place. Work is proceeding, in the time-honored manner, on reaching some consensus. This will take time. Whether there is enough time, however, is a moot point, for there is one great handicap which is inseparable from the group decision making, the national sense of solidarity, and the willingness of Japanese to line up and work to achieve one clear goal: It is very hard to change the signals, once they are set.

13. March 1973.

Decisions on important matters should not be made by one person alone. They should be discussed with many. It is only in the discussion of weighty affairs, when there is a suspicion that they may miscarry, that one should arrange matters in concert with others, so as to arrive at the right conclusion.

—The Constitution of Prince Shotoku

The Japanese is the world's committeeman par excellence. In the tradition of the group decision of the gods in the bed of the tranquil river, when the eight hundred myriad deities took counsel as to how they could coax the sun goddess Amaterasu Omikami out of her cave, the Japanese have been discussing and discussing before resolving almost any problem. Their deliberations are time-consuming, wordy, and, to an outsider, consistently exasperating. But they generally end up with the kind of true consensus that is an envy to us all. The Japanese variety of group decision, for all its defects, can be compared with other memorable national inventions such as the Macedonian phalanx, the Viking longboat, the Lancashire spinning machine, and the Empire chaise-longue, as one of those inspired devices which has gained prominence for its nation at a particular point in history.

Americans make poor committeemen. Despite this, we have abandoned most of our own decision making to committees. Although it is beginning to dawn on many of us that something is lacking in this arrangement, the situation will probably not change. And since so much of our technological future seems fated to be decided by meetings and relatively little, any more, by crusades, individual divine inspiration, earth-shaking dissenters, or other classic furnishings of the West's past history, we would do well to examine the success of the Japanese committee in action.

I have had a unique chance to enjoy (and suffer) the skill of Japanese committeemanship while serving as president of two Japanese companies and consorting with people from many others. Not only did I have my own companies' meetings to cope with, but I had to participate in a wide variety of discussions with scholars, outside editorial collaborators, other Japanese businessmen, and government officials. Although I still bear a few nicks and scars from these encounters, I learned a lot from them. There were lessons in patience, tact, and thoughtfulness, as well as how to steer a group toward a good decision, on occasion, and how to head them off from making a bad one.

When I went to Tokyo as a businessman in 1966, my experience in committees had been mostly foreign-policy seminars, editorial conferences on magazines, or congressional committees (the last when I worked as a consultant for the House Committee on Space and Astronautics in 1958 and 1959). The foreign policy conferences and seminars, although productive of occasionally interesting results, generally ended either in convinced disagreement or bland joint resolutions. Editorial meetings reflected the fact that most good journals are run as editorial dictatorships. ("I can't hear you too well," said the late Henry Robinson Luce in the course of a long-distance conversation with a *Life* editor. "Does it matter if you can't hear us, Harry," replied the editor of *Life*, "as long as we can hear you?") In the House of Representatives it seemed to me that a committee or subcommittee was just about as good or as bad as its chairman. The chairman staged the meeting, he asked the important questions, and he kept a hand on the rudder, for better or worse.

Instinctively, in my first meetings with my Japanese colleagues, I tended to adopt the strong-chairman approach. "Thank you for that proposal," I would say, "very interesting. And don't forget to send me a memo about it, so we can see it all on paper. Meanwhile, I think we'd better strike out along the lines of the plan which I've sent you. Please let me know if you have any questions. Does anyone have any questions now, by the way?" Of course there were none. The Japanese bow to no one in their reluctance to respond to the phrase "Any questions?"

This is certainly an approved procedure in some American companies where the memo is the medium, and whole squads of accountants and special assistants may be employed researching why various subordinate executives had failed to comply with the president's memo of August 14, 1967, or his final warning in the decree of September 19, 1972, etc. The Japanese, however, hate paper as a means of communication, however much they prize it as reading material. Although some final matters must necessarily be put down in memo form, the exchange of ideas is something they prefer to do face-to-face. Indeed, the oral commitment often counts far more than the written contract. ("I don't care what you write in those memos," one of our Japanese executives cautioned me before one labor negotiation meeting, "But please be careful what you say to them.")[1]

It worked out that our strong-chairman meetings grew shorter and shorter. I noticed a rising concern among the department heads. One of our partners finally took me aside and imparted some good advice: "At least listen to everyone's opinion before you give yours—even if you have made up your mind," he said. "People like to be heard. And who knows, you might hear something interesting."

I finally recognized (as many of my fellow Americans in business here have yet to do) that my tactics were breaking every rule of the Japanese committee. The committee in Japan is neither a debating forum nor rubber stamp. It exists to exchange different views and to achieve in the process a meeting of minds, that kind of comforting harmony—the *wa* ("harmony") principle—which means almost everything to Japanese community living. As a friend of mine in Japanese politics explained, "When we meet in a committee we are all groping. We put out our antennae and we try to see what the other man is thinking. We want to get a line on everybody's thoughts before we move, so we'll know whether there is any really serious obstacle to agreement."

In business or in politics, it would seem, the good Japanese chairman is seen rather than heard for most of the meeting. His presence should be felt, but not urged. He is the presiding power, and, for that very reason, he should wait until everyone has been heard, instead of trying to stimulate their thinking. After all, they are supposed to be stimulating his. For a while, therefore, I sat back and listened, intervening only when I thought the meeting was running down. At the end I would deliver my summation.

Uncritical listening, however, proved disastrous. The Japanese

1. Part of the reason for this is mechanical. A Japanese typewriter is actually a mini printing press and it takes at least six months to master. Typing the simplest memo can be a major production. The language itself, however, is so vague that clever explanations and attempted niceties of expression may cause more communication problems than they solve. Most Japanese company typewriters, when available, are reserved for unambiguous declarations like financial reports and price lists.

meeting is full of hidden pitfalls. For example, it is normal protocol in Japanese committeemanship to indicate one's sympathy with an opposing point of view, before attacking it. Expressions such as "*goi shimasu*" ("we are in mutual agreement") or "*dokan desu*" ("I concur") are almost sure-fire signs of impending trouble. When a man approves of your idea "*gensoku teki ni*" ("in principle"), it is time to take to the trenches. Such statements of apparent accord with an opposed viewpoint are essentially made out of consideration for one's colleague's feelings in a society that likes to avoid abrasive exchanges, especially if they are unexpected. One learns to hear out these apparently affirmative sentiments, ears turned for the inevitable "*keredomo*" ("but") that is bound to follow.

The crafty committeeman will get the chairman's or general approval on the bulk of a proposal, then he will slip in a curve with an extra provision or two at the last minute, hoping that it too will receive approval. A few shrewd committeemen, acting in concert, can engender a sense of urgency which transforms a discussion into a decision which others were not quite ready for. In this situation, the prudent chairman will retire to the safe high ground of "*kangaemasu*" ("I'll think about this"), while emphasizing that one must never be too precipitate in launching a project or making a decision, without weighing all factors involved. Such a statement is bound to find favor with a certain number of those around the table, especially the ones who have been sleeping or not saying much.

All too often, however, overwhelmed by tons of honorifics and the bright faces of the men with the winning idea, one would say "yes" and give a decision then and there. This could prove disastrous. Inevitably, several days would follow of meetings with people on the wrong side of the decision who explained (often validly) just how badly one had sold out the store. In my own case, I would often write voluminous memos in English tortuously revising my decision, confident that my memo language was just as confusing to my Japanese associates as their long-winded oral explanations had been to me.

After a while I got the hang of the thing and accepted the Japanese committee for what it is: a device for bringing people's views out into the open, stimulating different opinions, and, in the process, resolving them. If serious disagreements remain, they are ironed out in private so that, when the moment for final decision comes, everyone knows which way the decision will go and has adjusted his mind and mood to it.

There is something marvelously stimulating, yet soothing, to the Japanese committeeman's mind about that magic phrase "*kentō shimasho*" ("let's make some estimates"). It means that nothing at the conference has been really decided, but that the time has come for a final weighing of views before any concrete steps are taken toward solving the problem. *Kentō shimasho* is the finest moment in the Japanese consultative process. It means that finally the powers of the particular committee are getting ready to consider all plans that have been put forth, however clear it may already be from the course of the discussion which way they are going. When a decision comes, if it is a good decision and the committeemen are masters of their art, it will emerge from the *kentō* process long days later, rather like a camellia coming out of its bud —a beautiful but inexorable and completely predictable result.

This committee system also has its defects. It takes an awful lot of time. It is ill-fitted to cope with crises. Sudden, unexpected developments are overwhelming to the Japanese committeeman. Often committeemanship results in homogenized, compromise decisions instead of bold ones and all too often it is the refuge of incompetents. I have watched grinning, fluent mediocrities trample on the views of better men who, because of lower seniority or lack of forensic ability, were forced to keep silent.

Yet the gain in solidarity, all things considered, is well worth it (to a Japanese, at least). So is the feeling of participation that everyone has, after going through the decision-making process. It may have taken a long time, but one has a feeling that the eight hundred myriad deities in the tranquil river bed each could count on getting a hearing.

This sense of committee solidarity is not unknown in the United States. In fact, in recent years the American committee has tended to drift away from the strong-chairman system and abandon itself to a process which the Yale psychologist Irving Janis in a recent study labeled, after Orwell, "group-think." Taking his starting point from a study of various bad committee decisions of the United States government, for instance, the Bay of Pigs, Janis calls this "a deterioration of mental efficiency, reality testing and moral judgment that results from in-group pressures." In other words, members of the committee were more interested in keeping the approval of their fellow conferees than in coming up with strong opinions. "I was surprised," he wrote, "by the extent to which the groups involved in these fiascoes adhered to group norms and pressures towards uniformity, even when their policy was working badly. . . . Members consider loyalty to the group the highest form of morality. That loyalty requires each member to avoid raising controversial issues, questioning weak arguments or calling a halt to soft-headed thinking."

Somehow this does not happen in Japan. Perhaps it is because the Japanese have used the committee system so much longer that they can distinguish between loyalty to the group and differences of opinion. They can express controversial views without feeling that their loyalty to the group is thereby called into question. Perhaps the American, hearkening to a still dimly felt Jeffersonian past, has to justify a group decision by pretending that it is really his individual decision, too. Whereas the Japanese, even when overruled, is confident that he is making his contribution to the group. More importantly, he cheerfully goes along with the group once the decision is taken. That is, perhaps, what loyalty means.

Chapter IX

Japanese and American Businessmen
Or How They Got Ahead

Renounce desires and pursue profits whole-heartedly. But you should never enjoy profits. You should, on the contrary, work for the good of others.

—Suzuki Shozan, Zen master

It was only 7:45 A.M., but the early-morning Shinkansen express —familiarly known to foreign tourists as the bullet train—was already well out of the Tokyo-Yokohama-Kawasaki industrial belt, purring along at 140 kilometers per hour like a huge mechanical cat, past a few patches of farmland, distant mountains, and a tiny blur of sea on its way to the massed smokestacks and flat factory buildings outside Nagoya. There was just room for two more in the window seats at the end of the buffet car. The businessmen who sat down, one Japanese and one American, had already begun a familiar dialogue.

"Profits," said the visiting vice-president from Benchmark Manufacturing, "are the name of the game. You know that as well as I do. Fuji, look, you get your return on investment calculations, right, correct that plant amortization figure, then you set your yearly profit objectives and prepare your forecast. Stick to that forecast. We'll trim costs if we have to—so sales go down in a year. As long as we're flexible enough with our costs and keep trimming when we have to, that's no problem. The big thing is to make that profit objective. If we don't, I may be sending my résumé around next October, and so will you, Fuji.

"Now I know you don't like to lose heads in a Japanese business, but that's the easiest way to cut. And quite frankly, a lot of the people I've seen here in the last week we could do without. . . .

hose girls bringing the tea—and who's that guy they call a sta-
onary auditor?" [1]

Fujimoto, the Japanese general manager at Benchmark's new
'okyo branch, is worried. Ever since Calvin Knox flew in from
ndianapolis, barring a few man-hours lost to golf and evening
ntertainment, he has been pounding at the same thing—it is as if
Knox were back at the Des Moines plant, where Fujimoto served his
pprenticeship and which he still remembers with a shudder.

"But Mr. Knox," he begins what has become a refrain, "Japan is
ifferent. We have different ideas about business here. We like prof-
s, too. But we build a business around our people. We just can't
re the way you fire in the States. We need all our people. They
uild our market for us. The market is what counts here. I can tell
ou, we have to get that corner of the consumer market, or we'll
ever grow. By the way, you've seen the plans for the new build-
ng and the R & D lab. What do you . . ."

"Fuji, you must be crazy. An R & D lab? Why, we don't have
ne at the home office in Indianapolis. And let's not start thinking
bout a building when we're only getting 3.2 per cent pretax on the
usiness we're doing. Get that profit up to 15 per cent, where our
rofit goals should be and then we'll talk. You can't tell me that
apan is all that different."

"Look, Mr. Knox . . ."

"Cal, please. . . . Remember, you're Fuji, I'm Cal."

"Well, ah, Cal, would you mind reading a little article I have
ere. A friend of mine in business wrote it up for a management
rientation course at Sophia University."

"You've got those management courses here, too, haven't you?
haven't been in a classroom since I got my M.B.A."

"If you wouldn't mind, please take a look at it when we go back
o our seats in the green car." [2]

1. Japanese companies are required
y law to have *statutory* auditors,
'ho sit on the board of directors,
ut whose only duty is to certify the
ompany's financial statements. Un-
l recently, at least, they may or may
ot have been accountants and the
'ork they were expected to do is
inimal.

2. Taking his place in history along
de the United States airline states-

man who substituted "economy" class
for "tourist," someone at Japanese
National Railways decided several
years ago that terms like first class
and third class were undemocratic.
So the first class type of accommodation was christened green car, with
green cloverleaves painted on the
cars. Critics and all concerned were
happy and the service has not
changed.

"O.K., Fuji, old buddy, you think I can get some orientation by reading an article, I'll read it. The one called, 'Why the Japanese are Ahead'? It starts, 'In 1976 the United States and Japan . . .'"

"Yes, Mr. Cal, that's the one."

In 1976 the United States and Japan did more than $25 billion worth of business between them—$15.5 billion in Japanese exports to the United States, $10.1 billion in United States exports to Japan. Following the example of American investment in Japan, Japanese businesses have been steadily moving into the dollar market. By 1976, Japan's direct investment in the United States was about $5 billion, one fifth of total Japanese investment overseas. The trade between these two countries is one of the largest between two trading partners anywhere and it represents one of the free capitalist economies' better hopes for the future. If these two well-developed countries, with their interlocking trade, cannot solve the problems of trade and exchange for the future, one had better start thinking about either Marxist planification or conservative capitalist autarchy.

Yet the people who do the actual trading, Japanese and American, remain notably unsophisticated in their approach to mutual problems. The American businessman has spent considerable time complaining about competition from Japanese products, but relatively little time studying the philosophy of business, not to mention the people behind the tactics that make such competition effective. The Japanese businessman has devoted a great deal of academic study to the behavior of American business, not to mention the American consumer. When he is in America or Europe, he is constantly reminding himself of the differences between them. But in Japan, especially, surrounded by all the reassuring externals of his own tight island society, he is honestly perplexed when the foreigners behave so completely differently from himself.

Superficially, the Japanese and the American businessman have a lot in common. "Both of our peoples are hustlers," a former ambassador to Washington once said, "We like to move fast and get things done. We Japanese get that feeling in the United States always. In addition, you are so outgoing. I feel I could talk to Americans. In Europe, it's quite different. They are a closed society, more like ours, and they do things in very strange ways. We are used to the American way, at least."

Japanese businessmen are even more obsessed than Americans

with the cult of newness. As in the United States, the line between technical innovation and gadgetry, between advanced ideas and sloganizing, is very thin. Almost every new wave that washes over American business carries on across the Pacific. Massive public relations campaigns, management consultants, head-hunters, direct selling, computerized decision making, negative options, software and the information revolution—you name it and the Japanese have it. The *Harvard Business Review* regularly has articles reprinted in Japanese. Both *Fortune* and *Business Week* have got together with Japanese counterparts to give Japanese businessmen the latest word on the newest trends, in the popular magazines *President* and *Nikkei Business*.

Yet the many obvious similarities between Japanese business life and American probably serve only to obscure the differences between them. We would do better to look at the premises of both parties: what motivates them, what kind of security do they have, and what kind do they seek; their views on markets and profits; their feelings for contracts and commitments; the way their companies grow; the way they make decisions. Then we may see not only how vastly different the Japanese and the American businessmen really are, but how much the very structure of Japanese business relates to its success.

The American businessman is taught early that profits are the name of the game. His yearly profit and loss record is his Bible and also his report card; his success with the profit percentage is, by and large, what will determine his utility to his next employer. (In this sense, at least, Americans think very far ahead.) The Japanese businessman is primarily interested in the growth inside his particular business community. He is anxious for profitability and realizes as well as the next man that no company can operate indefinitely at a loss. But profitability is not his only yardstick of achievement. His measure is the security and growth of his company. The American first demands efficiency of a business. The Japanese wants permanence.

For all the group-think at work in our society, the individual American still sees himself as a free agent in the market place. The Japanese does not. Once he has entered a company, he rarely contemplates any changes. He has entered a family. Americans who boast about the variety of companies they have worked for shock their Japanese counterparts. For management consultants scouting job candidates for their clients, Japan is a head-hunter's nightmare.[3]

The American businessman lives in a society of written proposals, comments and guarantees. He dotes on memos, countermemos, and multiple-copy Xeroxing. In American business, people often seem to be judged by the amount of words and figures they can combine on paper. This often leads to excesses like the late comedian Fred Allen's storied molehill man, generally found in the advertising business, who could start with a paper molehill on his desk at 9:00 A.M. and work it skillfully into a documentary mountain by 5:00 P.M. There are molehill men in Japan, too, hundreds of thousands of them, but they are more apt to perpetrate their inefficiencies by word of mouth—which at least saves on paper and duplicating charges. The Japanese businessman puts very little down on paper (although he is indecently fond of graphs, charts, and other forms of visual or statistical presentation).

The American businessman, newly arrived in Japan, will out of habit go to the file cabinet and ask for all the memos his key Japanese subordinates have written. He is generally surprised to find the memo output so small. (I can count on the fingers of one hand the times Japanese executives who worked with me have written memos on their own initiative, other than to prepare fixed reports when they are due.) Conversely, the Japanese taking over his firm's office in the United States or elsewhere would be apt to gather his new subordinates and talk to them, to see what kind of people they were and how they thought. This is in character. The Japanese, obsessed by the need to develop people and get along with them, sees even a foreign office as part of his company's village. The American, by contrast, tends to think of his place of work as a nine-to-five (or nine-to-nine) arena, where he does gladiatorial combat with all comers, wary of strangers and constantly checking his figures. An American banker friend of mine sadly testified, after ten years of Tokyo business life, "In Japan business is founded on trust. At home business is founded on—well, distrust. I am sorry to go back."

3. When Japanese do switch, it is generally well arranged in advance— and generally because a younger man wants more money than the stuffy seniority promotion system allows. Recently, when the president of an American company in Japan sought to hire a young banker, the banker first had to clear things with his branch manager, to see if the bank would permit his leaving in good order. Ultimately, the American had to visit the president of the bank, who insisted that his young employee must have a guarantee of long-standing employment, before the bank would release him.

It is to be expected that those two veteran American business gladiators, the lawyer and the accountant, are less important in Tokyo than they are in Chicago or New York. But the degree of difference between these two business worlds in this respect is little short of astonishing. The approximately 200 million Americans enjoy the services of roughly 400,000 lawyers. The 60 million West Germans have only 20,000 lawyers. But for 114 million Japanese (as of September 1977) there were some 11,000 lawyers. One would think that this would literally pose serious traffic problems outside the courts, with Japanese litigants fighting to get in for their ultimate judgment.[4] But the Japanese are not a litigious people, at least in the sense that they prefer compromise agreements to an open and shut decision. Farmers who sued other farmers, eschewing the time-honored methods of compromise and negotiation, have been ostracized from their villages. And, while urban businessmen are not that extreme in their feelings, they have a great hesitation about turning to the law.

In the generalists' world of Japanese business, the phenomenon of the dynamic attorney who becomes company counsel and goes on to the presidency is virtually unknown. There are, in fact, almost no company counsels. Most companies employ lawyers sparingly and use their outside lawyers only when they have to. To the Japanese mind, there is something slightly sneaky about having a lawyer pore over a contract, once a basic agreement in principle has been concluded. Americans, who have been taught to check everything out with the company lawyer, are surprised to find that in Japan having the company lawyer join the meeting may be more of a liability than an asset.

Accountants are equally suspect, especially the American accountant who insists on doing everything the way the book says. (As recent United States business history shows, many large accounting firms may pass on very dubious sets of assets and liabilities with equanimity, as long as all the entries are in the right place.) Doing things by the book, while obligatory in filling out the nu-

4. With the accent on the word *ultimate*. In normal cases, one can feel confident about getting a verdict in four and one half years. If a tricky case goes up to the Supreme Court, as in recent famous cases, it can run as long as twenty years. The Matsukawa Case was not untypical. In January 1951, the case was brought before the court in Fukushima, with Communist party members accused of derailing a JNR train. After successive appeals, reversals, and reviews in the lower courts, the Supreme Court finally judged the defendants not guilty for lack of conclusive evidence in 1973!

merous forms and registrations at the bottom of the Japanese bureaucracy, is not useful in negotiating real problems at the top, in either business or government. Originally a great many American accounting firms came to Tokyo to assist the many branches and joint ventures of American firms there. But even non-Japanese found that the necessarily cut-and-dried formalism of United States accounting procedures and carefully hedged recommendations were having little effect when it came to dealing with Japanese government ministries, as well as with one's own opposite numbers in business. One Japanese accounting firm appeared quite suddenly in the late sixties, both well versed in international accounting standards and wise in the ways of Japanese habits of compromise and negotiation. This firm is now doing a land-office business; foreign accountants have had to do some serious stock-taking.

In sum, most Japanese are intensely disturbed, not only by the table-pounding of American business confrontations, but also by the analytical approach that brings every business problem down to a matter of numbers. The keen-eyed controller, with his insistence on cost control and manpower efficiency ("The first thing we have to do in the Shipping Department is lose some heads.") may be a hero in the United States, but he spells only trouble to the Japanese. If the American feels his primary obligation is to save the company's money, it is a major concern of the Japanese to save the company's people.

All of the classic social characteristics of the Japanese village which we have noted—the dedication to the group goal, the sense of hierarchy, the dependence on superiors, the formalism—are intensified in the modern Japanese corporation. It is just this sense of community within the individual Japanese company that has made the Japanese worker and the Japanese businessman such formidable international competitors.

The Japanese executive is rather poorly paid, by American standards, and more heavily taxed. His exploits with the company expense accounts are justly famed. Yet there are not many $100,000 a year executives in Tokyo; stock options and various similar sweeteners are rare. Company cars and houses abound, as do bonuses; but not many "salarymen" can get rich quickly in Japan.

The incentives given the Japanese executive, like those given the factory workers under him, are those that go with permanence. Both he and the company have made their choice carefully and are prepared to live with it. Unless he is picked as a director or leading department head, he will retire at about fifty-five, helped with a small retirement allowance and maybe the gift of a company house; but he has saved for that day. He may even come back into the company or a subsidiary to work as a temporary employee, in a less critical job.

Not for the Japanese salaryman is the desperate frolic of extending overrides, stock options, the ever-more expensive houses, the constant shopping around for future job openings, of the American executive as he claws his way up his peculiar ladder. The Japanese executive's life, if humbler, is far more stable. He has the leisure to plan for his future, and the company's as one indissoluble unity. "Why do you think Japanese research is getting better than ours at a quarter of the cost in time and effort?" an American technical man in Japan asked. "Because the Japanese research people have continuity. They have the time to think. They are in their jobs for life and they plan accordingly, while their opposite numbers in the States spend half their time typing up résumés to find better jobs."

The Japanese company can afford to have constant training programs, because the man whose training they invest in, they rarely lose. They train consistently. The security of long-term employment and constant on-the-job education are among the major reasons for Japan's high productivity rate. People make productivity. And in Japan, productivity and market share are paramount.

For while the American businessman watches his yearly percentages, the Japanese is heeding the territorial imperative. He is out to get his share of the market. This is partly out of instinct, partly because competition in Japan's tightly packed society, if orderly, is almost unbelievably intense. Once he gets his corner of the market, and begins further expansion, he will start worrying about increasing the profit slice. He may, indeed, shamelessly pile up profits, as the large trading companies demonstrated in 1972 and 1973. But even such profiteering was incidental to their primary objective of cornering the market.

Morita Akio, president of the Sony Corporation and the man who practically singlehandedly made Sony a power in the international market,[5] summarized the difference of view in a 1973 talk to foreign correspondents in Tokyo:

"Because of this keen competition in Japan, the gaining of market share often takes precedence over profits. This has meant at times that

5. Sony's original investment was ¥190,000—about $500 at that time. In 1973 total sales exceeded $1 billion.

Japanese enterprises have had to operate at a lower margin than enterprises of other countries. Therefore, the oriental concept of face having priority over money, appears in enterprise operations. Japanese enterprises have invested tremendous amounts of effort, intangible as it may be, in countries abroad. I would say that this type of investment, in one sense, brings back far greater returns than the investment of money, in the long run.

"An American enterprise, which is concerned generally with short-term gains, would not be able to pursue such methods. An American businessman would immediately drop a project if it does not show up soon as a good return on the financial report. Also the person in charge of such a project would be fired. Fortunately Japan has a lifetime employment system, which encourages the long-range view even among lower and middle management levels. For example, a member of our company may be stationed in some far-off land, struggling to learn in a country with entirely different customs and characteristics. But he realizes that, with the knowledge he has gained in five years or so, he might become chief of the department in our head offices that deals with this area, and that in ten years he may become director in charge of our international operations, and later have the chance of becoming a top executive of our company. He, therefore, is keenly interested in how strong the company will be in five or ten years from now, at the same time that he gives his attention to the business at hand. He is thus not only working constantly to achieve today's objectives but also paying close attention to what should be accumulated over the years ahead.

"As a result, Japanese enterprises steadily move ahead, and even if progress is slow, it is very rare to see their projects backslide. American companies are constantly concerned with figures, and if rapid returns are not produced, the rating of the company drops. Except for very large corporations, therefore, I wonder whether American companies are willing to embark on world-wide marketing ventures that require long-term investments."

It would not be hard for an American company president to explain away Mr. Morita's criticism. Unblessed by the cozy Japanese banking system, the American businessman prefers to finance his expansion out of profits, if possible. He dislikes the idea of being perpetually in hock to his local banking institution, at formidable interest rates. Stock ownership in the United States is far more widespread than in Japan and the value of a public company's stock tends to fluctuate predictably based on its earnings reports. Thus a bad annual score in the profit and loss statement is always embarrassing, and can be disastrous to the whole structure of a business.

The unsuccessful American executive will then doubtless be forced to call it quits, receive his settlement, and sit back in his Carribbean condominium hoping for a new job offer. The successful American president will make some changes in a hurry when profits sag. Or he will, in some cases, pump up his fourth-quarter sales figures, do a little constructive juggling with the inventory, and get out on the crest with a better job offer, before the stockholders realize that there are some serious cracks in the company ceiling. This kind of behavior is hardly conducive to sound growth. However the system came to be as it is, the fact is that much of American business has become a numbers game, in the worst sense of the word. Three relatively booming postwar decades, helped by the vagaries of the tax laws and the inflated performances of an overvalued stock market, turned many from the honest pursuit of profits to the manipulation of figures for profit, which is an entirely different thing.

In such a system, production and productivity become quite secondary. Companies exist not to be built, but to be traded. The American conglomerate, so called, is all too apt to be merely an accretion of generally unrelated companies which have been acquired either for their attractive assets or for their equally attractive tax losses, or because they can be readily dressed up, traded, and sold at a profit. What happens to the employees in the course of these transactions is of secondary concern.

The Japanese, by contrast, rarely buy companies—at least in Japan.[6] Mergers they have plenty, but conglomerates, as the Americans know them, are rare. The various groupings of Japanese businesses, whether middle-sized like Furukawa or giant-sized like Mitsui or Mitsubishi, generally start as a composition of related industries, very much at one in their business philosophy and approach. Even when they diversify, they are kept firmly in harness through the group's central bank or trading company.

The reason subsidiaries proliferate in Japanese business is not, of course, wholly the result of altruism and a desire for organic growth among Japanese businessmen. Until recently Japanese tax law did not require consolidated balance sheets for parent and sub-

6. Overseas it is a different matter. The increasing volume of Japanese investment in the United States and Europe, as well as the underdeveloped countries, includes frequent acquisitions of foreign companies or, at least, a substantial share interest in them. Except for the ubiquitous trading companies, the Japanese are still nervous about starting branches on their own in the middle of alien urban cultures.

sidiaries. With subsidiaries and parent company all reporting their results separately—frequently with different fiscal years—the possibilities for juggling profits, overvaluing inventories, and disguising losses are almost infinite. Given a chance to confuse the government and the shareholders (to say nothing of the consumer) Japanese businessmen can prove to be as wily as any American or European counterparts. In addition, they have a notorious flair for backroom maneuvering.

Yet the way Japanese subsidiaries evolve is also a revealing commentary on their methods of doing business. Most Japanese subsidiaries are created in the interests of logical expansion. For instance, the Dai-Ichi Tractor Company which manufactures small tractors for rice paddies will decide, after a look at the market, that it would be advisable to go into home lawnmowers. To do this within the framework of Dai-Ichi itself is often impossible. The production department chief has his own goals to meet and he is opposed to the whole idea of lawnmowers. The personnel director shakes his head over the problems with the union that will result from such a shift in the company's traditional goals. The eighty-two-year-old chairman has been making ominous grumblings from his seaside estate. After all, he started the company forty years ago to make tractors for rice paddies and he is not about to see his tractors turned into lawn mowers. Although he has no formal power, his personal influence could cause great trouble.

In a situation like this, facing a possible showdown, the average American company would either buy a lawnmower company or set up one as a fait accompli. The chief executive who has control of the votes and the stockholders' confidence would simply give the order to go ahead. The production man could be retired early with a fine cash settlement; the chairman might be sent off on a cruise; the personnel director would be impelled to use his good labor relations record for a connection with a rival firm. In sum, the decision would be announced and dictated from the top.

The Japanese firm rarely does this. Despite the many advantages of strong direction, even the most dynamic company president would hesitate to risk the rending and tearing of the social fabric within his company that a drastic and unpopular decision might force. Does he abandon his decision? By no means. He discusses the problem with his board of directors and the department heads involved, then founds a subsidiary company. Thus he solves a multitude of problems. A subsidiary company will include all the executives who are really in favor of the lawnmower plan, without the disadvantage of having any of the old tractor men around. It will start out from scratch with its own union, thus no built-in labor problems. It has all the advantages of the back-up, support, and financing of the major company, without any of the social problems. In addition, suppose the lawnmower business does not work? Nothing happens to the parent company. Indeed, the problem may barely appear on the major company's balance sheet. Japanese corporate accountants work in myriad ways, their wonders to perform.

The subsidiary is also a nice device for bringing in foreign capital or technical help without in any way damaging the basic purity of the parent company. The Acme Lawnmower Company of Cleveland can readily be invited to join the new lawnmower company, preferably as a 49 per cent partner. This will enable Dai-Ichi Tractor to take advantage of Acme's management expertise, as well as the technical advances in the field, including several Acme patents which Dai-Ichi's planning division has been after for years. At the same time, the Acme people will not get too close to the main action, which is back in the headquarters of the parent company.

A successful company which grows one or more subsidiaries will control them most loosely, by American standards. The manager of a Japanese subsidiary is given considerable license, as long as he has the self-confidence to use it. Once he has his job, he and his executives are released to run their company by themselves. All they have to worry about is periodic reports to the parent company's directors, plus occasional inquiries from the parent company's planning board to see how they are doing. As long as they are doing well, they will not be interfered with. The subsidiary company is not regarded as a chattel, no matter how junior. At the same time, they are kept in line with general company objectives much in the way that village elders used to keep the junior households in line through the *honke-bunke* relationship. As there are periodic family councils in the village, there are periodic meetings of the managers in a given group—where policy is somehow decided on and coordinated. The mind boggles at the problems of coordinating them, but somehow there is always the Monday Club or the Friday Club.

The process of subsidiary making continues indefinitely, like that of biological reproduction. Dai-Ichi Lawnmower may, at one point, indeed, find the lawnmower market sagging a bit, but they may discover a wonderful opportunity in surgical instruments. Then you will have the Dai-Ichi Surgical Instrument Company, a subsidiary of Dai-Ichi Lawnmower which is, in turn, a subsidiary of Dai-Ichi Tractors. On and on they go.

In measuring the American and the Japanese businessman we should remember to weight the scales a bit. The Japanese economy has not been handicapped by heavy armament expenditures and the appalling costs of the Vietnam War. The postwar Japanese investment in new plant and equipment plays a heavy role, too, in Japan's relative increase in productivity. And if some apologists for Japanese protectionism claim that American industries, like the computer businesses, were fostered by government defense contracts, it is more than equally true that too much dependence on defense industry did more harm than good to many United States businesses. As the chairman of the Sony Corporation, Ibuka Masaharu noted years ago, "The transistor was an American invention, not a Japanese one. But you developed your transistor industry hand in hand with the military, and got used to the big prices they would pay for equipment. In Japan, we had to start out on the civilian market. We had to make a product that people could afford and still give us a profit. So we had to learn economies and short cuts, to survive."

The United States has been quite correct in criticizing Japanese protection of key industries at home. Japan's long-standing reluctance to liberalize imports and foreign investments would be understandable in a developing country, but not in the world's number two economic power. However, recent quotas on Japanese goods coming into the United States and complete liberalization of Japan's imports provided only part of the solution to the trade imbalance between the two countries. As Okita Saburo, former president of the Japan Economic Research Center, remarked in a 1973 seminar in Tokyo, "The productivity of American industries has not been advancing as fast as wages have risen, with the result that the export competitiveness of American products [is poor] in comparison to imported products. . . . Though opinion has often been voiced that the recent deterioration of America's international payments position is caused by Japan's inappropriate economic policy, basically it is attributable to factors, economic and social, in the United States." Mr. Okita, like all good economists, gave other cogent reasons as well for the trade imbalance which came to a head in 1971 and 1972.[7] By 1974, in fact, the extreme imbalance was corrected, only to re-occur again in the "exports" crisis of 1977–78.

American business, however, has yet to develop Japanese markets with its old vigor. Although there are significant successful exceptions to this, in far too many cases the potential of steady growth was sacrificed in favor of immediate return. There are two ways to view the emergence of another industrial power with a huge consumer potential: (1) as a new market, to be welcomed, or (2) as a competitor, to be feared and fended off. For all our professions of being internationalists and free enterprisers, Americans seem now to favor the latter view of Japan. They tend to avoid the effort of going into markets which may have strong local peculiarities. Sony's President Morita is fond of telling the anecdote about the two shoe salesmen who ended up in an underdeveloped country in Africa. The first cabled back to the home office, "No market, natives are all barefoot." The second cabled back, "No one wears shoes here. We can monopolize the market. Please send all available stock." One assumes that Morita's man was the latter.

A question of attitude is involved here. It is not that the Japanese system is better. Few foreigners would like to work permanently for a Japanese firm, with its network of codified loyalties and observances, stiff protocol, seniority rules, and submarine in-fighting. But, aside from the matter of national characteristics and local peculiarities, it is obvious that the success of Japanese business, its long-term planning, its unified decision making, its social incentives, has much of value for us all. This is especially true in a world where, inescapably, the collective plan is replacing the individual insight—a tendency that can only be accelerated by the dwindling of the world's resources. In the process of studying Japanese business techniques, Americans would do well also to restudy the Japanese market, in which there is still a great and expanding

---

7. Among them: (1) Almost two-thirds of Japanese imports from the United States are in farm produce and raw materials, for which demand growth is less responsive to increases in income; (2) the source of Japanese imports of raw materials and farm products has been diversified in terms of region, i.e., dependence on the United States has decreased; (3) "the adjustment of the foreign exchange rate by Japan and the United States was delayed, as was import liberalization by Japan."

opportunity for American products. The opportunity is yet to be fully exploited.

By now it was 9:40 A.M. Well past the eel farms of Lake Hamana and the ancillary factories of Hamamatsu and Toyohashi, the Shinkansen morning express was nosing its way into the Nagoya station. Calvin Knox finished the article and handed it back to Mr. Fujimoto.

"That's some pretty interesting sociology, Fuji," he said. "The man may have a point here and there, but I don't think he's altogether fair. Anyway, it doesn't have anything to do with our business. Just remember what I told you when you came to Benchmark first: It's all numbers. You can talk about growth all you want—we're all for growth. But this year it's those numbers on the bottom line that count—for you, for me, for everybody."

I am writing these paragraphs sitting at my desk in the offices of a rising Japanese publishing company, of which I am the president. It is 1972. The corridors outside my room are packed with about sixty young people with red armbands and red *hachimaki*. They are members of our union. Most of them carry placards, a few, flags. A flying wedge of militants, led by their Tokyo University-educated chairman, have forced themselves into the general manager's office next door to mine, protesting some alleged violation of labor rights (i.e., someone in management had taken down one of the thousand-odd paper slogans with which they have ornamented the editorial working area). They are trying to wear him down by repeated accusations and shouts, from "son-of-a-bitch" to "colonialism." The young women among our militants are even more abusive and vocal than the men.

My secretary has locked the doors of my room. I cannot leave to join the battle outside (at last, the real adversary situation which every good American businessman waits for!) because our Japanese executives have convinced me not to. The greatest hope of the unionists would be a confrontation in which I, a foreigner, could be pictured as jostling some Japanese workers. Nothing could be juicier meat, either for the newspapers or the pending case at the Tokyo Metropolitan Labor Arbitration Board hearing. So I am reduced to writing and reflecting on the strange turns of fate that brought me to this pass.

Our loyal employees are technically not on strike. They are merely conducting a slowdown and partial strike (an hour one day, two hours the day after) which has had the effect of wrecking the company's production schedule. Their form of militancy includes taking over the company rooms by force for pep rallies, distributing leaflets to passers-by on the street outside, which denounced the company's allegedly crooked sales practices, and tacking up signs labeled "die" on the office door of our executive editor, whose only offense is that he is almost seventy. Under Japanese law they cannot easily be prevented from doing such things, as long as they are not caught in the act of attempting physical violence or actually destroying property. Such things they tend to avoid.

The scene shifts. It is about a year and half later. The loyal editorial employees of the company, which include the same people who were screaming obscenities outside my office door not so long ago, are gathering in their offices for an informal party, to celebrate the publication of our latest volume. Desks, file cabinets, and the inevitable personal lockers, not to mention a mass of tea trays, cups, slippers, doilies, and other homey touches typical of the Japanese office, have turned their space into an office planner's nightmare. But work is now being done there, with some efficiency. For the party, desks have been cleared and quantities of beer, whisky, sake, and hors d'oeuvres imported. The ladies of the copy department are wearing their better pants suit ensembles and even the dignified retired journalists who check the articles have abandoned their game of *Go* for the occasion.

Accompanied by a flanking force of department heads, I venture down to open the festivities; the president of a Japanese company is always throwing out the ball in some sort of season's opener. The executive editor gives a toast by way of introduction. I make my speech, as planned—praise for past achievement, with a justified reminder of even harder challenges to come. Then I circulate. A few words to every department, but not too many. None forgotten. A drink with every division, if possible every group, and a little conversation over hors d'oeuvres. The cumulative effect is personally impressive. After an hour or two I begin to feel like a peanut floating in a sea of Kirin beer.

Nothing substantial, however, has been said. The president makes his appearance to sustain the sense of solidarity, but any direct, pointed message from him would be regarded by many as counterproductive. If the president tells his junior employees that their ideas are wanted—and something new for a change, this could be taken as a reflection on some of their immediate superiors, who might not be at all receptive. If he tells one or two of the people in an outlying section what a good job they are doing, this might be taken to mean some lack of confidence in their courteous depart-

ment head, whose seniority and gentle Japanese manners toward
his superiors make him unbudgeable. Can a Japanese chief executive
lay it on the line and tell them all that the planned new projects
will come quicker if they do the work they are supposed to do
better now? Too delicate a situation, the majority of his directors
and department heads would doubtless agree. Cooler heads would
then suggest that the subject might be better taken up at the next
executive committee meeting.

The single most frustrating thing for an American working in a
Japanese company is the apparent need to work through and
around people, rather than direct them. One can push the company,
at the cost of some effort, to soaring heights of comradely efficiency.
But the man who tries to pull it generally finds himself standing
alone on the top of the hill with a broken rope.

If it is a smoothly managed company, the key figures are not the
president or directors, but the department heads, the acting department heads, the division chiefs—who run any well-managed
Japanese firm. Almost every foreigner who blunders into a Japanese business is, at some point, led aside to hear a gentle lecture on
the virtues of bottom-up initiative, as opposed to top-down direction. The Japanese are well aware that in the American company, it
is the president or the executive vice-president who plans, who
initiates, who drives the company forward, in the manner of a
profit-conscious engineer pouring on the coals to pull a reluctant
train. When this top-down management happens in Japan, it is resolutely disguised. The department heads, generally, are the people
who work out the tactical plans. They participate in the innumerable conferences and agree in the concurred decisions. In most
Japanese companies, in fact, major decisions have to be recorded by
the extraordinary process which the Japanese call *ringisho* (literally, "decisions in a circle by document"). In *ringi*, after some
preliminary tallies, a memorandum setting forth the details of a
plan is prepared by someone on the planning staff or a junior department head. Generally, basic plans are prepared far down in the
organization. It is then passed from department head to department
head up the ladder to secure their opinions and agreement. Its passage is, of course, preceded by long conferences. Abridgements
and corrections are made, so that by the time the final paper is cir-
culated, everyone's ideas are pretty well on the table, if not in the
last draft. Everyone wants to see what is written on this basic *rin-*

gisho, which is one reason for the scarcity of copies of Japanese memoranda. (The fewer copies, also, the less room for considered objections.)

Ultimately, the *ringisho* gets up to the managing director level. Many seals are already upon it and it is close to final acceptance. The director concerned may well have been informed of its progress. In fact, he may have initiated it; although, if he is a good manager, he will have taken pains to keep up the appearance of the initiative having come from the lower ranks.

By the time the document reaches the president, all that is needed is his seal. The president, too, will have been kept informed of its existence, if not its progress. He generally has no reluctance about signing it, but it is his responsibility. If something goes terribly wrong with the plan that has been outlined, it is not the senior assistant in planning who is ultimately responsible. It is the president's resignation that must be in the chairman's office the next week.

Similarities between military decision making and Japanese business decision making are striking, especially to someone who learned Japanese in a wartime atmosphere. Staff officers, often at quite a junior level, work out major strategies to be initialed by the chief of staff, and then submitted to the commander, who in most cases, does little more than initial them himself. Indeed, the *ringisho* system is more understandable to a military-staff trained mind in whatever nation. Pentagon veterans would easily adjust to it. The late Dwight Eisenhower, America's masterful military adjudicator, is one of the few American executives I can think of who would have flourished in such an atmosphere. With most Americans, however, the mechanics of group decision, the oblique confrontations, and the seniority system become oppressive.

Policy differences and my personal feelings apart, however, I find much to admire in the Japanese way of doing business. The system of bottom-up decision making, for example, although frustrating to Americans, is by no means bad. It is, indeed, as the Japanese are fond of saying, more humanistic. We are all on this world for a limited span, one should remember. Viewed *sub specie aeternitatis*, it is a good system that gives everyone his say and his dignity, whether he is individually able or not.

In addition, the fact that relatively low-ranking people in a company can make plans and have them taken right to the top gives them both personal recognition and the beginnings of a confidence that will really pay off when they get to the critical department head or director level. They may not always have authority, but they grow familiar with its uses.

Recent history has not been kind to the cherished American principle of giving the boss his head. The excesses of the postwar decades —from the everyday arrogance of generals and company presidents to apocalyptic pretensions in the White House—may yet make us aware as a people of the strong streak of authoritarianism we have. If our genuine democracy still works, the egalitarian principles on which we founded it are visible only on a clear day.

The current role of the American company president seems increasingly less like that of an entrepreneur and more like the storied Priest of Nemi, made memorable in Fraser's *The Golden Bough*. The proprietor of Diana's sacred Italian grove was all-powerful in his sphere, but he had his problems. As Sir James Fraser wrote, "In his hand he kept a drawn sword, and he kept peering warily about him as if at every instant he expected to be set upon by an enemy. He was a priest and a murderer; and the man for whom he looked was sooner or later to murder him and hold the priesthood in his stead. Such was the rule of the sanctuary. A candidate for the priesthood could only succeed to office by slaying the priest, and having slain him, he retained office until he was himself slain by a stronger or a craftier.

"The post which he held by this precarious tenure carried with it the title of king; but surely no crowned head ever lay uneasier or was visited by more evil dreams than his. For year in, year out, in summer and winter, in fair weather and foul, he had to keep his lonely watch."

In the American corporation the keenness of the president's drawn sword is generally represented by the quarterly earnings report of the company. If the company is doing well, so is the president. He is lionized, rewarded with extra stock, and (at least in the business storybooks) told to write his own ticket by a grateful board. If the earnings start to slip, however, he is in trouble. There he sits in his palatial Lake Nemi executive suite, nervously peering around him for that skulking figure who will someday displace him. The best way he can defend his position is by increasing the earnings and showing the board and the stockholders more cheerful figures on the bottom line.

The Japanese president, by contrast, has reached his job in a more leisurely process. He is not immune to palace revolutions or dethronement. Sometimes an apparently sleeping chairman can resurrect himself and throw the president out. (This happened in the sixties in Tokyo, when the progressive and hard-working president of a prominent bank decided, disastrously, to start a merger with another institution without receiving his chairman's approval.) But, in contrast to life in the American company, the Japanese chief executive can generally assume that the succession up the ladder is quite regular, that the man who succeeds him will likely as not be someone he has picked himself.

If the Japanese president is deposed, for a real or fancied failing, for incompetence or for low profitability (in the end the Japanese worry about profits, too) he does not face the corporate obliteration that the American president awaits. The chances are he will stay on the board of his own or an affiliated company, or be sent off to manage a tranquil subsidiary. One Japanese executive I heard of, after having successively wrecked a trading subsidiary, a computer manufacturer, and a good portion of the home office, finally found a rewarding and comfortable niche as president of the company bowling alley. At the least, a president who has worked for a while in his job will become what the Japanese call a *sodanyaku*, a sort of senior management consultant who draws a steady but reasonable income and can be called on for occasional painless consultations, but generally is not.

Which type of organization is better? It is easy to say that neither can be duplicated in the other environment. But each can learn from the other. As a foreigner in Japanese business, I was from the first impatient at the hierarchical posturings, the cloying formality of announcements, the excessive use of titles. I still am. The years have not eased the ennui of three-hour conferences spent discussing the pros and cons of a tiny transportation allowance to employees or the transfer of two half-able department heads to replace one hopeless incompetent or the patient hearing of masterful golf stories (life as a businessman in Japan has confirmed me in a lifelong devotion to tennis). But I am grateful to my Japanese colleagues for having given me a glimpse of the solidarity and the sense of corporate egalitarianism which this system produces.

The spirit is catching. When I returned to Japan to live in 1966, after not having lived there since the end of 1950, I was naïve and condescending about the strength of the Japanese business society, even though I had written about it before as a reporter. If the employees in my company wanted a company song, so much the better, I thought. But I smiled privately at the thought of this kind of ingenuous school spirit. If they wanted to have their outing together, instead of going on vacation trips, and needed money for the office ski club, the office hiking club, or whatever, that was probably a vaguely good thing for company morale. Harmless, but who would have thought of doing it seriously? My children and I used to play a little game counting businessmen and buttons, to see if we could recognize the different organizations faithfully bannered on the lapels of Japanese businessmen.

It is now a decade later and I fret when I cannot find my company button before going to work. I have developed a habit, half-conscious, of looking to see whether anyone in the office is not wearing his. If no one is organizing a company club or no one wants an outing, I worry about spirit in the company and may mention the fact at our next executive committee meeting. When I have a meeting of the company staff, or a party at home, the invitations are by strict order of rank. I have long since given up trying to mix people on different levels. And when I meet someone from outside, I instinctively reach for my calling card and preface my identification, so he will know where I stand and whom I represent.

It is not a world I would have made myself. But how am I going to explain to the politely attentive faces in front of me at the Hotel Okura reception that I want to be known as myself, apart from my organization? How can I say to the man at Immigration asking my occupation, that I am a journalist by trade. "But what is your newspaper," he would ask, "If you don't belong to a newspaper or a magazine, how can you be a journalist?"

So I give up and write down "company president." And here, of course, is my button and my card.

Chapter X

The Mandarins
Japan's Singular Intelligentsia

The basic function of any intelligentsia includes both the icono-
clastic and the pedagogic, the destructive and the constructive element.

—Arthur Koestler

At the nerve center of Japanese society, commenting, informing,
exhorting, teaching and denouncing, talk several hundred thousand
members of the Japanese intelligentsia: the professors, scholars,
writers, critics and journalists who do their best to justify both ends
of Koestler's definition. In that they represent a definite segment of
the population, are articulately aware of the differences between
themselves and the other people, and certainly claim for themselves
the guiding role of an "intellectual, social, and political vanguard,"
the Japanese intelligentsia would fit Webster's dictionary definition
of this European term. They claim to be much more than this, and
they are.

Thanks to the traditional Japanese respect for learning, virtually
codified by the self-educating surge of the Meiji Restoration, their
prestige is high. For more than a century, except for a period of
relative silence in the thirties, their collective voice has carried as
far as mass communication can amplify it. In fact, given the cohesive nature of Japanese society and its relative remoteness from
the international conversation, the broodings, spats, and causes of
this intelligentsia are often magnified out of all proportion—like
people muttering to themselves over a public-address system. Yet,
partly because of history, they do not play the more active, aggressive roles in the national life that most European or American intellectuals do. They are of the establishment, but not quite with it.

They are self-consciously isolated from and generally disdainful of the society which, to put it crudely, feeds them. Too much tied to the Japanese order of things to become an active force for revolution, they are too perceptive and idealistic to accept the political and social realities they see around them. Yet their comments and criticisms are less like a modern vanguard than an ancient chanting *utae* "chorus" of a *No* play. They warn, moralize, and criticize from the far side of the stage.

They are called many things in Japanese. *Bunkajin* means simply "a cultured person"; [1] *Chishikiso* means "intellectual class"; *interi* is short for "intellectual." They are unlike even the European intelligentsias in the fierceness of their caste consciousness and the extraordinary ordering in their ranking systems. Many of them (especially the journalists), do attain their positions by examinations, and keep them (especially the professors) by rigid rules of seniority. Most self-conscious intelligentsias tend to share tastes and attitudes in Europe and the United States as well as Japan: *for* labor, free expression, and the people; *against* business, censorship, and the police; *for* string quartets, rough sweaters, and long hair; *against* popular classics, crew-cuts, and cutaways; and so on. The Japanese *interi* are this way, but more so.

I use the word *mandarins* to describe them all, even though the free-lance painters and essayists among them may superficially seem worlds apart from the original mandarins of the old Chinese civil service. Their general conformity and sense of being elite merit the term. The mandarins comprise the professors in Japanese universities and a good number of school teachers, the editors and journalists of the world's largest newspapers and most active magazines. They include the novelists, playwrights, actors, and poets of the world's most changeable literary society and one with probably the world's widest popular base.

If there were some kind of comparative I.Q. testing of various national intelligentsias, I suspect that the Japanese would win. Like the general population's literacy level, their level of perception, sensibility, and general academic competence is close to the world's highest. But, in common with other segments of Japanese society, the intellectuals tend to travel in groups. Although the artist is sup-

---

1. In a stage play, the Japanese dramatist Fukuda Koson satirized this view by spelling out this word—*bun* as in *bumbukuchagama, ka* as in *bakemono* and *jin* as in *ninpinin.* In English, "I for idiot, N for no good, T for troublemaker," and so on.

posed to be a man who is different, Japanese society does not allow him to be too different. No one likes to travel too far from the mainstream. Lonely genius is really lonely and uncomfortable. The record of suicides among Japanese literary men—from the volatile Akutagawa Ryunosuke of the twenties to modern literary monuments like Kawabata and Mishima—is an inverse index of their conformity to the general intellectual pattern. It is no accident either that Mishima and Kawabata both tended to the right rather than the predominant, "progressive" left of their fellows. It is no accident that the conductor Ozawa Seiji was first hounded out of Japan by jealous associates and had to make it in the United States and Europe before returning in triumph to Tokyo; nor that the films of Kurosawa Akira were cheered as masterpieces by foreign art movie audiences before they were much appreciated in Japan; nor that the most recent Japanese Nobel Prize winner, Ezaki Leona, has chosen to live and work in the United States.

The market for the mandarins' wares is almost unbelievably vast. A captive audience in their compact islands, the Japanese have no equals in the unified field of education-improvement-information. As ex-Premier Tanaka, not an intellectual himself, once said, "That is the one area where our statistics are ahead of the Americans'." Japan's close to 2 million university and college students (including 300,000 in junior colleges) rank just behind the United States and the Soviet Union in sheer student-body size. In 1974 about one out of four people in the eighteen to twenty-two age bracket was in college. This was more than double the figure ten years ago.

The average Japanese, in or out of college, spends between three and four hours a day watching television—roughly the same as the U.S. figure; but he also reads. As of 1977, Japan's total general daily newspaper circulation was about 43 million—58 million if you added trade and other specialized papers. The total daily circulation in the United States is slightly more than 62 million. Just about every Japanese household (1.25) has a newspaper, on average. Given the population differences, this means that newspaper diffusion in Japan is almost double that of the United States.

The purchase of books is no less impressive. Some 30,000 titles are published in Japan each year, slightly under the United States figure (but the amount of fiction published in Japan is double). Statistics on Japan's magazine industry are enough to bring tears to the eyes of the former editors of Colliers, the Saturday Evening Post, Life, and the pre-1966 Show. Out of a total just under 10,000 periodicals—including quarterlies, government publications, etc., there are 1,400 monthlies with a total average monthly circulation of 90 million and 56 weeklies with a total circulation above 20 million per week. Even allowing for newsstand returns, it is rather staggering. There are magazines on current affairs, knitting, arts, philosophy, medicine, sports, entertainment, and, inevitably, golf, as well as a huge group of children's magazines. Circulations range from the 600,000 plus bracket for the big sex-and-scandal weeklies (most of them run by the newspapers to publish all the news that the dailies do not see fit to print) to the few thousands for really specialized intellectual fare. Yet a monthly like *Bungei Shunju*, with a level well above that of *Harper's* and the *Atlantic Monthly* in the United States, can bring in almost 600,000 circulation on its own.

In America, the combination of a slack reading public, hidebound publishing practices, and overactive direct-mail merchants ("Unless we receive your answer in the negative within two weeks, you will receive free of charge, the first trial volume of our splendid, lavishly illustrated, 20-volume set "Gun Fighters of the Old West.") have all but destroyed the bookstore. It is, by contrast, an exhilarating experience to walk into almost any bookstore in Japan. Past the crowded piles of magazines, people are standing reading at the shelves, picking up volumes and discarding them, exchanging comments and, inevitably, marching toward the cash registers ringing up their unending sales. In the Shibuya section of Tokyo, one establishment, Taiseido, bills itself as the "book department store." It houses eight packed floors of books on sale and does business like a prosperous discount house. Given Japan's reading public, its success is not surprising.

From historical novels ("Read the book that inspired NHK's current TV series") to etiquette manuals, the trade in best-sellers is tremendous. Japanese writers are almost indecently prolific. Popular novels, translations, poems, and essays (driven out of almost every other literature, the discursive literary essay remains solidly popular in Japan) can easily run into sales of hundreds of thousands. Sets come off the presses as fast as new Toyotas off the assembly line. *Masterpieces of World Literature, Great Works of Japanese Literature, The Complete Works of Mishima, The Complete Works of Tanizaki*—their number is infinite. Encyclopaedias do well.[2]

A really successful *bunkajin* can stimulate his book market with

2. The new *Britannica International Encyclopedia* in Japanese, finally completed in 1972, had sold 300,000 sets by mid-1978, despite the fact that its cost is roughly $1,000. Other reference books also prosper.

a spate of magazine articles, a few commercials, and series endorsements ("The new Masterpieces of Japanese Poetry series, in sixty-seven volumes, is truly a necessity for any cultured household") on the ubiquitous seminars and talk shows, as well as by participation in the almost unending parade of newspaper and television pundit discussions, round tables, and lectures. Fees for both television one-shots and endorsements are rather modest. Appearance on a talk show will rarely bring more than ¥100,000 ($335), while the average endorsement for a publisher's works can be had for less than ¥50,000. However, the market is a large one and the fees do add up.

The round of television and newspaper symposia never ceases. The slightest crisis, not to mention steady topics like United States-Japanese relations, the new China, Are the Japanese Really Economic Animals? and Whither the Working Girl—will bring familiar groups of mandarins into a studio, working over a topic in terms of mutual corroboration rather than confrontation. In television symposiums, as in business, Japanese hate the adversary method, much preferring to talk round and round in circles, gradually zeroing in on their topic, as the hour progresses, to the accompaniment of the obligatory rounds of "Of course . . . ," "Absolutely true . . . ," "I agree in principle, but I'd just like to add something to what you just said now . . ."

The deep-felt need for talking over problems within Japanese society makes the profusion of symposia (*zadankai*) inevitable and, not incidentally, fosters the hold of the mandarins on their public. As Japan's society of villagers moved into the cities, it made various adjustments to its new environment, in order to satisfy its old needs. The Japanese public, no longer able to get the steady guidance of the village elders, readily grabs at the bottled advice and admonition offered them by their huge communications industry. *Sensei*—that all-embracing word used to address a doctor, professor, diet member, teacher, or, indeed, any superior authority—has a magic ring to a society which believes in education as a kind of lifetime supreme value and goal. If the *sensei* says something in the newspapers, it is something to listen to with respect, if not complete agreement.

At its best this constant flood of advice and communication from the mandarins strengthens the consciousness of the one great family society and guides it toward a consensus of opinion on various subjects. At its worst it fosters an intense variety of the spectatorism that modern technology threatens almost every society with. "One hundred million people and all of them commentators," ("*ichi oku so hyoronka*") the late historian Oya Soichi, himself one of the few commentators worthy of the name, used to say of Japanese society. He would add, with bitter special reference to television, "A hundred million people and all of them idiots."

Behind the huge information society audience which Oya criticized—with its exasperating combination of literacy, conformity, credulity, and educated unconcern—lies the Japanese system of education which forms it. It is a unique and curious structure. Its extraordinary surges of growth, first in the Meiji Restoration, next after World War I, and finally the more recent postwar period, have been, if anything, more sensational than the corresponding surges in the Japanese economy. Yet, like the Japanese economy, the wondrous growth of education has brought with it problems of quality and content far deeper than the reassuring growth statistics. The schools and universities of Japan explain much about the mandarins' hold on their audience ("The only country I ever lived in," an American international banker said, "where you can look at a ditch digger relaxing from work at his lunch break with the local version of the *New York Times*."). They also say much about the mandarins themselves.

Until the end of senior high school, which is now close to universal (about 85 per cent of the population), Japanese education offers (outside of the Soviet Union) the world's most intensive and rigorously demanding system of mass instruction. Through their long disciplining in language, mathematics and history, and the backbreaking preparation for successive school and college entrance examinations, past the desperate cramming (and not so occasional breakdowns and suicides of their "examination hells"), Japanese children learn. They learn despite conflicts between Ministry of Education bureaucrats intent on restoring a kind of iron-fisted paternalism to the nation's schools and a dominant Japan Teachers' Union whose consistently strong left-socialist and communist leadership elements have negated a once legitimate concern for teaching independence.

They have to learn. The relaxed indulgence of the society toward babies and preschoolers gradually gives way, as the bonds of responsibility and accountability for failure are riveted. The merit system is faithfully observed. The elitist high schools like the famous First Higher School (Ichi Ko) in Tokyo, which used to insure its graduates for admission to the then Tokyo Imperial University the way Groton used to prepare for Harvard in the 1920s, have long since been changed.[3]

Once a person has arrived in the university, standards of marks and attendance are, to put it mildly, lax.[4] The core of what Japanese children learn, college graduates included, is largely absorbed by the time they are eighteen. Even assuming the defects of the university system, however, the school system as a whole produces good learners for future life. They are more skilled in the popular use of their language than people in any other large, modern country (this even despite a recent drop in standards). They possess a considerable, if undigested knowledge of Western culture, as well as a thorough grounding in Japanese (and an almost hopelessly academic acquaintance with the English language, which is generally taught as a problem in grammar, not communications). They tend to have orderly study habits. They read a great deal.

It is not an education that develops imagination and creativity, however. On this subject, it is worthwhile to note the observations of Dr. Ezaki Leona, the United States-based Nobel Prize scientist, on returning to Japan in 1974 for a brief visit. Thanks to the rigid system of college entrance examinations, Ezaki stressed, the Japanese mind is strong at executive ability, but low on creativity. The observation seems a fair one. The typical Tokyo University graduate is apt to make a talented executive, with a flair for constructive relationships and a born political sense. He likes to work through channels. At innovation, he is weak. He dislikes being singled out, even for praise, until he is sure he is on the right track, preferably with plenty of company. Innovators rarely make their mark, unless they are skillful enough to disguise their talents.

This criticism aside, the Japanese schools are not to be slighted. All too many United States educators, by contrast, have stripped down their scholastic disciplines in the interest of "creativity" and

3. More like the old-fashioned German gymnasia than the lower-level United States high schools, Ichi Ko and the other preparatory schools were designed for the former imperial universities. Ichi Ko itself was ultimately amalgamated, postwar, into the first two years of Tokyo University's Liberal Arts College.

4. There are some striking exceptions to this in Japan today. Sophia University in Tokyo among others, has the temerity to dismiss students who do not pass their courses. The students, not surprisingly, respect this indication of faculty concern.

have ended with students who had their own opinions, but could read and write and cipher only with abject difficulty.

That Japan's schools have succeeded so well is a tribute to the men who ordered up a new system of mass education by fiat in the Meiji era. The greatest was undoubtedly Mori Arinari, the extraordinary traditionalist modernizer who told his fellow samurai to stop wearing swords after his first trip abroad, suggested abolishing the cumbersome Japanese language in favor of a Japanized English, democratically drew up a contract marriage (sadly, it failed), but at the same time saw Japan's new school system and its pedagogues as a kind of samurai priesthood dedicated largerly to chronological learning.

Mori and the other Meiji reformers—Ito Hirobumi, Fukuzawa Yukichi, Okuma Shigenobu among them—set about producing a modern educational system which was undoubtedly elitist. As Ito and Mori saw it, the products of the new schools and universities were to be wedded to the service of the state, or in some cases, business.

The universities grew up fast, but expanded slowly. By the time World War II ended there were eight imperial universities, the apex of the educational system, and forty others, public and private, with a total of no more than 113,000 students, all male. It was a tiny foreshadowing of what was to come. Although the elitist mood was high upon them, a basic argument had already been in progress, over several decades, beginning as a dispute between Mori and Fukuzawa. Mori had wanted the scholar to serve the state. Fukuzawa, who founded Keio University and did a lot more for Japanese life as well, wanted Japanese scholars to stand aloof from the main current of society, true to their dedication as a separate, self-conscious intelligentsia.

The dispute, as well as the whole problem of mass versus elite, was thrown wide open when the United States occupation released a small army of well-meaning American education specialists, determined to reform their Japanese counterparts. As noted earlier, they had considerable success, at least in the quantitative measure. The prewar multichannel structure [5] of Japanese education, a complex of higher primary schools, nationalist teachers' colleges, college-

5. The multichannel and single-channel comparisons are borrowed from Herbert Passin's *Society and* *Education in Japan,* previously noted, an excellent study of the growth and problems of the Japanese schools.

level technical institutes (*senmon gakko*), as well as the universities was, indeed, organized on rather elitist lines. But it also had the virtues of separating children with different attributes, so that a future mechanic, for instance, could get an excellent education without either he or his parents feeling slighted that he had not been allowed to go on to law school.

After the United States occupation educationalists finished with the Japanese schools, some basic changes emerged. The schools were temporarily put into the hands of local school boards, many of which were quickly captured by Communist lobbies. Coeducation was made mandatory as a presumed premise for any real democracy. Most importantly, the entire school grade system was reorganized on the American 6-3-3 basis, with all the traffic ending at the gates of the new four-year university or, more precisely, at the room where they held the university entrance examinations. The resultant traffic jam was and is almost incredible.

As Japan's affluence increased, just about every Japanese mother made up her mind to get her children into the university at any cost. The *kyoiku mama* ("education mother") became a figure of postwar Japanese folklore, capable of almost anything, including stealing examination papers (such cases happened), if she could get her son or daughter higher-educated. Universities proliferated even more widely than in the postwar United States. By 1960 there were 245 universities with 711,618 students. In 1973 there were 404 universities with 1,597,282 students; of the universities, 75 are national, 33 public (metropolitan, etc.), and 290 private. In addition there were 500 junior colleges, with 309,824 students.

No institution could have been less flattered by the attention it received or less prepared to handle the new influx of customers. The Japanese university was, in 1945, and largely remains today a stronghold of aloofness. The great number of its professors, the core of the mandarinate, think of the university largely in terms of an old-fashioned ivory tower. It is no coincidence that the term *ivory tower* has such currency in modern Japanese discussion. Despite the hordes of mass-education postulants invading their campuses,[6] they have retained the stiff-collar German university traditions they developed in the late Meiji period, as well as the interlocking system of castes and precedences built up in the im-

6. Tokyo University, at the top of the system, has more than 18,000 students; Nihon University, the largest, has 83,000.

perial universities before World War II. After the wartime military leadership (*gunbatsu*) was destroyed by the defeat, the United States occupation tried next, with at least temporary success, to break the power of the old prewar politicians and their allied big money business circles (*zaibatsu*). Yet, despite all the lower school reforms, the professional caste (*gakubatsu*) of the universities was left untouched.

This comment on the university hierarchy is not meant as a reflection on Japanese scholarship. In the course of the greater part of a decade spent working with Japanese scholars, I have been most impressed by both the depth and breadth of their knowledge. In the liberal arts as well as the sciences it is basically only the language problem, plus a native reticence, which keeps Japanese scholarship from being a far greater force in the international academic world than it is today. High scholarship, however, does not necessarily translate into good teaching. For the most part, Japanese university students are taught, if they care to learn, in large, uninspiring lecture courses. Almost 15 per cent of university classes have five hundred or more students in them. At the major public universities which, on the average, can spend three times as much per student as private institutions, there can be seminars and some tutorial instruction. But they are exceptions, not the general rule.

Graduate study is generally weak and hopelessly tied to an apprentice system that long predated the university in Japanese life. The aspiring student becomes a graduate assistant, then a lecturer, working for a professor in his department, then rises successively to assistant and finally full professor, having done much of his learned predecessor's heavy scholarly work for him during his younger years. Since he must work within the same narrowing pyramid, promotion is slow. Transfers outside his particular department are very rare.

The professors are grouped within faculties. Each faculty is a power unto itself, virtually independent in its own administration. There is very little lateral communication, except in formal faculty councils, which are cumbersome and hopelessly inadequate instruments for administering modern university affairs. Since presidents are elected by faculties, administrative ability is rarely demanded of the candidates. The president has little real authority.

Thinking of the faculties, one immediately recalls the *dozoku* system of the old Japanese village, with its leading and subsidiary households, linked by their own peculiar semifamilial laws. Inside, the faculties jealously guard their preserves. A nongraduate, at

least in the most prestigious universities, has little chance of breaking into the faculty, even if he is a distinguished scholar. "No
outsiders need apply" is the unwritten code. Outside the particular
university the pecking order among the universities is rigidly enforced, in an even larger village-household system. Tokyo University is at the summit, with Kyoto University at its heels, and a few
of the private universities like Keio and Waseda also near the top.
The Tokyo University graduate is still widely assumed to be the
best material for government and academe, not to mention business. That an astonishingly high percentage of Japan's postwar
cabinet ministers are Todai graduates is hardly surprising.

Given this background, the spotty outbreaks of student unrest
since the war, culminating in the really bad period of 1968 and
1969, hardly seem surprising. What is surprising is that the explosion took so long to detonate and that it subsided so quietly. Watching the square tower of Tokyo University's Yasuda auditorium
shrouded in smoke, as several companies of mobile police units
stormed the militant student radicals holed up there, television
viewers in the nation of spectators had the impression of an assault
on a medieval castle. In view of the problems of Japanese university education, the impression was in character.

The government has given little constructive help to solving the
university problem. Private industry, ever-busy putting its money
into new factories, has given almost none. Until the mid-seventies,
at least, there was no place for American-type foundations in Japan's
socioeconomic structure (tax breaks are almost nonexistent). Private
universities which take the bulk of the student load, have to subsist
on tuition fees, with virtually no endowments; they receive almost
no help from either national or local governments. Professors' salaries
are wretched. In a 1974 survey, salaries of professors at state universities ranged from ¥150,000 to ¥210,000 per month; lecturers generally received under ¥100,000. Private universities generally pay
worse. It is safe to generalize that in 1974 the majority of Japan's
professors earned less than $6,000 a year.

Because of the low wages, outside part-time work is almost essential, for professors as well as students. As Nagai Michio pointed
out in a recent book,[7] fully 60 per cent of national university professors and assistant professors are dependent on additional sources
of income; in the private universities the figure is much higher.

7. *Higher Education in Japan.*

The need for constant outside work, most readily found in writing articles and giving outside lectures, incidentally, explains why so much of the professorial commentary in Japanese magazines is hasty, ill-thought, and verbose (they are paid by the page).

A powerful amount of discontent is brewing within the university system, particularly among the younger scholars, who do not take so kindly to the old apprentice system. Almost everyone recognizes that the impersonality of the university, the gulf between students and teachers, the lowering of standards are probably getting worse faster in Japan than in Europe or the United States, thanks to their swollen growth. The government is worried. Business is worried. The more thoughtful of the academic mandarins are themselves terribly concerned about the future of the Japanese university. But until now all these sectors of the society have been examining the problem separately. While government and business have been slow to avail themselves of scholars' services in solving their problems, the scholars cling to their ideal of magnificent isolation. Instead of trying to reform their existing society, they tend to look at it as through a glass, like a detached and rather distasteful laboratory specimen. When they do take a stand about the vibrant, intrusive, powerful, business-directed society around them, it is largely to set themselves against it. They tend to see only its warts. Much of the real value of Japanese academics' comment on their society is dissipated by a tendency to see it in the Marxian struggle terms of a "regime" somehow forcing capitalism on an exploited public, and doomed to fall of its own weight, through the wondrous workings of dialectic.

The mandarins' strong sympathies for the Marxist left, as well as a corollary petulance with the United States has a long history. Its roots lie far deeper than the current fascination with Peking, the current excesses of Japanese business profiteers, or reaction to American attempts at political and/or economic dictation. They come partly from the character of the intelligentsia itself, partly as the result of some wholesale intellectual importing from Germany and Russia over the past century.

In the 1890s, just as the Japanese university system was taking shape, it was inundated by a wave of European philosophy, especially German, from which it has not yet fully dried out. Hegel, Kant, and Schopenhauer were brought into the new lecture halls in Tokyo and given to the students rather uncritically, as something next to Holy Writ. The old song "DeKanSho"—for Descartes, Kant, and Schopenhauer—which has become something of a "Gaudeamus igitur" for old grads in Japan, fairly suggests the continental emphasis of Japanese university education over at least half a century. Marxism followed. The intelligentsia of this most intuitive of peoples thus began their long romance with systems of thinking which dote on deductive laws and inflexibly ordered processes.

Later, during and after World War I and the Russian Revolution, Japan's swiftly growing intelligentsia was swamped again, and profoundly affected, by a wash of proletarian and revolutionary literature, which seemed particularly appropriate to the unsettled economic conditions of that time. From Hegel to Marx to Lenin was little more than a hop-skip-and-a-jump. It quickly became fashionable in Japanese universities to think proletarian and to mount the barricades, at student demonstrations at least, before one went out into the world and started working for the Mitsui Trading Company or joined the army.

There was—and still is—something very basic in this university radicalism and the refusal of so many university professors either to come to terms with the society around them or to make any effort to alter it themselves. It had other sources than the most obvious one: the real poverty in which many students studied and their understandable, if exaggerated, contempt for the materialistic business society growing up around them.

The Japanese university was founded in the aftermath of the Meiji Restoration principally as an interpreter of the West to Japan. Even though many great universities of Europe started out, like the University of Tokyo, as a translation agency—in their case translating the rediscovered classical literature of the Renaissance —their roots were far deeper. So too, the American university has grown up with its society, from Yale and Harvard in the early eighteenth century to the land-grant colleges in the nineteenth. Japanese critics have argued, with much cogency, that, in contrast, the Japanese university has never looked the society around it in the face. For however much the hierarchical structure has followed ancient Japanese patterns, the matter of learning has been largely Western. Both the way of thinking and the content of the texts studied are often not only different from the native tradition the student has grown up with, but often sharply opposed to them, for instance, in matters like the relative importance of individual, fam-

ily, and class. A natural reaction is to take an extreme revolutionary position and conclude that the only ideal way to deal with the
society outside the university walls is to subvert or overthrow it.
This is especially true if the political science professor says so, too,
in his lectures.

The steady growth of education, the basic Japanese respect for
the scholar and literary man, and the vigorous activity of Japan's
printing presses propelled the university and its professors to center stage. The *sensei* began to be asked for his opinions. The columns of the newspapers and magazines began to fill with scholarly
exhortation as well as explanation. Much of the social indignation
in this period was justly caused. The professor in particular, could
not merely respond emotionally to the plight of the Japanese
farmer or the newly proletarized factory worker, he also had the
books to show that their plight was not unique. It was, after all,
just one stage in Marx's triumphant intellectual process. That, at
least, was the way he interpreted it.

The coming of military rule to Japan changed matters, but the
change was by no means as suppressive or explosive as many postwar accounts picture it. With very few exceptions, there were no
Thomas Manns or Benedetto Croces in Japan to make a protest as
in Europe. Almost no one apart from the Communist party core
went into exile in protest against his country's policies. Those who
stayed at home in silent protest were not numerous. There were
some, though. The great novelist Nagai Kafu refused to write during the Pacific War and confined his bitter reflections to his diary.
The novelist Ishikawa Tatsuzo, who saw the barbarities at Nanking in 1937, was jailed for writing his courageously realistic novel
Ikiteiru Heitai (*Living Soldiers*). Professor Yanaihara Tadao at
Tokyo University publicly criticized the annexation policy in
Manchuria in 1932 and continued to oppose Japan's militarist policy in China until he was forced to resign from the university in
1937. But these were lonely voices.

On the whole the Japanese intellectual establishment hedged its
bets. Although many of the intelligentsia had obviously guilty consciences over the successive incursions of the Japanese Army into
Manchuria and then China itself, the fact that this could be interpreted as a move against Western imperialism in China gave them
a face-saving out. There were also ample instances of European or
American dollar diplomacy or racial arrogance—the United States
Exclusion Act of 1924 not the least—which could distract a liberal's

indignation. The political scientist Royama Masamichi developed the idea that Chinese nationalism was merely a tool of Western imperialism. Japanese "regionalism," he argued, was something quite different. The Japanese, in his view, were genuinely trying to construct a cooperative regional order in Asia. Like the young army officers from farming villages whose socialistic militarism included a hatred of big Japanese capitalism, university intellectuals could pronounce Marxist anathema at the imperialism of the Western capitalistic powers, principally Britain and the United States. In the process it was not hard to look indulgently on Japan's own incursions into China. Its beginning, at least, seemed no worse than an Asian Monroe Doctrine.

They had some right on their side. In the perspective of forty years after, it is next to impossible for Americans to re-create the easy moral indignation of people like Cordell Hull denouncing Japanese aggression in China. Stupidities as well as atrocities in Vietnam have torn the consciences of the American intelligentsia in the most public way; and the parallel to Japan's China War has not been unmentioned. As for the hired intellectuals of China and the Soviet Union, if one reflects on the Great Purges, the Cultural Revolutions, and the interventions in the name of assistance committed by those countries in the last few decades, the sight of Peking and Moscow spokesmen denouncing prewar Japanese militarism is high irony. Yet, not merely in America during the Vietnam War, but also in the physically repressive world of the European or Asian police states (North Vietnam conspicuously included) there were many intellectuals who had the courage to oppose their government's policy publicly. In Japan of the nineteen thirties there were very few.

When war finally came in 1941, left and right in Japan united against the American enemy.[8] The left merely dwelt on the capitalism and the racial arrogance of Americans in Asia, leaving the right to beat the drum for the more mystical side of Japanese imperial patriotism. But the entire mandarinate, including several squadrons of postwar antifascist critics, lined up behind Japan's semifascist new order.

Donald Keene, the American authority on Japanese literature, once commented on this in the course of a thoughtful essay,

8. The Communists, on the orders of their Moscow superiors, were anti-American during the Soviet-Germany nonaggression pact period, but supported the Allies after Hitler's attack on Russia.

Contrary to the impression by Japanese who are anxious to gloss over the war years, the members of the literary profession were at the outbreak of war almost solidly united behind the militarists. They exulted in the triumphs of the first year and urged redoubled efforts when the ominous signs of reverses appeared. Only when Japan's defeat became imminent did some writers lose their enthusiasm, although others whipped themselves up to an even more frenzied patriotism. There was no resistance to the militarists save for the negative action of a few authors, mainly older men, who refrained from publishing. Some left-wing writers were rounded up by the government immediately after the outbreak of war in December 1941; this fact is well-publicized, although it is less commonly known that, unlike their counterparts in certain European countries, Japanese writers were not executed or left to die of maltreatment in prison. With a few exceptions, those imprisoned in December 1941 were released shortly afterwards on their avowal that they had changed their views.[9]

At the end of the war, the occupation restored freedom of speech and the press [10] to a Japan which had not known either of these for a decade. The disillusionment of the intelligentsia with the wartime Japanese government, its false promises, and bloated expectations, was massive. Partly because of this disenchantment, partly out of relief at the chance to express thoughts long withheld, partly out of expediency, the Japanese intelligentsia did a disciplined left face. At first grateful to the Americans for ventilating the atmosphere, the mandarins quickly developed a gymnastic facility for praising the occupation reforms and defending the new occupation-imposed constitution, while at the same time becoming increasingly critical of the United States.

Marx once again blossomed in the universities. The contradictions inherent in capitalism were held to be at work just as inexorably as Marx had said they were. Attacks on crude American imperialism became the parlor game of the intellectual left.

The Americans, it must be admitted, did a lot to further this trend. The sudden switch in United States policy at the time of the Korean War, however understandable its causes, was most paradoxical. The country which had set Japan up with an antiwar constitution as the "Switzerland of the Pacific" (one of General MacArthur's less happy phrases) set out to assist Japanese rearm-

9. From "Japanese Writers and the Greater East Asia War," *Journal of Asian Studies*, 1964.

10. Except for censoring any opinion critical of the occupation, one of the many contradictions of this effort to impose democracy by military fiat.

ament. The calumnies of the McCarthy era in the United States were vividly reported in Japan and seized on by the mandarins to show how easily the capitalistic Americans could drift into fascism. Long after Joe McCarthy's raspy voice was stilled, the Vietnam involvement revived the whole Marxist cant about American ruling circles and their "inexorable" drift toward fascism. The domestic horrors later exhibited to the world by the Nixon administration did little to soften the intellectuals' distaste.

One could not fault the intellectuals for being almost instinctively hostile to the United States, as an arrogant superpower, still endowed with much of the race prejudice and national nearsightedness that had served to antagonize the Japanese before World War II. Yet, at the same time, there was a considerable amount of good will for a country which had acted magnanimously, restored freedoms, and, not so incidentally, given out a great number of fellowships and scholarships to Japanese scholars who had been half-suffocated by the imposed isolation of the Showa dictatorship of the thirties. The average Japanese professor or newsman can not be considered anti-American as such. One of his problems, indeed, is that he criticizes the Americans as much from the standpoint of a colleague or a relative as an outsider and he secretly expects their standards to be higher than he will publicly admit.

For it is important not to overstate the degree of the intellectuals' political commitment to the Marxist left. With many, it is more or less a matter of professional equipment. Like the box lunches at the *sumo* wrestling matches, the doilied neck and headrests on automobiles, the huge gold-lettered characters announcing the names of noodle shops, and the prevalence of Lieder records on the FM stations,[11] the Marxist sympathies of so many intellectuals are canonized by long habit. It is a lonely *bunkajin* who is not *kakushin-teki* (literally, "progressively inclined"). One follows the crowd. The degree of commitment is extremely superficial. It rarely manifests itself in any serious attacks on the society around one, with which one may be privately quite satisfied. Barring the core communists, the average Japanese progressive is the socialist equivalent of a "rice Christian." That is, he is quite vocal about his

---

11. It is difficult fully to explain the extraordinary popularity of German Lieder in these media, where Fischer-Dieskau has the popularity of a pop singer. Possibly some Lieder were played in the early days of Japanese radio and kept on the programming by force of habit, like so much else in Japan.

leftism when being interviewed by the press, but it rarely intrudes into his actual work and almost never into his life, which tends to patterns of a hierarchy and class consciousness generally more rigid than those of unenlightened businessmen or working folk. It is probably no accident that Jean Paul Sartre is so popular among Japan's *interi;* rarely has any self-professed champion of the proletariat been so far removed from the people he claims to represent.

A bigger problem, particularly in the intellectuals' evaluation of American postwar politics, was the isolation of Japan from the world in the thirties and forties. This continued, in fact, until the occupation officially ended in the fifties. The Moscow purge trials, the perfidies of the Nazi-Soviet nonaggression pact, the subservience of the European and American Communist parties in that period, the armed interventions by the Soviet Army in postwar Europe, not to mention wartime activities like the virtual genocide against several ethnic minorities in the Soviet Union and the Katyn massacre of ten thousand Polish officers—such facts had been ground into the experience of European and American liberals and socialists. Somehow, they never made much of a dent on their Japanese counterparts. They were simply not there. Similarly, the differences between the curious American synthesis of individualistic welfare-state capitalism and the textbook Marxist definitions of capitalism never really came clear in most Japanese intellectuals' consciousness.

The professors who had had to lock up their copies of *Das Kapital* in the early thirties brought them out again in the late forties and resumed lecturing, as if nothing had changed in the process. Most continued to preach to their students the evils of capitalism and insisted that the only socialism was the doctrinaire Marxist variety. Modern British socialism, modern German socialism were largely ignored. Never having experienced the real struggles of the left in Europe and the United States during the thirties, the average Japanese mandarin is little more than a parlor Marxist theoretician. The kind of ideal Marxist society he says he is for has never existed and it is highly doubtful that one will emerge. He would be the last to enjoy it.

Yet in the middle of the world's most successful welfare-capitalist state, the mandarins opted out. Any Japanese government not run by socialists was automatically classed as reactionary. And to cooperate with the reactionary Japanese government was unthinkable. Therefore, they would just criticize, until Utopia somehow materialized. The utterly oppositionist attitude of the mandarins, of course, served to strengthen the really reactionary elements in the Japanese right and weaken the genuinely democratic forces in both right and center. By thus helping to polarize Japanese politics, the mandarins worked to create the same kind of evil they said they were opposed to.

The American intellectual, particularly over the postwar decades, has been nothing if not engaged. Find a cause, from antibomb to zebra preservation and the scholars, artists, playwrights, novelists, and assorted professional thinkers will mobilize, with press releases, buttons and slogans, telethons, rallies, and protest meetings. The Yale and Harvard faculties will jostle each other in an effort to be first with the signed one-page manifesto in the *New York Times*. Prominent New York conductors, playwrights, book-reviewers, and the casts of off-Broadway shows were old reliables for the rallies of the late sixties, especially. Battling them to support law and order one has found smaller but equally strongly motivated peers from the current intellectual establishment, be it McGeorge Bundy, Herman Kahn, Henry Kissinger, or whatever group of academic economists which happens to be posted to Washington at the particular moment.

The public debates and demonstrations of these years only served to publicize the network of interlocking ties that binds American academe to the counting house and city hall in a thousand different ways. A Ph.D. degree is an automatic asset for promotion in the United States Civil Service; its possessors are called doctor in government offices more than they are in the universities. Historians as well as chemists from the universities have found good jobs with business and the foundations, the latter of which are indeed almost an academic preserve. Newspaper editors are used to giving public affairs speeches, and are not adverse to breaking bread with favored clients of the advertising department. The average American professor, by the time he becomes a department head, if not before, is as well versed as many businessmen in the mechanics of budget making and personnel relations, not to mention fund raising.

For better or for worse, the ivory tower has gone from America. Like the American clergyman who long since decided that the social gospel of active help for the living was far more rewarding to his temperament than prayers for the dead, the American intellectual has merged himself into the mainstream of his society. The merger did not come overnight. It has had to survive occasional Know-Nothing outbursts like McCarthyism and the antiegghead movements of the fifties; but such activities have been, increasingly, anachronisms. The tradition of intellectuals participating in the American community and doing their best to lead it is as old as the founding fathers. It took only the postwar era of mass communications, scientific discovery, and theoretical policy planning to destroy the ivory tower, perhaps forever.

Much has been lost in the process. One misses the aloof dedication of the old-fashioned teacher and the determination of past generations to use education as a means to form the minds of students, instead of training them to be citizens, racially aware, and good drivers. This is especially true of the university. Adam Ulam, in his recent book, *The Fall of the American University* rightly deplores the "politicization and bureaucratization" of that institution, at the expense of its basic function of educating students.[12] Yet, in one sense, the gain for the United States may have outweighed the losses. At least the mental discipline of the intellectual, his ideas, his penchant for thinking more in terms of principles and ideals than practice and techniques, has been merged in the national bloodstream. He is a part of his community.

The Japanese mandarin is, but he is not. Nothing more illumines the differences between Japan and America than the resolutely ivory tower attitude of Japan's intellectual establishment. It is not even a real ivory tower that the *bunkajin* have fashioned for themselves. They are ready enough to shout preachments from its windows, offering a barrage of not-quite-free advice to the Japanese community on almost every conceivable subject. But with few exceptions, they have no stomach either for defending their ideas or helping to put them in practice in the market place outside. The whole intellectual establishment of Japan has, in effect, opted out

12. "Over the last twenty years there has grown a belief, as erroneous as it is basically incompatible with democracy, that the university can and ought to instruct society on how to conduct its affairs; that those who instruct and administer it possess, quite apart from their professional qualifications, some special wisdom unavailable to the ordinary citizen and which entitles them, in fact puts them under a special obligation, to prescribe cures for social ills, solutions for foreign policy dilemmas, and the like. . . . For the American university the beginning of wisdom and reform must lie in acknowledging that there are clear limits to what education can do for the individual and society, and that it has been the attempt to transgress those limits which has been responsible for most of our troubles with education."

of the mainstream of its society, like the professors on smashed and burning campuses who refused to allow the police and firemen entry because they might represent government interference. Government and business circles, it should be noted, have not been very active in requesting the mandarins' services either. Nevertheless, their isolation is basically self-imposed, like that of a small Tokugawa Shogunate within a modern country. It is his own mandarinate that best deserves Oya Soichi's accusation of a "nation of spectators."

In a short and perceptive essay,[13] Okochi Kazuo, former president of Tokyo University, definitively described the problem. It had its origins partly in the cult fostered by Japan's great Meiji Era novelist Natsume Soseki, who set a Proustian premium on the "Spectator" as opposed to the "Man of Action," partly out of the intellectual's disgust with the prewar rightist drift of Japanese politics. "The Japanese intelligentsia," Okochi writes, "instinctively opposed to the mood that was developing, assumed from the first a negative attitude and showed a reluctance to act. As a result the intelligentsia itself evolved into something like a weird basket-case, with a huge cranium and shriveled arms and legs. Granted that their critical cast of mind, their work of emendation and explanation, their explaining of interrelated cause-and-effect were ill-suited to a life of action. But far from saying what should be done, they lost even the inclination to speak out. At this juncture, nothing was left but to be "spectators." This kind of approach to life came to be highly esteemed among intellectuals. To view any constructive program rather negatively, to look at only what was put before you, to add your explanation to what had happened, of course to criticize it—this became not only a way of life. It was thought to encompass the intellectual's social duties."

Consistent with this aloofness within their own country, the intelligentsia of Japan exhibit a sort of isolationism in international affairs that ranges from hesitation to outright xenophobia. Most scholars and journalists in the United States and Europe are consciously part of an international community. However high the barriers of politics and economics grow, the presses and the screens have tumbled most of the walls between the world's intellectuals. Not only do they meet and correspond on the level of international conferences and exchanges—the symposium circuit has done almost as much for the airlines' business as the guided tour. But they

13. Part of a collection of essays and commentaries, *Jibun de Kan-* *gaeru*, by this distinguished political economist.

are almost automatically exposed to each others' works, both in the original and the relatively simple translations between European languages.

Not so the Japanese. Several thousand books and major papers will be translated from foreign languages into Japanese in one year, hardly more than fifty will emerge from the Japanese. Most of these will be translations into English. Few foreigners—barring the handful who can read Japanese—are exposed to the bulk of the comment, criticism, and original literature that rolls off the presses in such fascinating volume. Thus comment and criticism from the outside are denied most Japanese writers. Like their own tidal Inland Sea, the waves of new thinking, old thinking, and argument wash up endlessly on Japanese shores and are thrown back into Japanese waters.

The isolation is most marked among novelists, not to mention poets and essayists, who pride themselves on their Japanese style. This is a subjective matter on which a foreigner is reluctant to venture judgment. It is generally true, however, that the kind of Japanese writing most esteemed in literary circles is elliptical in meaning, rich in obscure or cleverly used readings or combinations of Chinese characters and a sort of stream-of-mood sense that triumphantly frustrates any kind of grammatical analysis. (Since my own thought patterns are irretrievably Western, I have adopted as a rule of thumb that any prose which I can read easily must be seriously wanting in literary graces.) Many novelists or other *bunkajin* deliberately shun any contact with foreign languages, especially Western ones, on the theory that their style will be infected by foreignisms and hopelessly lost in the process. A few, like the late Mishima Yukio, would enjoy the companionship of foreigners, but generally kept them segregated. At one of his foreign evenings you met largely fellow foreigners for dinner. By contrast, he dug ever deeper into difficult and traditional Japanese in his writing. He was a most exceptional artist.

For the most part, however, the vogue to avoid the contamination of foreign languages or literatures is more often the device of mediocrity than genius. I have too often heard the argument "that is not the Japanese way" or "we Japanese simply have to write or edit things this way" advanced by smiling incompetents in an attempt to disguise their individual inefficiencies as group nationalism.

Many Japanese intellectuals now worry greatly over this international literary isolation—the more particularly at a time when most Japanese are coming to feel that an international approach is a necessary virtue. In recent years, notably with the founding of the Japan Fund in 1972, universities and the government, over the trench between them, have made some efforts to internationalize Japanese scholarship; in other words, to get more Japanese scholars circulating overseas and more than the present trickle of foreign scholars and students into Japan. The support for the new United Nations University which the Japanese have given is one very heartening sign. Yet, in the main, the isolation of the mandarins has resulted in a defensiveness and a kind of passive, waspish nationalism that sees the world more and more through Japanese-made filters. There is so little meaningful outside contact. Most of the polite symposia and international meetings have little relevance. So many of the mandarins turn in upon themselves. This modern drift toward a kind of cultural isolationism is an ironic comment on Mori, Fukuzawa, and the great outgoing pioneers of modern Japanese culture in the Meiji Period. (Even Natsume Soseki went to Paris to learn about novel writing.) It also clashes with the genuine feeling for internationalism among Japan's youth. In the great energetic impulse of the Meiji Restoration, Japan's scholars and teachers sailed out to meet the world and begin a world conversation. Their descendants *speak* of meeting the world. There is quite a difference.

Chapter XI

Mandarins Continued
The Press

"In editing a newspaper I encourage reporters to write bravely and freely. I have no objection to severe criticism or extreme statements, but I warn them that they must limit their statements to what they would be willing to say to the victim face to face. Otherwise, they are what I would call *kage-benkei*--fighting from the shadows, i.e., attacking from the security of their columns. It is very easy for *kage-benkei* to fall into mean abuses and the irresponsible invective which is the shame of the writer's profession."
—Fukuzawa Yukichi, *Autobiography*

The three great Japanese national dailies, *Mainichi*, *Asahi*, and *Yomiuri*, trample the bush of world journalism like huge dinosaurs, each surrounded by an attendant swarm of weeklies, monthlies, quarterlies, and book publishers, camera magazines, journals for the intelligentsia, low-brow weeklies, high-brow publications on economic affairs, lavish yearbooks, jiffy dictionaries, sports papers, and specials for the children and the ladies. They are about the last press dinosaurs left. Their circulations are something that most of the world's editors (apart from the compulsorily heeded didacts of *Pravda* and *Izvestia*) can only imagine with difficulty. As of this writing, the *Asahi* remains the most prominent, with a total daily circulation of 7.3 million. *The Mainichi* lags with 4.4 million, while *Yomiuri*, which used to be number three, has pushed up over the 7.4 million mark and threatens to enlarge its gains against the former front runner.[1] All three put out separate evening and morning edi-

---

1. These figures are based on Newspaper Publishers Association estimates for 1977. By comparison, the London *Daily Mirror* has a circulation of 4 million, the *New York Daily News* over 2 million, the *New York Times* under 900,000.

tions, and have distinctive local editions for most regions of Japan, centering on the large cities.

The satellite publications run to scale. *Asahi*, for example, has three weekly magazines, including the resolutely high-brow weekly *Asahi Journal*, with six-figure circulations, four specialized monthlies such as *Science Asahi*, one quarterly, four annuals, a book publishing firm which puts out 200 titles annually, a year-book, and two dictionaries. *Mainichi*, with its *Sunday Mainichi* weekly, the more cerebral *Economist*, and others, has a similar range, as does *Yomiuri*.

All three are distributed almost everywhere in Japan, with local headquarters at Osaka, Fukuoka, Sapporo, and Nagoya. A few strong local papers challenge them in certain areas like the *Chubu Nippon Shimbun* in Nagoya (morning circulation, 2,167,000), or *Nishi Nippon Shimbun* in Kyushu (morning circulation 629,000). One business and finance-oriented newspaper, the *Nihon Keizai Shimbun* has its own national network and a group of profitable subsidiaries. Although the morning paper's circulation is nationally only 1,450,000, it is highly esteemed among the business circles which support it and frequently feed it information in advance of the other dailies. But generally the three big dailies reign supreme. Along with NHK, (the Japan Broadcasting System) and Kyodo Tsushin, the national wire service, they control the news, information, and opinion given to the Japanese people more thoroughly than any similar press in the nontotalitarian world. The only conceivable parallel, the onetime domination of British journalism by the *Times*, the *Daily Mail*, and the *Daily Express* groups, pales by comparison.

Despite their undoubted influence and the superficial prosperity of big circulation, however, the dinosaurs are running into ecological problems. They have been actively competing since their founding in the late Meiji Era, and their size has swelled enormously—almost ten thousand employees at *Asahi*, slightly under eight thousand apiece at *Yomiuri* and *Mainichi*. Their distribution networks carry the news into each Japanese neighborhood, elbowing each other aside in the process, for there are relatively few newsstands in Japan. Yet the bulk of their circulation is home-delivered by an antiquated and underpaid network of newsboys, [2]

2. The papers are troubled by this, particularly since the distributors have been making angry noises lately about unfair labor treatment. Newsboys are, indeed, something of an anachronism in a full-employment society. They

fortified by a system of premiums and inducements to keep subscribing. (Try to cancel a subscription to a Japanese newspaper in the middle of a subscription drive.)

The present cost of a single newspaper, since the latest price increase, is 50–60 yen, but subscribers can get both morning and evening editions for a special monthly rate. For this the subscriber gets a daily reading diet which is intellectually vastly superior, both in writing and content, to all but two or three American papers and about half a dozen in Europe. Complex financial problems and educational reform bills are given front-page treatment, where in the United States, they would be lucky to get into the average paper at all. The world news network of all three papers is formidable. The foreign datelines are as regularly spotted as those in the *New York Times*. Art and culture sections are big, with a wide range of Japanese and Western topics. The editorials and the opinion columns cannot be read lightly. Even the serials are apt to be highbrow. Sword murders and similar disasters beloved by reporters everywhere are heavily played up, but generally (unless they happen to be really big) on the last two pages, which are given over to local news. Different back pages are printed for different localities.

There is one great editorial peculiarity about the Japanese papers. They read much the same. Where other countries have rightwing papers, left-wing papers, Democratic papers, Republican papers, Catholic papers, or socialist papers, all the big three in Japan —familiarly called by one word *ChoMaiYomi*, are indistinguishable except in matters of degree and emphasis. The *Asahi*, for example, is far more obviously antigovernment, further left, and generally more antiforeign than the others. The *Yomiuri* is a bit middle-low brow in its language. The *Mainichi* is promotion-conscious to a fault; for instance, an inordinate amount of space is given to news of *Mainichi*-sponsored art exhibits, and the like. But, on the whole, its news columns are fairer and more objective. These are nuances, however. The general tone and news handling of the three are very similar.

One obvious reason for this is the manner in which Japanese news is gathered. It is a far cry from enterprise journalism. Government ministries and other important foci of news are covered,

are desperately courted by their employers with scholarships, prizes, and overseas vacation premiums; but they are increasingly hard to retain. There is talk of recruiting Korean students in Japan, among other possibilities, to take over some of their chores.

if that word may be used recklessly, by groups of reporters from the major Japanese newspapers organized into reporters' clubs (*kisha kurabu*). Each club has its own officers and operating regulations, stated more or less formally, and serves as the sole channel for news emanating from the agency involved. Membership is restricted to reporters from the major Japanese newspapers. No free lancers are tolerated and, with one or two exceptions, foreign correspondents are rigidly excluded.

There are often several members from one newspaper in the same reporters' club, thanks largely to the swollen numbers of journalists allowed house room on the staffs of Japanese papers. (The *Asahi* has fifteen hundred reporters, for example.) As might be expected, seniority rules. When a piece of news is given to the club, the senior members of the group informally but inexorably decide how to play it; and they call the turn. Once more one recalls the village society in action. Although some newspapers may on occasion get a bit of news before others (*Asahi*, for example, often seems to get announcements of imperial visits ahead of time), there is rarely a question of one man scooping another. Once the importance of the story has been discussed, members of the club telephone or trot back to their respective city desks at the same time (and at the same pace) to hand in their stories. Not surprisingly, the stories turn out about the same—facts, slant, and all.

The same is true of editorials, most opinion articles, and the general placement of news. Each morning as a hobby, I pick up the three leading Japanese dailies and compare the front pages. With rare exceptions, the contents are strikingly similar, down to the pictures used and the choice of headlines. No country, outside of the Communist world, can boast the terrifying homogeneity of opinion that the mandarins of the press put into Japan's newspapers. (The word *opinion* is used advisedly. Helped by the general vagueness of the Japanese language, reporters manage to make fact and opinion in a news story almost indistinguishable.) The closest parallel to Japanese news reporting is the kind of opinion-reporting which the national affairs and foreign news departments of *Time* magazine made so unforgettable in the United States during the forties and fifties. As with *Time* in those days, the opinion enters into the placement of news; an account of American and Japanese cooperation in the so-called Kissinger security plan, for example, will be run on page one of *Asahi* directly above a huge photograph of an American nuclear submarine entering Yokosuka harbor for, as it happened, the hundredth time (September 17, 1973). Socialist or other progressive victories in municipal elections will get a huge, front-page splash, while conservative victories traditionally receive more modest treatment.

In an article in *Bungei Shunju,* the critic Omae Masaomi once called attention to the choice of words used by the *Asahi* in reporting the victory of "progressive forces" in this case during the Nagoya city election in April 22, 1973:

RIGHTEOUS INDIGNATION ELECTS A PROGRESSIVE MAYOR—COMMODITY PRICES AND POLLUTION PUSH THEM TO THE LIMIT—A WEAK POPULACE RISES UP: "WE WON'T ALLOW THEM TO IGNORE THE PUBLIC INTEREST . . ."

The progressive Mr. Yamamoto wins. The night of the 22nd, the conservative Mr. Sugito was finally defeated and ousted from the Nagoya Mayor's office that he has been occupying for the last twelve years. "Throw out the Liberal Democrats." "Make the Public Welfare our No. 1 priority." "Let's judge carefully." Slogans like that reflected the the new image of the winner. After a long time spent in preparation, the Communist and Socialist parties finally got together and threw all their resources into the fight. They caught up in the last month. Behind all the furore was a real surge of popular indignation, because of the rise in commodity prices and pollution. It produced a definite trend against the Liberal Democrats. A real ground swell!

Crowds of young people and housewives rushed through the streets in the big drive to strike a blow at the kind of conservative politics, inexorably allied to big business, that ignores the people's interests. The "greens," i.e., the color symbol for the progressive forces, finally came out on top all over the city. The conservatives were decisively routed. Housewives rushed up and down stairways visiting people in apartment buildings, cheerfully conducting meetings and distributing pamphlets. They carried on this work far into the campaign. Some of the women brought their children along—even pregnant women joined the fight. In one housing development, you could see the figures of these women with their children breathing hard as they rushed from door to door spreading the good word.

As Mr. Omae commented, "This impressionistic report about the Nagoya elections and the victory of Mr. Yamamoto, the progressive candidate was not taken from *Akahata* (the official Communist paper) but from *Asahi Shimbun.* The fact, for example, that the Communist Youth Group organization threw thirty thousand of its national membership into the local campaign was not mentioned. They were just young people, and so on." Although the article quoted above may be an extreme, it is by no means untypical. It should be added that an *Asahi* article in Japanese which criticizes the government in the usual circuitous way will seem subjective enough in that language, but when literally headlined into English, a more forensic tongue, by the *Asahi*'s English-language daily as "New Tanaka Cabinet Trying to Hoodwink the People," it has an even more heavily biased ring to it.[3]

More serious than the vagueness of Japanese is the looseness with which reporters and desks in Japanese papers check their stories. I was once in the middle of one minor newspaper controversy about alleged bad sales practices at an American-owned company. Except for one perfunctory meeting with two of the company officers, from which their remarks were printed out of context, the Japanese papers relied exclusively on the handouts given them by one relatively small consumers' league, which was trying to make capital out of the incident. Stories would appear whenever the consumer league spokesman wanted to hand them out. No one from any of the Tokyo papers bothered to cross-check with the principals involved. An offer was finally made by the principal offending paper to interview one of the company executives to get his side of the story, but only after three months of determined mudslinging in that newspaper's columns.

The extreme projection of this is the technique of the weekly magazines whose favorite tactic is to call the target of their story for a statement "just as we are going to press." As might be gathered, Japanese libel laws are hopelessly ineffective. The sex-and-scandal weeklies, carrying on a regrettably long tradition in Japanese journalism, take full advantage of this fact; it is only on rare occasions that someone, usually an entertainment personality maligned about

3. The English-language press in Tokyo occupies a very special corner of journalism. Running heavily to United States domestic news, international and United States sports, English-language movie reviews, and society columns, it supports itself partly by subsidies from parent newspapers or government organizations, partly by advertising revenue gained through top-heavy shipping and transportation news. The *Japan Times*, the relic of the once independent prewar *Japan Advertiser* preserves a polite Foreign Office tone that reflects its connections; the *Yomiuri* is distinguished by good United States columns, the *Asahi Evening News* by its heavily editorialized news. The *Mainichi*'s English-language daily is the best. Perhaps inevitably, the view of life in Japan afforded by the English-language dailies is heavily filtered.

marriage problems, will go to the trouble of suing. Weeklies like *Shukan Gendai* and *Shukan Post*, not to mention the weekly *Asahi*, *Mainichi*, and *Yomiuri* magazines buzz over the big stories covered by the dailies like flies feasting on leftovers. A whole race of tipsters called *topiya* (for "top stories," presumably) peddle ideas or pieces of evidence to the weeklies and, likely as not, write the stories themselves. Quotes are at best taken out of context, at worst invented, in this kind of journalistic guerrilla fighting. About the only protection from the worst of the weeklies is money (one acquaintance of mine hired a special press consultant in his company to be sure that, whatever bad was written about the company, his own name was never mentioned). There is, also, the leverage of counterthreat. Fortunately for the harrassed celebrity or businessman, the advertising departments of the weeklies are generally quite approachable and friendly. It is a very competitive business and a few pages of advertising space may melt away a certain amount of editorial indignation.

Foreign firms are, of course, always fair game. When one American company representative pleaded with a scandal sheet editor to remove a particularly unpleasant (and inaccurate) story about his firm from the next issue, he was told, "Maybe we don't have to run it. If our reporters finish their exposé on Coca Cola in time, we'll run that instead." The barbed reporting is not restricted to the weeklies. It is still the fashion in Japanese dailies, for instance, to refer to a foreign firm's opening of a Japanese branch as a military-type landing. If, on the contrary, the foreign company closes down part of its Japan operation, the headlines read "withdrawal in defeat."

The United States, for partly understandable reasons, is a constant target, from criticisms of foreign policy to disapproving articles about babies who suffocated because their mothers had used American style sleeping methods (*Amerika-teki nekasekata*, i.e., they put them to bed face down). Except for one paper, *Sankei Shimbun*, which still reports both favorable and unfavorable news about China, the People's Republic of China is a totally indulged sacred cow. The Soviet Union is attackable, although the mandarins go to great lengths to avoid this. In the summer of 1973, Premier Tanaka was rudely rebuffed by the Russians in his efforts to secure return of the four Japanese northern islands which the USSR occupies and to secure freedom from arrest and imprisonment for Japanese fishermen in the area. Most of the Tokyo papers contented themselves with denouncing Tanaka for not having handled the negotiations properly.

Despite the tone of the big three's criticisms, the press mandarins are very consciously part of the establishment. Like the bureaucracy, the professoriat, and the business-statesmen, they consider themselves guardians of the state—paid critics, in a sense. They do not want to change the government so much as they want to persuade the government—bludgeon it, if need be—to adapt their policies. They are nationalist above all, as most United States papers are not. When they believe the national interest is threatened, they will blow an angry whistle. Thus the press started to ask pointed questions about the government's economic growth policies and GNP fixation several years before the energy crisis of 1973. From the beginning of the seventies, a year or two before the energy crisis broke, the newspapers were running articles about raw materials shortages. They attacked business profiteering and land-grabbing relentlessly, in the best tradition of the free press.

Yet, when matters affecting the national dignity are concerned, like postponement of an imperial visit or too severe criticism of a big Japanese company, one has a sense of punches being pulled. The old school tie tends to show. I have yet to see any really serious editorial criticism of Mitsui, Mitsubishi, or Sumitomo, for example. Even after all the big Japanese trading firms, including these former *zaibatsu* interests and others, had been accused of scandalous profiteering, press criticism of the big ones was very slow and restrained. It takes really extreme behavior, like an indictment, to get a major company's name unfavorably mentioned.

The imperial family, of course, is handled delicately, although problems do occur. In 1966 the emperor's son-in-law was found dead with a bar hostess in the hostess' apartment. (They had both forgotten to turn off the heater.) That was beyond suppression.

The *Sokagakkai*, Japan's crusading Buddhist religious sect, besides enjoying the loyalty of its ten million adherents, has virtual immunity from press criticism. In the late sixties, however, when members of the sect tried to suppress a critical book about them, the controversy was aired in the press. This, too, was too much to hide.

To confirm its editorial judgments, the press regularly invokes the authority of hordes of mandarin colleagues in the universities.

Almost whatever is reported in Japan, from an Arab oil boycott to a dispute over Tokyo garbage disposal, will be garnished with the comments of various professors whose opinions tend to reflect the general intellectual line. These professors and other commentators —the equivalent word *hyōronka* in Japanese has a more magisterial connotation—occupy a unique position. Many specialists among them tend to stick to their lasts—art *hyōronka* will comment on art subjects, engineering *hyōronka* on engineering, and so *ad infinitum*. Yet the enjoyment of a full professorship, particularly in the social sciences, apparently gives a man authority to pronounce on any subject whatsoever. This phenomenon can be observed in the United States or Europe; but in Japan the rocks thrown from the shelter of the ivory tower are extraordinarily alike in size and texture.

Not surprisingly, the constant din of homogenized opinion has its effect. The generally favorable view of the *Zengakuren* student rioters by the populace in recent years (only their extreme violence was condemned) was hardly unconnected with the fact that the student side in a riot, whatever the cause, drew consistently favorable reporting in the press. The really bloody riots of April 1960, denouncing the Japanese-American security treaty, were intensified by a crescendo of press support, as well as the agitation of leading members of the university communities. This is not to condone the conduct of former Premier Kishi Nobosuke, who attempted to bull through the extension of the security pact on the eve of the abortive Eisenhower visit to Tokyo. His high-handed disregard of public opinion, not to mention political opposition, offered ample grounds to justify his resignation. But the dangerous intensity of the demonstrations and the mob violence that occurred were directly related to emotional and inflammatory newspaper reporting.

On June 15, 1960, the seven major Japanese papers jointly published a manifesto calling for the preservation of parliamentary government from violence. This was a remarkable occurrence. It had been made necessary by the bad behavior of the papers themselves and the free rein given by their editors to a combination of hot-headed reporters and hortatory commentators. The far-left Japan Journalists' Congress and the newspaper employees in Sohyo, the large left-socialist union, were particularly active in this self-styled struggle.

In fairness, it should be said that the press's readiness to criticize the government, like the chronically antigovernment attitude of the university establishment, has a great deal of history and justice behind it. Like most other modern Japanese institutions, the newspaper had its origin in the Meiji Restoration days, and grew out of the political struggles of that time. In the 1870s and 1880s newspapers sprouted as organs of particular leaders or political factions, exactly as they did in Europe and the United States at somewhat earlier periods. Although a few, like Prince Ito's *Nichi-Nichi*, were mouthpieces of the government, the bulk of the papers and the journalists were in opposition. Many of the samurai who disliked the new Meiji government went into journalism by choice or necessity, because it was about the only way to express opposition without resorting to open revolt.

Much of the reporting was scandal-mongering. Yellow journalism flourished in Japan as much as in Eruope and the United States. Some papers, however, developed a reputation for enterprise and integrity. Fukuzawa Yukichi's *Jiji Shimpo*, founded in 1882 was one of the best, as was the *Hochi*, later absorbed by *Yomiuri*.

The government was quick to move against press comment that displeased it. From the 1870s to the end of World War II, successive Japanese bureaucracies liberally used both formal censorship and occasionally police action in an effort to bring the newspapers into line. The *Yomiuri*, oldest of the big three, was in trouble with the government almost continually after its founding in 1873; for a while censorship was avoided by the simple device of staying away from political news. By the early twenties, *Yomiuri* had been crusading against militarism and censorship and, among other things, demanding home rule for the new Japanese colony in Formosa.

Both the *Mainichi* and the *Asahi* were founded in Osaka in the same decade; and only later expanded to Tokyo. Possibly because of their distance from the capital, they found it natural to assume the role of a loyal opposition. Some of the best Japanese newsmen came from Osaka. In their way, in fact, the early Osaka newspapers were the counterpart of the kind of twenties Chicago journalism made famous in the play *Front Page* and similar chronicles.

The *Asahi*, led by its founder Murayama Ryohei, came to take the lead in this kind of opposition journalism. In September 1918, Murayama was beaten up by right-wing gangsters for his opposition to Japan's Siberian intervention, a month after two of his reporters had been jailed for the same reason. His son-in-law was later attacked by rightists, as was *Asahi*'s editor. In the February 26 Incident of 1936, rebel soldiers smashed up part of the *Asahi* offices. With *Asahi*, as with the other newspapers, censor's suppression and suspension of publishing for antigovernment comments, real or fancied, became part of the job. Petty, exasperating, and crude though it was, it was something a newspaper had to live with. Some of the vagueness of Japanese press writing was not only linguistic, but arose from the efforts to avoid the censor's ban.[4]

As the militarist trend in Japan grew stronger, the papers had increasingly less latitude in which to oppose. By the outbreak of World War II they had all become, in effect, government mouthpieces. But if they suffered in silence, they had long memories.

When the war ended, the newspapers came out fighting. All the stored-up grudges they had held against the authoritarian governments of prewar Japan were merged in an attitude that was antigovernment almost as a reflex action. The problem was, however, that the postwar government was both weaker and, to put it mildly, less authoritarian than the prewar government, and often not deserving of such automatic hostility. Conversely, the free press was far less individualistic than it had once been. The enforced homogenization of the prewar and war period had done its work. There was quite a difference between the original, courageous free-lancers of Fukuzawa Yukichi's day and the new generation of newspapermen who, since the twenties, had been getting their jobs by competitive examination, like civil servants, in the best mandarin tradition. The average Japanese reporter, if he does not have the soul of a bureaucrat, possesses at the least the bureaucrat's strong survival instincts and his talent for protective coloration.

This is not to say that the postwar press fails to expose political or commercial wrongdoing or does not frequently act as the guardian of the public voter, or consumer interest. Whether reporting diplomatic news or local sewage system failures, however, it's guardian's role is completely self-defined. Two bits of recent history, one of them a continuing situation, illustrate the curious confidence of the Japanese press, not only that it knows what the public should read, but that, in addition, what the public doesn't read won't hurt it.

---

4. In 1921 the authorities forbade any mention of the fact that a prominent Korean in Japan had been murdered, worried about possible repercussions in their then colony. In his *Social Currents in Japan* (1927) Harry Wildes notes that *Hochi* got the news out by describing the incident as if the man were alive: "Bin Gen Shoku suddenly decided to return to Korea. . . . Escorted by the station-master, Mr. Bin entered a second-class compartment especially reserved for him, and decorated with wreaths."

On April 4, 1973, Tokyo police arrested two people on suspicion of violating the Public Service Code, which forbids civil servants from disclosing confidential government business to outsiders: Mrs. Hasumi Kikuko, secretary to the deputy foreign minister and Nishiyama Takichi, a reporter for *Mainichi* newspapers and a member of the Foreign Office Reporters' Club. The case was on its face a simple one. Nishiyama had made friends with Mrs. Hasumi, entertained her rather intensively, and secured from her copies of three secret Foreign Office cables.

The cables were most confidential. They had to do with a Japanese government promise to quietly repay the United States for compensating Okinawan landowners for damage done to their property by the United States military, as soon as Okinawa would revert to Japanese sovereignty. In the charged air of the Okinawan reversion controversy, they were bound to cause sparks. Getting the papers, however, and publishing them, would seem to be in the best traditions of newspaper enterprise. At almost the same time, indeed, the *New York Times* and the *Washington Post* were still energetically working over the Pentagon Papers exposé in the United States.

There were two things wrong with Mr. Nishiyama's scoop, however: (1) The elapsed time between his getting his hands on the cables and printing the news was no less than nine months. He got the copies from his good friend Mrs. Hasumi in June 1972. They were made public early in April 1973. (2) When they were made public, it was not in a by-lined story in *Mainichi*, but in a speech in the Diet by a socialist member, Yokomichi Takahiro, who had been given the contents of the cables by Nishiyama.

The controversy blew up in April 1973 after Nishiyama and Mrs. Hasumi were arrested for violating the secrecy provisions of the National Public Service Law, a rather mild Japanese version of Britain's rigorous Official Secrets Act. The newspapers were full of righteous indignation. With few exceptions, they denounced the government for interfering with the public's "right to know." Nishiyama, it was said, had thought about publishing the papers when he received them and discussed them with his editor at *Mainichi*. He even went so far as to hint at their presence in a story. But he and his editor both felt that there was no need to print the cables as they were. (For one thing, he would hopelessly compromise his source.)

So they sat on them for the greater part of a year. At length

Nishiyama felt that the matter ought to come out in public debate, since he judged that the government was deceptive in not telling the public about the secret agreement. On March 27 he delivered the cables to his friend in the Diet, who obligingly broke the story on the floor of the Lower House.

Behind the thundering and sanctimonious editorials, there were some indications that the press's collective conscience was bothering it—Nishiyama was allowed to resign, as was his editor. But almost nothing was said about the propriety of a newsman getting a story, holding it for nine months while he decided when it might be in the national interest to disclose it, then breaking it by handing his information to an opposition legislator.

In February 1974, the Tokyo District Court convicted Mrs. Hasumi, the tragic dupe, for betraying her trust as a public servant. Nishiyama was acquitted on grounds that his activities were justifiable as news gathering. The fact that he did not publish his curious scoop apparently had no bearing on this interpretation of the Japanese public's right to know.

The public, however, took its revenge on Mainichi in the market place. As a direct result of the credibility problem, circulation began to slip in 1974 and continued to the point of virtual bankruptcy. In 1976 Mainichi newspapers underwent a surgical reorganization, in an effort to turn them around. The stigma remained, however.

By contrast, the *New York Times* and the *Washington Post* were given the Pentagon Papers by Daniel Ellsberg and they decided that they should be published essentially because they were news. It was Mr. Ellsberg who broke his secrecy oath, but he was not seduced into doing so. The contrast between their swift publication and the nonpublication of Mr. Nishiyama's news is most instructive.

More serious than the Nishiyama case, in the sense that it represents a continuing malady rather than one strikingly ugly symptom, is the unitary attitude of the Japanese press toward China. "As soon as the figure of Chairman Mao came into view, a great shout of welcome rose and echoed through the hall. An unceasing round of applause, straight from the heart, went out: 'Long Live Our Great Chairman Mao.' Chairman Mao warmly waved his hand to the representatives."

This panegyric prose appeared, not in the official Peking press, but in the *Asahi*, part of a longer lyric written by *Asahi*'s then Peking correspondent, describing the Tenth Party Congress in August 1973. It is typical of the writing on China and Chairman Mao in the Japanese big three dailies and most of the other press. The utterly uncritical attitude of these papers toward China reflects not merely sentiment, but a calculated bargain entered into by the managements of all but one major Japanese newspaper. The bargain has yet to be mentioned to their uncomplaining readers.

Although Japanese correspondents have been reporting from Peking since 1964, the formal agreement to allow them there did not emerge until 1968. It has been amply confirmed since Japan and China resumed diplomatic relations in 1970. The agreement is built around three basic principles: (1) not to follow a hostile policy toward China; (2) not to participate in any plot to create two Chinas; (3) Not to obstruct the restoration of normal relations between Japan and China in any way.

Although these principles seem harmless enough, they have been interpreted by the Chinese as forbidding any criticism at all. Anything much less than the fulsome applause of the *Asahi*'s Peking correspondent is interpreted as critical. The agreement does not merely cover correspondents' writing. Any unfavorable comment by the parent newspaper or network in Japan is viewed as a violation, enough to expel that organization's correspondent, at the least. The track record is not a happy one. Five of the original nine media permitted Peking representatives in 1964 have had their men expelled. One correspondent was refused re-entry because his television network had run a program about Taiwan. Other correspondents were thrown out because their parent papers carried unspecified anti-Chinese cartoons. One correspondent of the *Nihon Keizai Shimbun* was imprisoned for more than a year on charges of espionage.

In September 1973, when two *Yomiuri* correspondents were ordered out of South Korea for reporting which that government objected to, all nine Tokyo newspapers immediately signed a joint protest to Seoul. Editorials rattled through the Tokyo press in righteous protest.[5] Not one of the protestors had spoken up pub-

<hr />

5. Interestingly, the Tokyo press, since 1972, refers to the government of the Republic of Korea as the "Park Regime," after President Park Chung Hee. The Communist government of Kim Il Sung, the Stalinist dictator in the North, is scrupulously noted as the "People's Democratic Republic." President Park's antidemocratic excesses are most justly criticized. At the same time, the positive achievements of South Korea are consistently down-played. By contrast North Korea, the most ruthlessly totalitarian of all the surviving Communist dictatorships, is generally depicted in the Japanese press as a happy country of cheerful gymnasts and burgeoning industry.

licly to help *Nihon Keizai*'s correspondent get out of his Peking jail; still less to protest any of the other arbitrary expulsions of Japanese correspondents by the friendly Chinese government. On the contrary, a stream of humble letters has gone to Peking from Japanese editors and company presidents begging admission for Japanese correspondents on terms that any self-respecting newsman might be expected to turn down. (As of this date, there are four correspondents from Japanese media in Peking.) As one Japanese critic put it, the major papers behave like "a cheap gangster whose degree of bravery is tailored to fit his opponent."

In their book, *The Slant in China Reporting*, [6] Professor Eto Shinkichi of Tokyo University and Miyoshi Osamu, the former editorial page editor of *Mainichi*, have analyzed this fascinating phenomenon. At some length they describe the extraordinary silence of the Japanese press about various upheavals in the People's Republic of China, the Lin Piao incident of 1971 among them, as contrasted with the abundant reporting of Western journalists, including correspondents in Peking. No word of their bargain to report only favorable news about China has yet been leaked to the papers' readers, however. As Eto and Miyoshi wrote in summarizing the problem, "The same newspapers who asserted 'the right to know' in the case of the Foreign Office's secret documents are themselves exercising 'the right not to tell' about their own secrets. If the newspapers, for the sake of their own commercial profit, cut off the free flow of information about diplomatic problems which have an important bearing on the nation's future, that would indeed be poisoning the well of public opinion. And this should be a matter of deep concern to every citizen. . . . In writing this book, we are worried about problems far deeper than the China question alone. In every free society there must be a balance of power between press, public, and government. It is only through such a balance that the free society can insure itself an active, healthy development, without falling into stagnation."

It is a fact, as Eto and Miyoshi note, that the profit motive played a part in the Japanese papers' "kow-tow diplomacy" in Peking. Given the intense competition between them, no editor wanted to see a Peking date line in a rival paper and not his own. That would mean not only loss of prestige, but possibly loss of circulation to the competition. Yet there is little doubt that the mandarinate's pro-

6. The translation is mine. The   *Henko o Tsuku.*
Japanese title is *Chukoku Hodo no*

Peking sympathies made their kow-towing less onerous than it might have been. The urge to make friends with the People's Republic, at whatever cost, has been very strong.

As of this writing, only the *Sankei Shimbun* has publicly stated that it will not submit to conditions laid down by Peking for admitting resident correspondents. The others, which have submitted, almost daily treat their readers to stories about China whose praise and enthusiasm recalls nothing so much as what the same papers used to say about the activities of their own country's armies in China in the thirties. No opposing views are permitted.

Japanese newspaper editors include some of the nicest people in the world. A good number of them are well-traveled, urbane, and knowledgeable. Speaking privately, their comments about what is going on are many cuts above most of their American counterparts and reveal an enviable balance and perspective. One is almost embarrassed to mention the wildly partisan reporting that slips into their pages. They nod with great understanding and say something about the bottom-up system of management and their reluctance to interfere with the autonomy of their reporters. It is in the best traditions of *amae*. An American is reminded of the wrinkled Indian statesman in the Grade B Hollywood Westerns, sadly admitting to Randolph Scott that some of the boys in the tribe have indeed been jumping the wagon trains again, "Young braves go on warpath. No listen to words of Chief."

In point of fact, most Japanese newspapers, like much else in Japan are organized like feudal kingdoms. Each department is a law into itself. The city desk (*shakai-bu*, or social affairs department) runs things its way; the foreign department has its own standards, and so on. With the exception of the *Nihon Keizai Shimbun*, which has a system of managing editors and assistant managing editors, in the manner of Western papers, almost no Japanese editor ever tries to edit his paper. He presides over it.

Sometimes delegation of authority reaches extremes, as in the case of *Asahi*. At *Asahi*, since the late sixties, the newsmen have been edited very little, particularly if they own stock. Partly this results from an intense struggle over *Asahi*'s management, climaxed in 1963 when the publisher and principal stockholder, Murayama Nagataka, was dethroned by a well-organized combination of employees working with the principal minority stockholder.[7]

Since so many editors and senior reporters threw their stock into

7. The confrontation was brought on by the highhanded activities of Mr. Murayama's wife, the granddaughter of *Asahi*'s founder in forc-

the common anti-Murayama fund, they became in a sense copro-prietors. They were not slow in conveying their sense of privilege to the front office. The influence of the desk in controlling *Asahi*'s reporting has steadily diminished. Management sometimes takes action against too extreme editorial behavior, such as when the Maoist editor of the *Asahi Journal* was removed in 1970. At least one *Asahi Shimbun* reporter was rebuked, during the 1968–69 demonstrations, for overenthusiastically joining the student riot he was supposed to be covering. In general, however, the reins are loosely held.

It is a back-handed tribute to the homogenizing tendencies of the Japanese that the views of so many editors and reporters stay in the same general mandarin mainstream, without much need for formal direction or indoctrination. In *Asahi*, as in most other Japanese newspapers, people with unorthodox opinions generally leave or end up in the research or the publishing department. It does not pay to be "different" in Japanese journalism—which includes attempting to be objective. The moderate publisher of one large Japanese monthly had editorial policy taken out of his hands and his leading editors forced out of the office by an energetic group of far-left staff members, who used their union connections liberally for this purpose.

The American newspapers, or most of them, have had their troubles in attempting to lead the consciences of their communities. Lacking the reputation—and the resources—of papers like the *Washington Post* and the *New York Times*, all too many American editors are reduced to handling the wire service copy, then watching the late news on television like everyone else. Most hopeful efforts at magazine formats, capsule news reports, new world reporting networks, and imaginative local reporting have had a tendency to die on the calculating machines in the controller's office or given way to the advertising department's need for yet greater department store space. Whether or not the counting house, not to mention the printers' union, has been to blame for the actual retrenchments and forced mergers in the American press is a moot point. But until recently, at least, the editor has had to fight a losing battle. It took the extraordinary investigative journalism of the *Post* in the Watergate exposé to dramatize the importance of the strong, critical newspaper, even in the society with the world's most massive television news coverage.

ing her views into the editorial pages. It also had some idealogical overtones, however, which dated from Mura-yama's belated efforts to curb the reckless reporting that helped provoke the 1960 riots.

In Japan the problem is opposite. With its big circulation dailies and docile reading public, Japan is an editors' dream. Television news has yet to make the serious challenge to the newspapers which has been made in the United States. (Although the semi-government Japan Broadcasting System [NHK] has some good news documentaries and awesomely comprehensive coverage, the content and presentation of most news programs on both NHK and the private television stations is far from dramatic.) Circulation still calls the turn. The advertising content of Japanese newspapers is far less than American; and the leading ad pages, reflecting both popular interest and the mandarinate's prestige, are for book publishing houses, not department stores. There is almost no one to dispute the newspaper's role as public-opinion molder.

The Japanese editors' response has been to ignore increasingly what the readers want to read or what they might think, and to concentrate on what the editors and their friends in the mandarinate want to tell them. What the reader gets, on any given day, is certainly not objective reporting. It is not even necessarily the opinions of *Asahi* or *Mainichi* or *Yomiuri* as recognizable journals with defined audiences, but the canned impressions of the reporters' clubs as filtered by the rewrite desks and the editorial boards.

For if sometimes an iconoclast (as it should be), the newspaper in Japan is more often a conscious pedagogue. It faithfully reports the press establishment's view of the world in its news and editorial pages. The serialized novels, heavy sports coverage, a few bits on cooking and embroidery, and the back-page blood from the latest mass accident or neighborhood murder are its principal concessions to the reader. And even they are virtually the same in all the papers.

Although the competition between papers is fierce, especially among the big three, their editorial staffs are so heavily manned and evenly matched—reporters' clubs and all—that one paper's beat over the others is rarely visible to the naked eye of the lay reader. How vital and necessary this seems to the average home newspaper buyer is a very good question indeed, especially in a day when families move more and are less rooted in their neighborhoods. If the thin line of badly paid newspaper distributors ever falters and householders begin to lose the newspaper habit, the great newspaper houses will be in for a heavy fall. It is on the steady subscriber that they depend.

The papers have heavy assets in real estate and other ventures to act as a certain financial cushion. The *Yomiuri* has been a leader in the lucrative real estate department, with its own amusement park, Yomiuriland, and other more prosaic ventures. It also owns the Yomiuri Giants, Japan's perennial baseball pennant winner. *Mainichi* runs its own direct selling company. *Asahi* produces the annual Osaka Festival of Music and the performing arts, in its own Festival Hall.

But the fundamental business of publishing is still the big three's major cause of concern. Profits have become paper thin and there have been loss years. Some executives talk of planned circulation cutbacks, recalling the Pompeian days of *Life, Look,* and the *Saturday Evening Post*. In December 1973, in a mood approaching panic, *Asahi*'s high command suddenly canceled plans for a huge projected *Asahi Encyclopedia*, ostensibly because of the paper shortage at that time. The real reason was the drain this effort would have made in *Asahi*'s not unlimited financial resources.

If any one of the big three falls—as *Mainichi* almost did—it would be ill-equipped for a comeback. Encumbered by thousands of excess employees, inefficient subsidiaries, and the same numbers-at-all-costs circulation philosophy that killed so much of the American press, a Japanese press empire could expire like a dinosaur in a swamp, dragging most of its ancillaries with it.

Could television and radio fill the void? Certainly they are better prepared to challenge the papers than they were a few years ago. The private broadcasting stations have all grown prodigiously since 1960—there are now more than thirteen hundred private television stations in Japan. NHK has grown along with them. There are almost thirty-five hundred NHK TV stations in the country (half of them educational). Virtually a government institution, NHK is supported largely by monthly broadcasting fees (the equivalent of $1.75 from an owner of a color television set, $1.25 for black and white). Its resources for news coverage are vastly greater than the private television stations. Only Tokyo Broadcasting System maintains a substantial correspondent network of its own. The private stations, which do very well from their commercials, concentrate on purely entertainment programming, where they are more contemporary and freer than NHK. Only lately have they been growing more venturesome with news documentaries.

NHK, officially Nippon Hoso Kyokai ("Japan Broadcasting Corporation"), has a stronger hold on its national audience than any other modern communications network, except the wholly propagandized services of the socialist countries. Part national conscience, part full-time instructor, yet necessarily committed to a policy of consensus entertainment, it sets out very consciously to act not only as an information channel, but as the executor of what the French used to call a *mission civilisatrice* on every level of Japanese society. Although it is formally organized as a private corporation, the government connection is close. Prominent establishment figures are invariably selected as its chairmen and trustees. It is operated as a national trust, much as the BBC and, indeed, more so. It is no coincidence that the NHK and BBC are, at least to my knowledge, the only stations in the world which have the time, money, and concern for a particular segment of two nature-loving populations to broadcast weekly programs of recorded bird calls.

As an educator, NHK does a superb job. Its educational programs, both on television and radio, are excellent and it has done more with instructional television than almost anyone else, anywhere. The network of instructional classroom broadcasts is formidable and authoritative. On its educational channel, channel three in Tokyo, viewers can be instructed in English, French, German, Chinese, and Russian; they can learn the fine points of *Go*, Confucian philosophy, and plumb further into the intricacies of their own language, as well as brush up on new automation techniques, European history, international politics, and the racial problem in the United States (which the uniracial Japanese view with the remote fascination of earthlings watching Martian canals explode).

In its popular programs, NHK tries to steer a course between entertainment and uplift, with varying degrees of success. It broadcasts *sumo* tournaments and seasonal high school baseball championships, as well as some major league games. Vignettes of history, local folk festivals, popular music roundups are heavily featured, as well as a certain amount of rock music and jazz. The popular programs feature what can only be described as the NHK musical—expense-be-damned panoramas of oddly costumed chorus lines prancing about in garish Hollywood-type costumes.

The historical serials are generally excellent. Epics like the *Shinheike Monogatari*, a fifty-episode, fictionalized re-enactment of the fall of the Heike chieftain, Taira Kiyomori, and *Kunitori Monogatari*, which covered the rise to power of Japan's sixteenth-century dictator, Oda Nobunaga, command the services of Japan's best television, movie, and Kabuki acting talent.[8] They are exciting enough drama to keep a good portion of the country, at whatever educational level, in front of their sets from 8 to 8:45 P.M. on Sunday evenings for an entire year, straining to make out the reconstructed eleventh-or sixteenth-century dialogue, whose soaring honorifics can be understood about as easily by modern Japanese as the Shakespearean declamations of Henry V by modern Americans.

With these, as well as the folk festivals and the *sumo*, there is rarely any doubt about NHK's conscious mission as the defenderpropagator of Japanese culture in a world full of foreign blandishments and temptations. Its news documentaries are complete and frequent foreign coverage is provided by the on-the-spot correspondents as widely distributed as those of the major dailies. Its news is determinedly objective, although, appropriately enough, NHK has been known to lose its cool when the national honor is at stake. Voices of the most professional announcer can break in happy excitement when, for example, Japan wins the ski jumping at the Sapporo Winter Olympics.

In its didacticism, its careful choice of subject matter and the whole of its commentary, NHK, too, shows the mandarin's hand. Its style is best typified by the announcer sitting authoritatively at his desk in the studio, reading the news comment against a background of still pictures of fleeting film clips or, most characteristically, talking on the telephone with some far-off correspondent giving half-audible descriptions of a hijacking or a world monetary conference, with the inevitable commentators waiting in the next studio to explain it all over again. There is no attempt at showmanship on NHK's antiquated television news broadcasts. Unlike the American newsman on television, he is not trying to get the public's attention. Like the faithful mandarin he is, he is telling the public what it ought to know.

8. The subjects are chosen with care, like everything NHK does. It would be tempting to infer some connection between the choice of aggressive military leaders like Taira and Oda for programs and the aggressive Japanese economic advances of the early seventies; whereas the subject of NHK's serial for 1974, a time of regrouping and international conciliation for Japan, was Katsu Kaishu, the statesmanlike seeker after foreign learning, who helped prevent civil war from widening in Meiji times. But this might be giving the NHK program department even more credit for omniscience than it deserves.

Chapter XII

Passive Politics at Work

"... the ruler is the least important."
—Mencius

The American and the Japanese democracies both work out their destinies in a man-made cave of winds. The contrary blasts of multiplying technologies, market shifts, inflation, pollution, welfare needs, competitive business and social pressures, compulsive consumer demand, vast profit incentives, and accelerative generation change have become commonplace to both. Both peoples are overcommunicated and yet underinformed by the constant noise of their own mass communications—press, polls, and picture tubes. They have, as a result, built up amazing tolerances for high-pressure living, rather like two heavy drinkers.

The strain of high living shows. Both the Americans and the Japanese began the second half of the seventies in a mood of frustration, confusion, and increasing discontent. Half-stimulated by affluence and half-surfeited by it, they are worriedly pondering the uses of wealth, the distribution of resources, the responsibilities of power. The strain shows most at democracy's most sensitive nerve: the process by which people make, or at least sanction, political decisions.

The political problems of decision making in both these complex societies are strikingly similar. The way they actually run their politics, however, is strikingly different. History, background, and present circumstance have made the Japanese approach to democratic politics almost the exact converse of our own.

The United States evolved in an attempt to restructure society as a mass free-enterprise democracy. Americans grew up confident of their abilities to assimilate people of any race or culture on the basis of equal opportunity. In a way, the American nation was its own national goal. This avowedly individualistic society is now, however, bound up in all sorts of group commitments and acquiescences, many of which it makes with ill grace. Its traditional melting pot is showing its age. Its democratically elected modern presidents have been uncritically entrusted with so much power that they threatened to become more potentates than public servants. The huge power concentrated in the modern American leadership has produced qualitative changes in the traditional American concept of what power is and how it is to be used.

Ideally, each American citizen is a public man or woman whose loyalty to both the principle and practice of the American democracy is assumed to supersede all other temporal allegiances. The successive shocks of the last fifteen years, from the half-hidden beginnings of the Vietnam disaster to the arrogance of the Watergate manipulations, are causing Americans, hopefully, to think back on the premise of their democracy: that the citizen is a public person whose concern and participation in government must, in the end, be the only warrant for its success. There is more than a little truth in the long-standing criticism by the Swedish economist Gunnar Myrdal that there is an "American pattern of individual leadership and mass passivity." The Americans, as Myrdal argued in his memorable racial study, *The American Dilemma*, have a great facility for developing leaders at the top—the executive, the boss, the community leader, whatever the title may be. At the same time, there have been few strong and continuing group movements from the mass of the population, as there have been in Europe. People have been content to pick their leaders (after all, any American can become one) and follow them, confident that the built-in network of checks and balances in the American system will protect them from most conceivable leadership abuses.

The Japanese have a different kind of political passivity problem. An ancient, sophisticated society cemented by social commitments, they still begin their thinking with the group, not the individual. Historically, their revolutions as well as their repressions were channeled and directed from above. The concept of a public person, whether leader or citizen, with his primary duty to a public, self-governing commonwealth, is not easy for them to contemplate. There is a general consciousness of one vast Japanese family, a nation-society. The Japanese democracy is strong in the sort of things that family members need and want, like free expression and protection from injustice. But being automatically part of a family is a different thing from being politically engaged or at least aware of one's responsibilities in a nation built on political commitment.

Where the American, too confident of his own authority and security as a citizen, too uncritically gave his proxy to his national leaders, the modern Japanese has been nervous and grudging about delegating his rights at all on a national level. One's company, one's university, one's neighborhood are a part of life in Japan. Japanese are comfortable with their private hierarchies. Yet they are suspicious of public leaders. It is one thing to line up behind the man-with-the-flag and depend on him within the structure of a company, a ministry, or a political ward. It is more complicated and far more unsettling to give one's allegiance to a national party or a national public figure. For most postwar Japanese, the national government is something really "other." It does not belong to them, nor they to it.

The same society that is instinctively collective in so many areas —whether sweeping out the neighborhood for the festival or working long hours at the office—is at best docile, at worst surly about supporting its national government. A charter member of the *amae* society, a Japanese expects to be taken care of. He is, in a sense, a family dependent of the government. As such he obeys the household regulations faithfully. He almost never crosses a street against the light, unquestioningly obeys the orders of police and other local officials, pays any number of taxes and assessments without prodding. But ask an individual to contribute to a political party, or even to enroll in one, or join in a national political movement, and the response will be very, very weak. Although the politically committed and the politically contributing can be found in Japan, their numbers—barring a few, dedicated groups on left and right—are quite small.

The whole idea of a party system is still weak in Japan. Question an American or an Englishman about his political beliefs and he will likely as not say, "I'm a Democrat," or "I'm a Conservative." A Japanese would merely say, if pressed, that he voted for the Liberal Democrats or the Socialists in the last election. He might add, alternatively, "I am a member of the progressive (or, more rarely, conservative) camp." But that is more an indication of ideological bias than political commitment.

Much of this "let-George-do-it" attitude toward the national government and its politics stems from the immediate experience of the Japanese before World War II, if not, indeed, since the Meiji

Restoration. The Japanese are a long-memoried people. Once an impression is set, it is hard to dislodge. And fully as much as the university professor or the newspaper editorialist, the man in the street was embittered both by the arrogance of his pre-1945 government and by its failure.

When the Meiji reformers built their new political house in the 1870s, they laid some basic foundations for a parliamentary democracy, but they anchored the whole edifice to a sort of central authoritarian control. The Meiji reformers were themselves of two minds on the issue, in that some wanted heavy popular participation in government. But being Japanese, they were not overly bothered by logical contradiction. In the end, despite the trappings of liberal parliamentary rule, the imperial institution was firmly set up as the basic authority in the nation. Enough veto power was built into the Meiji constitution to give the emperor, or those who spoke for him, an ultimate weapon against any popular reformers who might get out of hand.

The fight between democracy and authoritarianism in Japan was a long one; and the democrats got in some good blows. The so-called Taisho democracy of Japan in the early twenties (taking its name from the Taisho Era) was far from a fiction. Strong rural, and city workers and cooperative movements developed, including fast-growing Socialist and Farmer-Labor parties. The modernizing city population post-World War I had a good idea of its rights as citizens. But the militarists, able to ground their power in the emperor, finally, by threat and terror, extended the control of the central government bureaucracy to every area of Japanese life. The notorious peace preservation laws of 1925 in the end mopped up most dissent, real or potential.

Both the two largest political parties of the prewar period, the Seiyukai and the Minseito [1] were heavily wired into the ruling establishment. The former (and older) was bank-rolled by the Mitsui *zaibatsu*, the latter by Mitsubishi. Neither was distinguished for its grass-roots influences. Leaders of both parties were picked by the emperor and his advisors to form cabinets, generally without reference to elections.

As a result, except for a very few distinguished political leaders, the party governments were interested more in ruling than in rep-

1. Their full names were the Rikken Seiyukai ("Friends of Constitutional Government Association") and Rikken Minseito (Constitutional Democratic party").

resenting. Bribes were open and frequently passed between the parties and their supporters in the big business of that day. Regulations on profiteering were few, generally coming only in response to occasional public outcry, as in the nationwide riots over rice prices in August 1918. Even the most successful of the party premiers, Hara Kei, although known as the "commoner prime minister," tolerated wholesale corruption around him and fought a cautious delaying action against universal male suffrage. Until the militarists took over finally, in the thirties, Japan was actually run by its civil servants, the spiritual descendants of the men who made the Meiji Restoration by decree, led by a small group around the throne.

In the popular mind the major parties became little more than a side show. Their platforms were almost indistinguishable. The Diet, whose membership they supplied, was little more than a debating hall. In 1940 they were finally dissolved, with the Socialist Masses party (Shakai Taishu To) of Katayama Tetsu going first. The Seiyukai, the Minseito, and remnants of most of the smaller groups were all taken in under the umbrella of Premier Prince Konoye's Imperial Rule Assistance Association. They ceased to exist as units. They were not mourned, as the populace turned with general satisfaction to the pursuit of what most Japanese thought would be a winning war.

In 1946, along with the new antiwar constitution,[2] the United States occupation sponsored a return to parliamentary government in Japan, based for the first time on universal suffrage. In the 1947 elections (the first to be held under the new constitution), the Socialist party received a total of 143 seats in the Lower House, to become the largest single-party voting bloc. Although the total seats gained by the two new conservative parties was 281, the Socialists were able to put together a coalition government with the Democrats, led by the rehabilitated Katayama. With a moderate Socialist, that chronic dream of liberal Americans, in office (to make things even rosier, Katayama is a Christian) the young government experts of the occupation settled back in their chairs to watch the gradual emergence of a healthy multiparty system in Japan, which might even end up looking like the Democrats and the Republicans.

2. So-called from the famous Article Nine, by which the Japanese people "forever renounce war as a sovereign right of the nation," along with belligerent rights and the maintenance of armed forces.

It was not to be. The Katayama cabinet lasted less than a year; it was brought down when the left-wing socialists pulled out, going over to the opposition. The conservative Liberals and Progressives, benefiting from the heavy rural allocations of Diet seats as well as the generally conservative mood of the voters, began to increase their percentages of the total vote. They reached a high-water mark with 66 per cent of the popular vote in 1952. The Socialists, disunited and implicated in some unfortunate scandals, lost heavily in 1949. The Communists by contrast, gained an unprecedented 10 per cent of the total vote, but they lost this strength when they changed their popular-front policy for one of militant revolution in 1950–51. The Socialists, however, never got within striking distance of a majority or any kind of a plurality again. Through the sixties they generally held under a third of the total vote, reaching their high point in 1958, with 166 Lower House seats out of nearly 500, and 32.9 per cent of the popular vote. In 1969 they were scalped by the rising city strength of the other opposition parties, and their membership sank to 90. In 1972, Socialist strength returned to a stable level of 118 (22%), little changed after the 1976 election to 123 (20.7%) in the enlarged 511-man Lower House.[3]

The conservatives never once lost their Diet majorities through this period, in both houses. In 1955, when the left and right-wing Socialists came together for a time in a united party, the conservatives followed suit. They received 63.2 per cent of the popular vote that year and packed the Diet, emerging with close to 300 Lower House seats. Since then they have steadily slipped. In the 1972 elections, and the two preceding, they failed to get a majority of the popular vote. In 1976, their popular vote percentage sank to 41.8 and they were able to put together a majority only after several members elected as Independents joined the 249 Liberal Democrats.

The men who led the Liberal Democrats through more than a quarter-century of one-party government, taken as a group, have hardly shown spectacular qualities of leadership. Yoshida Shigeru, who called the turn for the others, was the only one to come on as a strong personality. He exercised great influence over his succes-

3. All the figures given are for the elections to the Lower House. The Upper House, the House of Councilors, is sharply restricted in its powers. Its members are selected on a national as well as a local basis. Although many prominent people are Upper House members, a working politician must almost invariably get himself elected to the Lower House, from a set constituency, before he can be seriously considered for higher things either in the party or the government.

sors long after his retirement. Tokyo wits used to comment that disciples of the so-called Yoshida School like Ikeda and Sato (whose eight-year premiership, 1964 to 1972, set a new longevity record in Japanese politics) spent half of their waking hours on the road to Oiso, the ex-premier's villa south of Tokyo, where they received counsel and instruction. Ikeda, the most straightforward as well as the most competent of them all, deliberately elected a low-posture policy to heal political dissension between right and left in Japan.

None of them, however, made much of a mark on the outside world. "Who was that transistor salesman?" was De Gaulle's typically cruel comment after Ikeda had visited him in Paris. With one or two exceptions, they came from a bureaucracy which disliked show and preferred to order things behind the scenes. Ikeda had worked his way up through the Finance Ministry, whose elite in Japan, as in France, are regarded as the bureaucracy's shock troops. Sato came from the National Railways. Kishi got his apprenticeship for politics in the prewar Ministry of Commerce and Industry. Fukuda Takeo also possessed a Finance Ministry background. Tanaka was the one businessman premier who managed (to his ultimate regret) to combine his contracting work with membership in the Diet. Miki Takeo is the only one of all the recent prime ministers who was a politician by trade, since entering the Diet pre-war, and he has been the most liberal. All of them have been long on experience and —barring Tanaka and for different reasons Miki—even longer on caution. They generally played the game by the rules.

In terms of its accomplishments the Liberal Democrats' administration would have to be judged a success. They have presided over Japan's extraordinary rise from destitution to the status of an economic superpower. Wise, long-range government planning and, as we have noted, the constant interaction of government, business, and finance in their decision making were prime factors in the success of Japan, Inc. Graft has reared its head, but on nothing like the prewar scale. Government is orderly. Although riots and demonstrations have echoed through the streets of Tokyo, over the years they have been, with few exceptions, well-handled. Indeed, they rarely represented anything more than a small, if loud, minority of the Japanese people, despite the occasional scare headlines in the foreign press.

The government, the courts, and the police have all been careful of human rights, and often forebearing in the face of great provocation. (As a Japanese friend commented to me once, after returning from the 1968 Democratic Convention in Chicago, "When I watched your police in Chicago brutally clubbing those people, I was really proud of our own Japanese police. That sort of thing could never have happened in Tokyo.")

Japan's foreign policy has certainly been satisfying to the United States. A steady adherence to Free-World politics and a continuing reliance on the American nuclear umbrella gave way to more neutralistic tendencies only in the early seventies—and this thanks principally to the adversary tactics used by the United States in its economic warfare and political pressurizing against Japan. The Liberal Democrats, for their part, regained Okinawa and the other Ryukyu Islands, resumed relations with mainland China without losing their profitable trade with Taiwan, and extended increasing amounts of judiciously given economic assistance to the developing nations. (The North-South problem, as they call the gap between developed and underdeveloped, is a favorite concern among Japanese foreign policy experts.) For years their international diplomatic image was gray, dull, but respectable, rather like the drab, but spotless, background of striped-pants and cutaways and black-crested kimono at a fashionable Japanese wedding reception. Only with Japan's increasingly aggressive business tactics abroad has this neutral image mixed with the controversial figure of the "Ugly Japanese" trader.

The political climate inside Japan, however, has been cloudy since the late sixties. Except for pockets of enthusiastic minority politics, mostly Communist party cadres or the religiopolitical cohorts of the militant Buddhist Komeito ("Clean Government Party"), the mass of the Japanese people remains disappointed in politics and politicians. The disappointment centers on the ruling Liberal Democrats, but it is vented also on the Socialists and other minority parties. Only a few political leaders, like Yoshida, managed to get their heads above the crowd, and are remembered for it. Mostly, *politician* remains a dirty word.

To understand just why, we must take a look at the workings of the Liberal Democrats, their opponents, and the factors that have kept them in power. As many Japanese see it, the balance is surprisingly unstable. Much of what the outside world regards as keen planning and decision making looks to the Japanese themselves more like extraordinarily lucky coasting on the momentum of Ikeda's double-your-income policy.

The centerpiece of Japanese rural and neighborhood festivals is the *omikoshi*, a portable shrine of the local tutelary diety, which is hoisted on the shoulders of several dozen shouting, struggling young men and pushed back and forth, depending on the waxing and waning strength of the people on each side, as it rolls down the street to the real shrine and the booths where the sake and refreshments are. The analogy to the Liberal Democratic party is hard to resist. The political complexion of its Diet contingent ranges from far right-wing nationalist to moderate but non-Marxist left (not unlike the current span of American Democrats). It includes regional old politicians, rising businessmen taking a fling at politics, numerous former civil servants, and a few popular young actors and entertainers. Each group in the party, like the shouting young man around the *omikoshi*, does its best to push the whole unwieldly edifice its own way.

The party has a strong Taiwan lobby, although some of its factions paved the way for diplomatic relations with Peking. Some Liberal-Democrats would like to rearm to the point of possessing nuclear weapons. Others advocate a kind of neutralist Japan, not far from what the opposition is saying. There are some general areas of agreement, notably the maintenance of the free-enterprise economy and the utility of some security relationship with the United States. But even here, one sees strong shadings of opinion. Since the 1971 Nixon shock, particularly, doubt has been growing about the worth of the American connection.

The differences of opinion in the party reflect its origins. Three postwar parties (two of them with links to the prewar Seiyukai and Minseito) and numerous local political fiefs came together in 1955, principally for political survival. The differences are perpetuated by the division of the party into semipermanent factions, grouped around various party leaders. These have the character of fractious, semi-independent feudal baronies. In 1978 there were no less than nine factions in the Liberal Democratic party, of which even the closest are only loosely allied. The factions do their own fund raising and subsidize their own candidates as heavily as they can. Promising new men must be recruited, old backers must be rewarded, and voting unity be kept. In this real sense the Liberal Democrats are nine parties in one. For a prime minister (who is also the party's president) to keep them all in line involves the most constant sort of power brokerage. Each cabinet must represent some balance of factional strength, to keep the party united. It is a game of musical chairs that never ends.

The factions are not geographical. It is not unusual to find two big factional leaders from one area, like the Third District of Gumma prefecture represented by both Fukuda Takeo and Nakasone Yasuhiro (each of whom maintains his own national political organization as well as keeping up a strong local base). They grew up out of personal contacts and intraparty alliances and stretch across the country. Divisive as they are, the factions remain the chief source of most politicians' support. Business donors especially—and they are the lifeblood of the party—feel they can get more for their yen by giving money directly to factions rather than the party itself. Thus everyone resists doing away with them. The factions, in turn, polarize around leaders (when one dies, either the factions disintegrates or a trusted friend or deputy takes his place), so the party's grass-roots support, such as it is, centers on individual members. "Vote the local leader, not the party" is still the unspoken premise with most Japanese voters. A Diet member is generally as strong as the koenkai ("support association") he can put together.

The electoral system is an odd one, in which the top three, four, or five candidates will be elected from a given district out of a larger field. To get both party and faction endorsement for a campaign is critical and cutthroat. There may be five Liberal Democrats, four Socialists, three Komeito, and two each from the Minshatō (Social Democrats) and Communists contesting four seats—depending on the parties' estimate of their local strength. Two popular Liberal Democrats may end up taking votes from each other in this open-field running, with the Socialists profiting, or vice versa. The combinations are endless and intraparty competition often worse than that between rival parties.

Over the years, the Liberal Democrats have built up an increasingly strong central party organization, with its own bureaucracy in Tokyo. Nevertheless, the strength at the local level, where it exists, belongs to individual candidates and their factions. The strongest organizations exist in rural areas (a fact not unrelated to the LDP's consistent support for artificially pegged farm prices), but as urbanization continues they inevitably weaken. Individual contributions to the party are not abundant. Which means, paradoxically, that, while the conservatives need local participation desperately, they are still thrown back on big business for the money to fight and win national elections. It is not a healthy alliance. The deference the Liberal Democrats pay to business, in turn, is all too obvious and has turned away many potential Liberal Democratic voters in the cities. In the July 1974 elections for the House of Councilors, the notorious "business package-elections" (*kigyo gurumi senkyo*), the party went too far, even for some of its strongest supporters. Not content with running prominent TV and stage personalities for the national constituencies, the party persuaded various large companies to back their own Liberal Democratic candidates! Mitsubishi, Kajima Construction, even Shisei-do cosmetics publicly backed their own chosen politicians. This tactic was too much even for the patient Japanese voter. It turned a probable Liberal Democrat victory into a stand-off.

Why, indeed, have the Liberal Democrats lasted so long? There are three reasons: The first is the lopsided weighting of the voting districts in favor of rural and semi-urban populations. This has remained almost unchanged since the war, despite the prodigious shift of population from the countryside to the great megalopolitan belts along Japan's Pacific coast. Many Japanese rural voters have the same comfortable overrepresentation as, say South Dakotans in the United States Senate. As a *Mainichi* commentator once noted, "Such prefectures as Tottori and Shimane with very small populations have an apportionment of five seats, whereas the Second District in Osaka has an apportionment of four seats, although the population is several times greater. . . . In the general elections of December 1972, a Liberal Democratic candidate was successfully elected with a little more than 30,000 votes in the Third District of Gumma prefecture, whereas a Socialist candidate who obtained more than 140,000 votes was defeated in Tokyo's Seventh District."

The second reason is Japan's record of economic growth. Despite the party's lustreless leadership, recurrent scandals, and frequent cases of highhandedness, the voters have been prospering. Capitalism has been kind to Japan. A factory worker who is able to buy his own small car and has started thinking about a place in the country is not going to start reading Marx or attempt to overthrow the established order. Not until the beginning of the seventies, when inflation began to hit hard and the need for improvements in housing, transport, and other public services, became more urgent, did the voters start to protest. At that, the protest registered most sharply in municipal, not national, elections. It is no accident that the mayors of Tokyo, Osaka, Kyoto, and Nagoya were, as of

1974, all progressives, elected by joint opposition party efforts. By voting in a Socialist or Communist mayor, who has quite limited powers, the electorate can protest, without having to endorse a new system of government. (Nor are the city-dwellers' votes watered down by the weight of rural constituencies.)

The third major fault behind the unhealthy calcification of Japanese politics lies within the Socialist party itself. Founded on the privations, hopes, and distinguished traditions of Japan's persecuted prewar Socialists, it has never been able to make up its collective mind whether it is a revolutionary band fighting a fascist government or a modern parliamentary opposition inside a modern democracy. In showdowns within the party, the left has generally won. The more moderate Democratic Socialist party was founded in January 1960, by a supporter of parliamentary government and a non-Marxist, as a reaction to the growing domination of left-wing socialists and their insistence on Marxist revolution. For in the face of the highly workable free-enterprise society around them, the socialist leadership remains firmly Marxist, talking in terms of struggles and revolutionary zeal that seem more suited to Japan in the 1930s (or Europe in the 1870s) than Japan in the 1970s.

The Socialists' big power base is Sohyo (officially, the "General Council of Trade Unions"). Sohyo, the largest of four Japanese union confederations, draws its main support from the well-disciplined government employee unions, such as the teachers and the railwaymen, whose leadership is almost by definition perpetually at war with the government. Urged on (and fenced in) by the aging ideologues in the Sohyo hierarchy, the Socialists have continued to preach Marxist revolution to a prosperous capitalistic country whose people do not want one. Since their most popular leader, the late Asanuma Inejiro shifted his own stance from right to left in the 1950s, they have rejected any proposals that the party turn to a gradual, non-Marxist approach to socialism. The former secretary general, Eda Saburo was continually voted down for advocating the positive social democracy that has elected governments in Europe; before his death in 1977 he bolted the party.

For almost two decades the Socialists have been locked into this trap of their own making. On the one hand, their power base has not grown. Sohyo has had difficulty keeping its younger workers in line. More economically than politically oriented, they increasingly resent voting the way the union tells them. (The radicals among them are disposed to vote Communist.) On the other hand, the Socialist party bosses, with their own comfortable core of rural or semiurban supporters, are themselves the prey of factions so divided that the easiest way to unite them is by ritualistic appeals for revolutionary struggle and peoples' republics and against American imperialism. At that, the Socialists, who were strongly pro-Mao for years, were badly back-winded by the conservative government's successful rapprochement with Peking, after 1972.

In the growing public concern over rising prices, government slackness, and big business profiteering which broke in the early seventies, a Socialist like Willy Brandt might have expanded his party base and won an election. But the Japanese Socialists were slow to see their opportunity. Obsessed by their own denunciations of global imperialism, they were pushed aside in the big cities by the well-organized Communists, who more profitably spend their time among the voters, helping housewives boycott high-priced grocery stores or calling up the old folks, on occasion, to see if they need advice on medical care.

For almost three decades the Socialists have hung on to their safe barely one-third of the vote. It is just enough to prevent revision of the constitution, for example (which requires a two-thirds vote), but hardly enough to block legislation. In the process, they have built up all the paranoia and obstructionism of an opposition which knows in its heart that it can never govern. The Socialists use every device known to parliamentary experts, from mobbing the speaker's chair to deliberate slowdowns in voting to boycotting the Diet proceedings altogether, in efforts to blackmail the Liberal Democrats into making concessions which they could never obtain at the polls. Admittedly, the conservative majority is often negligent of minority rights. But the Socialists, although they demand that the government majority compromise constantly with them in the best *amae* tradition, have been rather uncompromising themselves.

In fomenting the divisive national demonstrations of 1960 over the United States Security Treaty, the Socialists were extreme enough to bring down on themselves the censure of the press, which had quite a hand in the agitation itself. Ryu Shintaro, *Asahi Shimbun*'s great editorialist, wrote not long after the 1960 rioting, "[Sohyo's and the Socialists'] insistence on 'absolute opposition,' in fact, was more like a blind obsession than anything else. Thanks to Sohyo's uncompromising campaign against the government, the question of the revision of the Security Treaty went far beyond the bounds of a normal political discussion in the Diet and became a struggle on a national scale—a struggle, moreover, which was less a genuine controversy than a battle against the government."

If the Socialists are a revolutionary party during the day, at night, Tokyo wiseacres note, they become a parliamentary party. That is to say, the Socialists, as long-standing members of the Diet, have their own need to keep on good terms with their fellow Diet members in the majority, as do minority Republican officeholders in Boston or the Bronx. When the fury of debate has passed and the press mandarins have approvingly noted the cogency of progressive protest against Liberal Democratic highhandedness, the partisan committee members involved might meet, say, at some quiet restaurant.[4] Money may even change hands. In the many political-business scandals of modern Japan, it is interesting to note that Socialist Diet members' names often come up along with those of Liberal Democrats—although their take is rarely as great.

Some day, the Liberal Democrats' government will be replaced by a coalition. Japanese political observers feel that an opposition coalition, if not a new party alignment, is increasingly practicable. But the stubborn adherence of the Socialists and their Sohyo leadership to Marxism has thus far insured that neither happen.

The Japanese Communists, ably led and zealously followed, have pursued with some success a new-image policy, concentrating on help for the consumer and wage earner domestically and a neutral, but nationalist Japan in foreign affairs. They seem open-minded even on the issue of national defense.[5] They are, in some senses, the most nationalistic of the Japanese parties. They can mobilize a considerable protest vote, but it is doubtful that they would ever actually get a Diet plurality. In the 1976 Lower House elections, their strength dropped to 17. The Japanese electorate remains anti-Communist in home affairs and anti-Soviet. It is suspicious of the party's ties with the USSR—despite the party's recent conversion to return of the Soviet-held northern islands to Japan, among other things. Although Communists and Socialists have teamed up to elect numerous mayors

4. In his excellent book, *How the Conservatives Rule Japan,* Nathaniel Thayer quotes from a *Sankei Shimbun* article, "During the day [Steering Committee members] exchange fierce and violent arguments and part without reaching agreement, but at night, meet secretly in a restaurant within the city and have another discussion. Generally they reach agreement on what the steps in the future should be."

5. Nozaka Sanzo, then the active head of the Communist party, did not oppose abridging the antiwar clause in the constitution, as far back as the fifties.

prefectural governors, in national elections the Socialists are wary of cooperating. Some Socialist leaders want a Socialist-Communist coalition. But it is clear to many others that the better organized Communists, with their active city cadres, might easily call the turn in any popular front.

Conversely, the Socialists' rejection of gradual reform makes it unlikely that the remaining opposition parties would join a Socialist coalition. The Komeito is an egalitarian, but not Socialist wing of the militant Buddhist Sokagakkai,[6] claiming about ten million zealous believers. Whatever tactical positions it takes, it is basically anti-Marxist. So is the smaller Democratic Socialist party, whose moderately collectivist thinking is probably closest of all the party platforms to what the average Japanese really believes, even though its lack of drive and flamboyance keeps losing at the ballot box. In 1974, notably during the local gubernatorial elections in Kyoto, the Socialists' national leadership's insistence on working with the Communists wrecked its chances of organizing a non-Communist opposition coalition and also further split the Socialists themselves.

There has often been talk of one or more segments of the Liberal Democratic party bolting to join with right-wing elements of the Socialists, the Komeito or the Social Democrats in a new coalition. This is conceivable. But no segment of the Liberal Democrats would ever consider joining Communists or Marxist Socialists. On the issue of free enterprise and human rights, the party stands firm. Here the numerous factions tugging at the *omikoshi* all pull in the same direction.

Whether as a one-party leadership or in a future coalition, how do Japanese premiers govern? Given the pulls of his own constantly warring factions, a semirevolutionary opposition, a chronically critical press, and a public that views the government with passive suspicion, is there any hope for a Japanese prime minister to get out in front and, as John F. Kennedy enjoyed saying to campaign audiences, "get things moving"?

Kennedy, as history shows, had enough trouble moving the Congress of the United States, let alone the country. Yet he had a huge

6. The Sokagakkai ("Society for the Creation of Values") was founded in 1937 by Makiguchi Tsunesaburo, a Tokyo school principal, as an outgrowth of the Nichiren sect of Japanese Buddhism. Responding to the need for identification and spiritual security among Japan's postwar city masses, it grew prodigiously in the sixties.

executive staff, an efficient national party network, labor support, direct control through his appointees of all government bureaus, a national television public forum, and a press that was disposed to be friendly. (It took Richard Nixon's peculiar talent for inspiring antagonism to set the American press almost unanimously against the White House.)

By contrast the Japanese prime minister has a very small executive staff and a weak national party organization. If he dares to go to the people on an issue, he runs the risk of antagonizing any number of the factions and special interests backing him who dislike public contradiction or repudiation. And, unless his talents are very special, too much visibility will lead to the damaging charges of "high posture" and "arrogance" which are visited by the Japanese on people who raise their heads too far above crowd level.

Generally, the Japanese public prefers a conciliator who can manage to keep the conflicting interests in the country in some balance, without rocking the boat. Eisenhower would have made a superb Japanese premier, as would Lyndon Johnson, in his Senate majority leader days. Obviously aggressive men like Tanaka can succeed, like Yoshida before him, but they run a proportionately greater risk of failure than their grayer fellows.

There is, however, one great asset in Japan which no American president enjoys: the Japanese bureaucracy. "The *kanryo* ('bureaucracy') and the Sohyo (typifying the unions) run Japan," is an old phrase which the country's postwar history has often proved true. Like the Sohyo, the *kanryo* is conscious of its power; but it exercises power with quieter confidence. With a sense of national dedication, at its best resembling that of the old British Civil Service, a feeling for progress gradually and sparingly announced, a disinclination to make outsiders privy to its secrets, a certain arrogance, and an undoubtedly high standard of efficiency, the bureaucracy drives Japan on its course. Its members are more respected than politicians, by far; and they know it. They plan the long-range programs for the cabinet, like the middle-level executives in a Japanese company who draw up the projects, while the president and directors merely initial them. Their pre-eminence has given to Japan's economic policy, at least, a sense of continuity that other countries, including the United States, sadly lack.

It is not easy for a minister to establish control over the bureaucrats in his ministry, other than by drastic means. ("How do you get control of a refractory ministry?" a political friend once reflected, "Ummmm. Well, you start by firing the administrative vice-minister. Then they take notice.") Yet under the Liberal Democrats, the path between the ministries and the Diet has become well-trodden by retired bureaucrats.

As a veteran party man, ex-Prime Minister Tanaka himself once estimated the problem of government in Japan as largely a matter of steering the bureaucracy. "As a party man," he said, "you have about 20 per cent leverage or maneuvering room. The other 80 per cent you must accomplish by moving the civil service. You have to alert the country to these problems, but only after you have lined up the civil service. Then you do your best to see that you don't get too much of a bad reaction."

"This is a small country. When everybody gets together and decides something, it moves. We can't afford the uproar you Americans have—the shootings, the fights, all the wear and tear of your society. We are too small—small, but very tight and homogeneous. It is surprising how quickly things can be decided once the government—that is, the bureaucracy, the politicians, and all, can get lined up in favor of something."

"There is a great deal that needs to be done in educating the average Japanese citizen to know what is in store for him. In one sense, you are supplying remedies all the time—like giving a sick person a pill. This will make him better, but you must at the same time concentrate on the area of preventive medicine, to see that these problems don't occur again."

Granted that a Japanese premier can move the bureaucracy and his own party, he is left with the problem of the opposition. In politics, too, the Japanese spirit yearns for the final unanimous decision—*manjo itchi* ("the whole house as one," literally). As with the *ringisho* in business, the impulse for consensus is strong. After all the pushing and tugging behind the scenes comes the gratifying group decision with everyone's seal on it.

This attitude conflicts directly with the principle of parliamentary government as Americans know it. A traditional glory of the Western parliamentary system, in a sense, is the opposition member with his hands folded across his chest, who defends his principle and position through the last vote, and lives to fight—and maybe win—another day. The minority will fight long and hard, but after the final vote is taken, switch its attention to another subject. It abides by the vote. That is the rule of the game. This is not true in Japan. There Diet stand-offs and protests against majority decisions are the rule, not the exception.

Anyone who has worked with the Congress of the United States knows, of course, that behind-the-scenes compromises are the stuff of government in Washington, as well as in Tokyo. Yet both party leaderships and individual senators and congressmen in the United States set great store on keeping their positions as sharp and well-defined as possible. There are limits, public and private, to the amount of compromising they can or will do. While individual filibusters are cheerfully allowed to impede the business of the Senate, no one in either house would think of boycotting proceedings as a party. To the American mind that would simply mean giving the whole game to the other team. If the Republicans stayed out and the Democrats could field a quorum, they would ram through any amount of legislation, with at least half the country approving.

By contrast, Americans can well envy the good points of Japan's consensus government. The American system is classically an adversary one, dependent on the continual abrasive action of parties and politicians, as well as its built-in checks and balances, to keep it in running order. The Japanese dislike such forensics. Once all parties to an issue are polled, talked to, and considered, however, once general support is gathered for a plan or proposal, the government of Japan moves with a swiftness wondrous to behold, just as Premier Tanaka said. Everyone has a sense of his interest satisfied, no one has been trampled on and the country moves forward. This may well be a better way of decision making in the postindustrial society than the tearing and rending that goes with the American system.

But what happens if a consensus cannot be reached? Or if the agreement finally arrived at is a watered-down version of a measure that should have been sudden and drastic, such as price and wage stabilization to control inflation? Is it better to have bitter partisan public debate, if a strong policy can be voted on and then followed? Is not the tyranny of an obstructive minority, in a system which sets such store on unanimity, as deadly as the tyranny of the overriding majority?

These questions have yet to be answered by Japan's politicians. In prewar days, it is true, the consensus sytem worked because it was essentially a private process, persuading leaders of various groups and interests to reconcile their views under the umbrella of the emperor's authority. Disaster followed in World War II. In post-war Japan the really hard political decisions have been avoided, or at best slipped through. The hard-line Ikeda policy on controlling inflation and rationalizing industry in 1949 was taken under the protection of the United States occupation, and at United States instigation. The equally hard decision to curtail the rice subsidy was postponed by the Korean War and never taken. Yoshida took one hard decision himself, when he negotiated the 1951 peace treaty at San Francisco, without the participation of either the Soviet Union or Communist China. Kishi tried to steamroller the extension of the United States-Japan Security Treaty in 1960 and arrange the Eisenhower visit at the same time. He failed disastrously.

No government, however, has had the heart (or quite the two-thirds majority) even to attempt revision of the constitution, which has resulted in the anomaly of armed forces (and the possible need for more) in a country formally pledged not to have any. Largely because of this, Japan has not had a foreign policy to speak of, since 1945—alternating only between dependence on the United States and subservient gestures to appease China, the Soviet Union, the Arab Nations or whoever seems likely to rock the boat.

No Japanese government has yet had the guts to tackle—or better yet, to anticipate—the terrible problems of rising prices, swollen land values, and profiteering by big business as well as the pressure tactics of big labor, which grew almost unchecked through the late sixties and early seventies. It does no good to say that the opposition is equally to blame with past governments. The fact remains that the postwar consensus system seems to work only when the economy is moving along at fast forward speed, with something for everybody temporarily. To have successfully avoided challenges over a long period of years is not the same thing as overcoming them.

The Izu Peninsula was one of the loveliest spots in the world. Jutting into the ocean south of Mount Fuji, between Sagami and Suruga Bays, a volcanic mountain spine running down its center, it is full of green hills, waterfalls, hot springs, and dusky glades. On east and west there are magnificent coastlines of bays, cliffs pitched high above the sea, vistas of pine and quiet sand beaches. The view from its old hills is unique in its serenities—the sea far below, the white cone of Fuji immanent in the background, the faint sound of rushing water, and the bells of distant temples. And its tip, the bay-girt city of Shimoda, is just about one hundred miles from Tokyo.

Riding along the cornichelike drive south between Atami and Ito or on the quieter west coast overlooking the fishing hamlets in Suruga Bay, one is reminded of the Côte d'Azur between Cannes and Menton, if not the trip up the California coast to San Francisco past Big Sur. Izu has those visual comparisons. But the feel and the mood are its own.

Izu is rich in history. In the hot-springs resort of Ito, the twelfth-century warrior Minamoto Yoritomo spent part of his exile from the vengeance of the rival Heike clan before he sprang back to become the all-powerful Shogun of Japan. Will Adams, the English shipwright, came to Ito in the seventeenth century and taught the Japanese how the Elizabethans built ships. Decades before, the scattered followers of the great warrior-baron Takeda Shingen fled to Izu, as others before them, to seek in its hill-locked valleys some security from the vengeance of a later Shogun, Oda Nobunaga. Shimoda itself still lives the memory of the entrance of Commodore Perry's ships and of Townsend Harris, the first American consul, who, depending on one's interpretations of social history, did or did not have a full life with some of the local girls while fulfilling President Franklin Pierce's mandate to open Japan to trade with the outside world.

As late as 1964 the Izu Peninsula was still primarily a farming and fishing community. To go from Atami to Shimoda by the coast road was a rutted, pitted adventure, but good for the senses. Good land could be had on the hills overlooking Shimoda for ¥5,000 to ¥8,000 a *tsubo* and there were not too many takers. When the Shimoda Tokyu Hotel was opened in 1965, it was thought quite daring that someone would start a hotel there. Most of the Japanese tourists stuck to Atami up north, which was even then metamorphosed into a mountainside version of Coney Island and Miami Beach, or else they went down the middle of Izu, following the paths of Kawabata Yasunari's famous dancing girl,[1] where the hot springs have spawned a series of classic Japanese inns stretching their many tiers up the sides of the mountains, along groves of high pines.

The drive through the Izu in 1974, barely ten years after, was a saddening, if still visually arresting, experience. The gentle old mountains remain. So do the spectacular views of Fuji, but wherever one goes there are also the telltale marks of the bulldozer snaking ribbons of new concrete roads out of the blasted mountainsides. Driving through with some friends one day, I asked casually about the real estate possibilities. The conversation ran like this:

"That mountain there is absolutely beautiful, isn't it? Is there anything . . ."

"It all belongs to Mitsubishi."

"What about the land above the tunnel, overlooking the port?"

"Mitsui Real Estate has that, and the shorefront next to it was taken by Tokyu several years ago."

"Is this the road to the mountain?"

"Yes you can get there by following the signs to Orange Town. They've bought most of the slope, but Sumitomo has the good land on the other side."

"Is there any land left for people to buy?"

"How about this hill above the road. There's two thousand *tsubo* of that left. Of course, there's no water."

The big trading companies and developers have literally bought

---

1. *The Izu Dancer*, his great novella of the 1920s is about the encounter of a sensitive student who romanticizes a little girl in a dancing troupe. It has about the same currency in Japan as epics like *The Diamond as Big as the Ritz*, or *The Snows of Kilimanjaro* have in the United States.

up most of the good land around the peninsula at prices which were once ridiculously low and which, one can rest assured, will be ridiculously high when they are set for the leisure-bent, would-be homeowners who stream out to Izu to find a few hundred square feet of buildable land. Land that could have been bought in Izu for ¥8,000 a *tsubo* in 1964 goes for ¥50,000 to ¥80,000 or more in 1974. Around Shimoda the price is up to ¥100,000. New towns, gardens, leisure lands are announcing openings almost every day. Tokyo newspaper supplements are full of glossy advertisements showing how happy families can escape the heat and smog of the cities by packing themselves into happy new developments, along with the thousands of other happy families who are willing to pay the prices the trading companies ask.

Down the length of the coasts and high on the plateau and the hills in the center the developers have left their mark. Signs announce several varieties of sky towns, garden mansions, and Fuji views, as the road builders hew out their highways over the hills to the neat squares of concrete foundations. Where there is no construction, the mark is still there. "It is reserved," the signs and the people at the real estate office tell you, "But buy now. Or reserve space. It's going to be more next year."

The blight of Izu is merely one among thousands of examples of how the trading companies have been acquiring and hoarding land and commodities to bid up the prices. It is far from the most serious example; but its beauty makes the spoliation all the worse.

Between them, the six great trading companies—Mitsubishi, Mitsui, Marubeni, C. Itoh, Sumitomo, Nissho-Iwai—are estimated to own about a quarter of the natural resources of the entire country, in land, raw materials, finished goods, and stored commodities. Buying up vacation land is only a tiny part of their activities. For the last decade they have been engaged in a relentless effort to buy land, commodities, natural resources, to fulfill their function as the indispensible middlemen between the consumers and the producers in the Japanese economy. The trading companies market Japanese and foreign products in Japan. They sell Japanese products overseas and buy raw materials that go into Japanese factories. Originally created as the sales outlets for Japanese manufacturers (who like to keep production and marketing companies separate) they have come to stalk the world, as they do their own country, buying at the market's best price and selling when the market meets their price.

Between 1968 and 1972 their combined sales doubled, from ¥10,-000 billion to ¥21,000 billion—the equivalent of more than $70 billion. By 1972 their sales accounted for almost 20 per cent of all Japan's wholesale business. Their activities have been a prominent factor in accelerating Japan's inflation. They have been investigated by the Fair Trade Practices Commission, their presidents summoned before the Diet to answer charges of hoarding and speculative buying. Their officials have been indicted for price fixing. Because of their often preclusive buying, many more commodities than land have been bid up, almost out of sight.

The land grab in Japan has been executed with all the rapacity of the old land grabs in American history. The trading companies, at that, are staid and respectable, compared to the fringe kind of real estate speculator, whose recent activities recall the worst excesses of development gougers and real estate racketeers in the United States. There is one difference, however, between the Japanese experience and the American. In Japan there is not enough land to go around.

Everyone in Japan wants his own house or, if not a house, a small apartment that he owns. The urge to own a bit of land and one's roof, basic to everyone, is almost unimaginably strong in this nation of smallholders and village people. Yet in no country does the would-be homeowner have a rougher path to follow.

An extensive report on housing by the Mitsui Bank in 1974 illustrated the problem well. Between 1955 and 1972 prices of land in or near Japan's big cities rose about 2,557 per cent! This was six times the rise in city workers' disposable income. Put another way, between 1963 and 1972, it became more than twice as difficult for a resident of Tokyo to buy housing. (Rental apartments and the new cooperative mansions are almost as hard to secure at a reasonable price.) Behind these inflated prices lies the staggering increase in construction costs. A building job which cost ¥200,000 per *tsubo* in 1971 could require ¥550,000 yen per *tsubo* in 1974.

The problem was somewhat less in Osaka and much less in cities like Sapporo in Hokkaido, Japan's wide-open-space island, where the increase in city land was still manageable. But the average price for housing units in all big cities now exceeds $33,000 (¥10,000,-000). This is five times the annual income of even the well-paid Tokyo worker. The average price of the houses offered in the Tokyo area by major real estate companies is about $66,000 (¥20,-000,000). This gets the home buyer three rooms and a dinette-kitchen, an hour and a half's commute from Tokyo. Since 1972 the Japanese government has put large sums of money into public housing. A program to construct 9.5 million new housing units by 1975 is well underway. But the congestion in getting the program off the ground has been appalling. Even when housing can be had, financing is difficult. (Before 1965 conventional bank loans for housing to individuals were almost unheard of.) Over the whole growth surge, the private sector, as expenditures for people are euphemistically called, was nowhere more scandalously slighted than in housing.

The houses and apartments that people do live in, all over Japan, are likely to be so small and rickety as to be slums even before they are inhabited. The average wooden housing unit in Tokyo has only 1.4 rooms with about 125 square feet of area. Only 10 per cent of these have their own private bathrooms. The number of people who share bathrooms, kitchens, and washing space is large.

Business land prices also have gone up in scale—often in frightening scale. The classic example of inflated pricing was the sale of the NHK (Japan Broadcasting Company) building at Uchisaiwai-cho in Tokyo to the Mitsubishi Real Estate Company for ¥35,-163,127,000—or the staggering amount of ¥11,112,000 per 3.3 square meters.

At the worst, potential homeowners can fall into the hands of outright crooks. The Fair Trade Practices Commission, already overworked in other areas, makes periodic checks of developments with impressive lampposts that are not wired and neat-looking sewers that go nowhere. Some of them have been exposed quite effectively on television. But still the prices went up. Big companies and small, the reputable real estate broker and the fly-by-night speculator alike, escalated them almost to the point of absurdity.

In July 1973 Miyazawa Kiichi, formerly minister of International Trade and Industry and then head of the Liberal Democratic party's Policy Research Committee, wrote a front page editorial to his fellows in the government party in the party paper. Miyazawa is one of the most responsible political men in Japan, and his comments deserve quotation:

Recently the economic activities of the large trading companies have come into question. A great many people have felt that these companies have amassed indecent amounts of wealth in the course of their business activities, in such a magnitude that people can hardly imagine it, let alone stop it. It is widely believed that the conduct of a strong minority has given rise to injustices which the majority find it impossible to prevent. Apart from the question of rights and wrongs here—the fact is that in any democracy there are a small number of the strong and far larger numbers of the weak. If great numbers of the weak feel that the "system" exists to benefit a minority of the strong and the influential, then the ultimate result is that democracy will be destroyed. . . .

As the party in charge of the government, we should and must take whatever steps may be necessary to restore a sense of justice among the people of this country. We should begin with a redistribution of our people's wealth and earnings. When we say "redistribution" there is one thing to consider first. Many things which have traditionally been thought of as free for anyone's unlimited use actually have their limits. Commodities that have been thought of as private resources or wealth should be thought of as public or common property. Air and water are among these commodities. And when we think of our great cities we must count land in as well.

This is not a question of ideologies. We believe in a free economy. It is precisely to save that free economy that we must exercise some control over the use of unlimited private rights. We must look facts in the face. To achieve social justice, it is absolutely necessary that we look at this idea of "common public resources" from a new angle. To redistribute incomes we have not only to work through social insurance and the tax structure. To widen everybody's share we have to make the pie bigger—there is no need to apologize for this kind of growth thinking. But to preserve the free economic market in the process we must realize very keenly that our spheres and channels of distribution are defective. They constitute a serious liability. I cannot emphasize enough that they must be changed.

I believe that the wisdom of the people of this country resulted in our postwar prosperity. If those of us in political life can eliminate the feeling that social distrust and injustice is now abroad in our society, then the Japanese people can go on to rebuild a new sense of values. But we have very little time left to us.

Miyazawa's advice was not heeded by most of his colleagues in the ruling Liberal Democrats. The party's behavior in the land inflation crisis remained consistently unreassuring. Legislation to stop profiteering, when it finally was introduced in 1974, came too little and too late. But the warning should stand on the record as a danger signal to Japanese free enterprise, and one which cannot long be ignored.

Over the years the American free enterprise society has developed its own variety of checks and balances to avoid the exploitation of the weak majority by a few of the strong and influential, specifically, by the large corporation whose profit-seeking often works against the well-being of the society. Apart from the press, there are the courts, the Justice Department, and the Federal regulatory commissions who can and should act as the public's surrogate against the big economic interests. Senate or House investigating committees have repeatedly blown the whistle themselves when big businesss was thought to be stealing too much yardage. How effective such "whistle blowing" has been is still in question. But it is at least a part of the American adversary system of justice, founded on the old American principle that the cops must always protect the sucker from the sharper, at whatever level.

In Japan many of these formal and informal safeguards and hedges on the governance of the strong over the weak are either faulty or totally wanting. The press does a good job in calling attention to the errors of big business—at least when they get so excessive as to be obvious to everyone. The Diet has not exactly distinguished itself by investigating committees, which are not its bag. In land, as in the whole energy crisis, both the government and the opposition parties' role has been weak and disappointing, nor are there many strong regional interests in this tightly welded country.

The courts are not notably efficient. Although government agencies like the Fair Trade Practices Commission and the new Environment Agency have been on the alert, their power is limited, especially if pitted against the Ministry of International Trade and Industry (MITI) which is generally on the side of business. Popular indignation has been long in coming.

The reason for the lack of these adversary-type safeguards is of course the feeling of trust that is the hallmark of Japanese society. Implicitly, in return for his hard work and fidelity to the powers-that-be, the weak man has the Japanese feeling of *amaete iru* toward authority. They will take care of his interests, because he and they are part of the same group. Just as the weak man's voice is included in the consensus, his interests will be protected in the transaction—nobody is after too much. It is just this feeling of group trust that has kept Japan's modern society away from Marxism, except as a theory. The worst abuses have been corrected even by authoritarian prewar governments. And how can you join the proletariat when there is not any in your country—everybody is just Japanese.

have read the special section in the February issue of the popular intellectual monthly *Bungei Shunju*, called "The *Economic* Sinking of Japan." The finance minister, the vice-minister of the Economic Planning Board, and the most prominent union leader in the country contributed articles and the title fairly expressed a general fear. The sudden squeeze in oil supplies and prices had not happened by itself. It came to Japan on top of literally galloping inflation, a critical pollution and environment problem, a shrinking labor force, and a shortage of essential raw materials almost everywhere. The energy bombshell and the threat behind it were so total as to leave people almost numb, resigned to the point of seeming nonchalance.

From a national point of view, the news of the unexpected Arab oil war against Japan in 1973 was as damaging a blow as had been the Allied oil embargo that directly precipitated Pearl Harbor—except that this time there was no Japanese fleet to counter ultimatum with force. Nor were there trusted allies to depend on. The negative Kissinger diplomacy toward Japan was never so consistently displayed as in the official United States indifference to Japan's predicament, which was, after all, partially a by-product of Japan's support for America's pro-Israel policy. There was nobody. The Japanese people had the helpless feeling of being trapped in a speeding subway car by a burly madman with a knife. So the Foreign Office hastened to announce Japan's newly discovered sympathy for the Arab cause. Three cabinet ministers rushed off to the Middle East, their portfolios bulging with plant construction and technological assistance programs. It was easy to criticize this latest case of Japan's kow-tow diplomacy. It was easy to say that Japan's vulnerability to Arab oil and almost every other energy possessor or producer was the result of poor foresight. But more accurately stated, it was good foresight turned bad, one of the many tactical economic advantages that were becoming Japan's long-range strategic liabilities.

It was, some considered, the end of economic man written all over again, this time in Chinese characters.[1] Which is to say that an economic superpower, relying for both its prestige and its livelihood on an apolitical free exchange of goods was fated, given the

---

[1.] With apologies to Peter Drucker, whose original book *The End of Economic Man* was written in 1938, with the theme that the Nazi emergence in Europe had changed the nineteenth-century faith in civilized peoples living according to economic laws.

breakdown of the other armed superpowers' authority, to be at the mercy of every bully lurking at an international street corner.

This view may have been too pessimistic. But judged by any standard it was time for Japan to pay the piper. Slightly more than 99 per cent of Japan's oil supplies come from overseas; and by 1973 oil was supplying more than 80 per cent of Japan's energy needs. It was not always so. In the 1950s coal still furnished 60 per cent of Japan's energy. But it was bulky to use and Japan's own mines were becoming increasingly costly to work. By 1960, encouraged by the buyers' market in petroleum, Japan was consuming 84 million kiloliters of crude oil annually and looking for more. By 1972 oil consumption had risen to more than 300 million kiloliters annually, close to 10 per cent of the world's total; and it was heading up. Just before the energy crisis hit, Prime Minister Tanaka had buoyantly told the country that its oil consumption was scheduled to "triple in five or six years."

There were no immediate sources of energy that could fill the gap. Although television stations ran features on American houses heated by solar energy and the papers staged symposia about the use of chemicals from coal and shale oil, these were all far futures. The Atomic Energy Institute estimated that 60 million kilowatts of nuclear power could be generated by 1985, but commentators conceded that this would still leave Japan more than 70 per cent dependent on oil.

Nor were the shortages of raw materials restricted to oil. Logs, copper, and iron ore were all going up. The oil cutbacks only increased the pressure on the existing supplies. A bad farm year in many countries was cutting back agricultural exports to Japan or raising their prices. Assured sources of supply were turning fickle.

For not only does Japan live by imports. But its industry and trade pattern is so finely balanced that the slightest jar will cause shakings and tremors throughout the economy. Even without the Arab oil crisis, a dry year had resulted in 1973 water shortages throughout the country, just as industry was shifting into high gear to satisfy new demand and make up lost ground after the 1971 recession. As the water supply lessened, aggravated by Japan's own huge consumption, the steel mills had to curtail production, with a ripple effect on most sectors of the economy. Chemicals were as precariously situated. In 1973 there was a big explosion in Idemitsu's vinyl chloride plant at Tokuyama. This was enough to set back housing constructions throughout the country, for want of vinyl coverings. The entire production of ethylene was running behind schedule. Because plant capacity was pushed, as a result, more accidents happened. Of course, the oil shortages made normal deficiencies all the worse.

While the planners who set the course of the economy were devising emergency countermeasures to cushion the energy shock, pulling the familiar levers of credit cutoffs, production cutbacks, and pricing guidance to industry, the consumers were outside, so to speak, angrily banging on the doors of the laboratory. Before the energy crisis occurred, Japan was already caught in an inflation that had outrun the normal inflationary rises in the price structure. Consumer prices for 1973 went up more than 15 per cent over the previous year, in contrast to 1972's rise of 4.5 per cent. By late 1973 prices in basics like vegetables and fish, the staples of the Japanese diet were skyrocketing—often increasing each month. Land was the most horrible example. Between 1964 and 1974 urban land prices more than quadrupled. Temporary shortages in items like toilet paper and soap flakes brought waves of petty hoarding, which drove their prices up further. Restaurant prices went almost out of sight. In the spring of 1974, as members debated inflation safeguards, the seven low-cost restaurants in the national Diet raised prices 32.8 per cent.

Through twenty years the Japanese wholesale price index, as has been noted, stayed remarkably stable. This was the result of a capacity that could almost always take care of domestic demand. In 1972 this began to change. With money plentiful and the 1971 "Nixon shock" a thing of the past, the buyers finally began to out run the sellers. By March 1974, the wholesale price index was 147, against a 1970 base of 100. In the same month, statisticians in the prime minister's office reluctantly reported that savings had depreciated 25 per cent in value over the past fourteen months. It was the worst inflation since 1951.

Still the demand kept increasing, totally without discrimination. Staples, luxuries, construction, and dissipation all together, the Japanese consumer economy continued to gulp down goods and services like a thirsty animal. New products kept appearing to jostle old ones out of the market place. Department store sales jumped —a 19.2 per cent increase over the sales of a year before in the last quarter of 1972, a 27.8 per cent increase over the past year in the third quarter of 1973—and more girls were hired to polish the escalators. Through it all, socially active businessmen continued to order lavishly on the expense account circuit. The 1971 entertainment tab was $4.2 billion (¥1,260,000 million) and the totals for 1972 and 1973 rose accordingly. In 1973, while the papers ran economic crisis reports almost daily, an information-saturated public heavily increased its spending at the races.

More and more people, however, were getting hurt; and it was not the people who visited the Ginza night clubs. The *kakebo* ("household account ledgers") of modest families were running in the red. Far from adding to their savings, people were having difficulty paying for essentials. And on top of inflation, there were the cumulative effects of industrial pollution. If the Arab oil cut-off was something beyond the power of the average man's retaliation, the inflation and the pollution were not. There were demonstrations, symposia, and television panel shows. And through all the discussion the mood of the public grew darker and more frustrated.

Like the sudden widening of the geological faults and the seismic movements that caused the fictional sinking of Japan, the economic crisis of 1973–74 was largely a matter of timing and combination. If a certain number of things had not happened, and happened at the same time, the matter would not have seemed so serious, the problem not so pressing. The energy crisis, bad as it was, was more of a catalyst than a prime mover.

Since 1972 strange things had been happening to the storied advantages of Japanese economy: the high-growth government economic policies, the dynamic corporate expansion, the disciplined work force that had made Japan an economic superpower. Economists and the shrewder businessmen and politicians were aware of them. Like the scientists in *The Sinking of Japan*, who first spotted the strange movements on the ocean floor, they had reported their findings. But not many people listened. Even among those who knew, there was little interest in spreading the word. It is one of the marks of the Japanese establishment that its members like to work silently, until they can get the ducks lined up in their own exclusive consensus. Then it is time enough to spread the word to the country and start the national consensus grinding. Because of the suddenness of the energy crisis of 1973–74, the facts were all too quickly laid out on the table. There was no safety margin for cool considerations.

Through the 1960s the Japanese growth rate had become the eighth wonder of the economic world, as it shifted from the 1961–

65 level of 10 per cent to the 12 per cent plus rate of 1965–70. Although experts in government and business alike thought it should, and must, taper off, they were not taken too seriously. Steady growth allowed corporate managements to keep up their now familiar cycle: new plant improvement makes for better productivity which makes for profits which can be put back into new plant improvement. Reserves were kept to a minimum. The companies borrowed constantly from the obliging banking system. (The percentage of owned capital before the war was more than 60 per cent, by 1972 it had declined to less than 20). Constant growth helped pay off the interest on the loans.

The unions contributed to the buoyant mood. Profits should be passed on to the workers, they argued; the annual base-up raises in Japanese companies became higher and higher. They hit 19 per cent in the spring labor offensive of 1973—and some bumped 30 per cent in 1974. The companies paid this to keep the peace, reasoning that relatively unopposed wage increases were better than production losses through labor strife. The base-up could always be passed on to the consumer, and generally was.

By 1972 plant expansion was coming to a halt. The growing murmur about pollution in the country was becoming a shout. Company after company was being forced to cancel plans for new construction because of local residents' opposition. The reason for the sharp drop in nuclear power construction—in 1972 only 28 per cent of what was planned and in the next years little better—was public distrust of this particular source of power. The "nuclear superstitition" has yet to be exorcised from Japan.

In addition, the great surge in technology of the late fifties and sixties had been largely digested. There were few easy patents to be had from foreign companies, which had grown stricter about loaning them out. The research and development effort in Japan, although increasing on a geometric scale, was still not enough to take up the slack.

The markets were still there, nevertheless—Americans, Germans, Thais, and Brazilians waiting for their Mazdas, Panasonics, Nikons, and much more. But the cost of reaching them had increased. Prices were going up on a world level. The frantic preemptive buying of Japanese trading companies was beginning to back up on itself. Even before the actual energy crisis, it was obvious to more people than the Japanese that the heady growth rates were pushing against the ceiling. Given a steady 10 per cent increase in the economic growth rate, an MITI study calculated that by 1980 "Japan's resource consumption would comprise 30 per cent of the world's total imports!" A heady percentage to contemplate.

As the American troubles of 1971–72 had suggested, the raw material exporter was getting unhappy. Unlike the United States, which did it to a fault, and Britain, when the British had their superpower day, further, the Japanese were notably slack either about investing capital or putting up plants in foreign countries. Thus they appeared as simple exploiters, without any mitigating factors like local employment to soften the normal reaction against them.

What had happened to the disciplined labor force in the process? Two things. The first was simply a reanactment of the modern Everyman drama already playing to capacity audiences in the United States, that affluence begets its own problems. The more Japan's workers became a consumer class, the more they were concerned with how to increase and spend the remuneration for their labor and the less with the products of the labor itself. The country that had made its mark in the international markets with its apparently inexhaustible source of skilled manpower was running out.

Slight rises in the birth rate would not be apparent until late in the seventies. For the moment Japan was a country where skilled, young labor was at a premium. In 1970 the total labor force in the fifteen-to-twenty-four-year-old bracket was 11,190,000. By 1980 it is projected that the total 15-to-24-year-old force will be something under 7.5 million. As the majority of the workers edged up into the middle age brackets, the average wage increased automatically, thanks to the old seniority systems. The uninterrupted gains of the unions, in addition, had produced a feeling that the yearly base-up and the bonuses of two to six months' salary twice a year were built-in perquisites of the job. For this Japanese businessmen had no one to blame but themselves.

The productivity curve continued to slacken. The rise in raw material prices, the rising labor costs, the dwindling room for plant expansion and technological improvement all acted upon each other adversely in what the Japanese economists called the *akujunkan*, the vicious circle. It was by no means a hopeless situation. Japanese business and industry had shown themselves capable of getting out of tight spots before and turning defects into advantages. But fast action was needed.

Where now was the finely calibrated combination of government and business, working to adjust the economic indicators, which is at once the goad and the built-in safeguard of Japan's economic progress? It was there. The government economists were not asleep at the switch. As early as the late 1960s, planners in the economic ministries had been worrying about the high growth rate of Japan, as contrasted to the lower per capita income of the Japanese. They were painfully aware of the scanty social welfare and general construction expenditures in the public sector: the housing, the sewage, the transport which Japan's population needed as much as its industry, in the continual internal elevation of rising expectations within its own islands. The 1969 projection of the Economic Planning Agency was a most ambitious blueprint designed to fill these needs. It envisaged a shift of spending toward the public sector and continual infusions of money and talent toward internal betterment, with the accent on blind growth diminished. In effect, the 1969 Economic Plan was the successor to the double-your-income policy under which Japan had been running for so long.

A refined version of this and similar plans was embodied in the book *Building a New Japan* (literally, in Japanese, the Reconstruction of the Japanese Archipelago), published by the incoming Prime Minister Tanaka Kakuei in June 1972, a month before he took office. The plan was the result of work done by the Ministry of International Trade and Industry while Tanaka was its head. Although later subjected to heavy political attack, his book is in fact one of the most far-sighted pieces of national planning ever to be issued by the head of a government. Taking into account all the building problems caused by Japan's sudden economic growth—pollution, overconcentration (almost one-third of the population in 1 per cent of the land area), lack of investment in the social infrastructure—the MITI plan proposed a massive dispersal of the country's industry, in order to "reverse the tide of people, money, and goods to create a flow from urban concentrations back to outlying areas, using industrial relocation and the national communication and transportation networks as the main tools."

The planners realized that the huge inflow of people and resources that built the heavy industry and chemical base of Japan, Inc., could not be indefinitely projected, so they advocated switching to the knowledge-intensive kind of industry, for instance computers. They were by no means prepared to abandon the growth mentality, but they accepted the fact that the growth rate might have to come down. The quality of life was high on the priority last. Along with switching Japan's industrial weight from the hopelessly crowded Tokyo-Osaka megalopolitan belt, the 'Tanaka Plan' advocated new systems of allocating water and oil resources, green belts, antipollution measures, and the growth of small cities that could counteract the appalling centripetal rush of Japanese into Tokyo.

The book found instant acceptance. It sold one million copies within three months. But its effect boomeranged. Although the bureaucrats at MITI and the Economic Planning Agency had been thinking about this plan for many years, the plan came too late to arrest some of the worst overcrowding, pollution, and reckless plant expansion. At the same time, it was announced too early. While the concerned ministries were quietly setting up new laws to remodel the archipelago, in a set timetable, the speculators, trading companies, and land developers did not wait. Even before Tanaka took office, the book was being used as the vehicle for a surge of cut-throat land grabbing which no one, not even Tanaka, had anticipated. It was a classic case of what the late Lyndon B. Johnson used to call "frontlash", i.e., the announcement of a plan produced a reaction which ended by altering the plan itself.

Political and intellectual critics, rightly indignant at the land grabbing, tended to dismiss the whole plan. And when the energy crisis broke, Tanaka's political opponents seized the chance to point out how unsound, wasteful, and visionary it all was. The subsequent exposure of Tanaka's corruption served to underline their point. There was nothing wrong with Tanaka's vision, however. There was a lot wrong with the rather casual way he had proposed the most drastic kind of national transformation.

With the Tanaka plan, as with the whole handling of the 1973 crisis, the government made two serious underestimations. First, they underrated the volatility of their own free-enterprise economy. The demand-shift inflation had already been brought into being by the pell-mell rush of capital not only toward building in the public sectors directly, but also by way of anticipating the public sector by private building and expansion programs. By the middle of 1972 prices in concrete, land, timber, and other construction materials were already going out of sight. When Tanaka became prime minister, his book was taken as gospel. The speculators, good and bad, crooked and honest, turned toward the areas favored for development and industrial relocation in The Reconstruction of the Japanese Archipelago, bought up land, and started building. Before the government controllers could decide how, when, or if, to apply controls, wholesale prices were going through the roof.

There were other problems, which the land grab made worse. For most of the sixties the average Japanese had taken the growing pollution of his air, land, and water as an inevitable by-product of progress, the price you paid for a soaring growth rate. Toward the end of the decade, however, things began to get out of hand. As the carbon dioxide in the air built up and various chemical concentrations from factories caused local pollution problems, the people affected began to protest. The news media carried their stories throughout the country.

The widely publicized Minamata disease was the result of methyl mercury contaminating fish in Minamata Bay, in Kumamoto prefecture in Kyushu. The methyl mercury was drained into the sea from the plants of the Chisso Corporation. Over the years hundreds of people fell ill from eating the local fish. Many died. The resultant outcry against the company received wide publicity and was finally settled in 1973 after almost a decade of prolonged court action. Similar problems occurred in Niigata prefecture, with the mercury-filled drainage from the Kanose factory of Showa Denko Company, and in Toyama prefecture, where a painful bone disease was caused by cadmium from the Mitsui Mining and Smelting Company. These, too, were settled by court actions early in the seventies.

In 1969 the Ajinomoto Company, maker of Japan's most popular seasonings, was rocked to its foundations by an American discovery, widely promulgated in Japan, that monosodium glutamate in its seasonings was harmful to people's health. While the Ajinomoto battle was being fought, victims of air pollution in Yokkaichi, a city with a literally poisonous concentration of oil refineries, were suing the refineries for an indemnity, which the courts later awarded. Shortly after Prime Minister Tanaka took office, he went on the air in 1972 and justified government measures taken against the refineries. The new nuclear-powered merchant ship *Mutsu*, fitted out with its reactor in 1971, was thereafter immobilized in Mutsu Bay, in northern Japan, because aroused local fishermen contended that radioactivity would destroy their catch. For years it was impossible to secure their consent for even a trial run! Now berthed at Sasebo, the Mutsu is still a ship without a regular port.

By 1970 most of the beautifully scenic Inland Sea was hopelessly polluted by the so-called red tides of polluted waters from the factories on its shores. Smog warnings became regular and asthma sufferers began trekking to the hospitals. Regional complaints and petitions about pollution, about 20,000 in 1966, had risen to 76,000 in 1971. The amount of money companies spent on antipollution measures doubled from 1972 to a figure well in the billions of dollars in 1974. The automobile was, of course, as great an offender as the factory, in a country where the car density per acre is already more than ten times that of the United States. Certain areas in downtown Tokyo were temporarily judged literally uninhabitable during peak traffic hours.

In 1971 an Environment Agency was set up at cabinet level to try to grapple with the problems of large-scale pollution which each step in Japan's industrial growth was compounding. It was noted at the time that in 1959, for example, Japan had an economic density like Britain's (density, in this case, being the ratio of level land to a country's GNP). By 1970 the density was 3.2 times greater and getting worse. As people continued to pour into the narrow belts around the Tokyo-Nagoya-Osaka megalopolis, the problems of water pollution and simple garbage control intensified until they were almost national emergencies. A nation-wide Sewerage System Consolidation Program was put in force, from 1971 to 1975, with an expenditure of more than $8 billion. It will fall far short of what a modern country needs.

By 1973 and 1974 the local protests against pollution, real or fancied, had grown to a roar. In Osaka a court ordered the government to curb the noise of planes landing and taking off, after a long suit by local residents. Similar suits were pending in five other airport cities; and in Nagoya, residents' committees had submitted petitions for damages due to vibrations caused by the passing bullet trains of the high-speed Shinkansen express. Residents groups and consumers protesting pollution numbered almost a thousand.

In October 1973 the prime minister's secretariat took one of its polls, this time about pollution. Some 61 per cent of those polled had noticed the effects of pollution near them and no less than 45 per cent said they had experienced them in some form themselves. Generally they were nervous about new public facilities such as express highways, airports, or garbage disposal plants in their own areas. Only 13 per cent would support such construction without reservation; 41 per cent said they would only cooperate "gradually." Thirty-six per cent expressed themselves as unwilling to cooperate, in various degrees.

The fact that so many antipollution cases exploded in the courts in the seventies was an index not merely of the gravity of the problem, but also of the fact that neither government nor private industry had taken such complaints seriously for a long, long time before.

The second way in which the Japanese planners underestimated the complexity of their own economy was more serious than the first; in its effect it reinforced the first and finally dwarfed it. For the past two decades, while Japan climbed out of the reconstruction stage of the fifties and built itself into a world economic power, the planners had continued to draw their projections like a man playing catch against a wall. The ball could be expected to come back, depending on the speed and skill with which you threw it. There is much in the Japanese character that encourages this point of view. Armored by the approval of their society, cheered by success in the GNP predictions, and confident of their economy's ability to expand, the men in the ministries and their counterparts in the trading companies drew their expanding lines up and up and up.

There was little sense of limits in their projections. As completely as the businessmen from Tokyo who went on their way selling and buying throughout Southeast Asia, with scarcely a worry about the needs or sensibilities of the local populations in their path, the planners single-mindedly projected their country's added imports without paying too much heed to what the exporters were doing or thinking. In other respects, like the changing pattern of Japan's exports and its own domestic industry, their planning was very finely calibrated; it was doubly disturbing to find that the imports were becoming a problem. The prices of raw materials were soaring. Even traditional suppliers, furthermore, grew increasingly nervous about their own domestic markets. The fact that Japan planned on using 30 per cent of the world's exports by 1980 was increasingly less of a secret.

The rise in raw material prices, although part of the general world inflation, also reflected some scarcities. Through 1973 the prices rose, on the average, at a rate higher than 20 per cent annually. The trouble with the United States over such imports as soybeans, which Japan had always taken for granted, was only one straw in the wind. People in the timber country of Canada and the United States showed signs of cutting their huge exports of logs to Japan. Ores and minerals, like oil, were looking more and more exhaustible.

At the end of 1973 the Nomura Research Institute of Tokyo put out an interesting report on raw materials and goods shortages. Nomura's analysts noted that the increasing prosperity of developing countries—which now had to be figured into the equation—was putting even more of a strain on established sources of raw materials for manufacturing as these countries expanded their own consumer demand. Steel and ethylene Nomura cited as interesting cases where increasing demand would be increasingly harder to satisfy. World-wide production capacity of crude steel, for example, stood in 1973 at 698 million tons, with an average operating rate of about 90 per cent. The capacity was just about at a limit. Yet, Nomura reported, "due to the expansion of demand based on the improvement of standards of living in developing countries and Communist countries, potential world demand for crude steel is estimated to reach 1,130 to 1,210 million tons by 1980, while the estimated production capacity will be at most 930 million tons. Thus, the present tight demand for steel is likely to become more serious. Moreover, such a situation will not be changed at a single stroke because it usually takes around five years to construct a fully equipped steel production plant." Japan, as of 1974, was importing 100 million tons of iron ore annually. Under the circumstances it is hard to see how the Japanese planners could push their imports up to the estimated need of 217 million tons annually by 1980.

In short, Japan by the early seventies was right in the middle of the world of macroeconomics. The Japanese with their spectacular recovery had indeed helped create this new world, a world where no nation's needs could be considered without also considering those of one hundred others and the interrelating factors among them. Just as economics was no longer a problem of pure growth, imports and exports were no longer a problem of a simple competition between the United States, Europe, and Japan, with the rest of the world chipping in when asked. Nor was macroeconomics merely a matter of increasing the budget for foreign aid and making sober pronouncements about the North-South problem and the need for technical scholarships for deserving Southeast Asians—although all of these elements played a part. To continue playing its leading role on this big a stage, Japan needed a rare degree of cooperation between business, government, and the general public, even more closely worked out than the sort of cooperation that had created the economic success story.

For the near term, at least, the energy crisis posed huge demands on Japan's economy and its trade. Prices had to go up to meet the new costs, with the possibility that they would go up again. Yet to be internationally competitive, they could not go up too much. For the long run, policy and goals had to be adjusted. To keep the peace inside Japan there had to be some meeting of minds about resources and reasonable profits, people's ability to pay and the country's ability to earn and work together.

That spirit was not much in evidence. For the first time since Premier Ikeda Hayato's double-your-income policy took off, there was a threat of serious divisions between the domestic partners. Business, the consumer, the unions, the small enterprisers, were looking less and less like members of a team. Big business behaved worst of all. Under the pressure of hard times coming, companies grew more, not less, selfish. In another country such behavior would have been merely expected. But in Japan, where a peculiar kind of trust reposed in the business leadership, it came as something very disturbing.

Despite much talk about social responsibility and the welfare of the community, most businesses reacted to the inflation crisis in the most primitive way. They raised their prices, on the theory that the company had better grab while the grabbing was good. The price increases were far above the added costs. For example, according to an MITI survey covering the last half of 1973 cement prices increased 32 per cent, although the actual rise in costs was 10 per cent; plastics prices rose 65 per cent, against cost increases of 7.6 per cent; synthetic fibers, 52 per cent, against a cost increase of 9 per cent; rolled aluminum, 36 per cent, for a cost increase of 5 per cent.

In 1974 twelve major oil companies were formally denounced by the Fair Trade Practices Commission. Documents were sent to the public prosecutor on scandalous price fixing, in the wake of the energy crisis.[2] The prices were raised five times between May

2. They were not alone. By the spring of the same year investigations were underway in France and Italy over huge cartelized price increases; and there was enough concern about the behavior of the oil companies in the United States that some far-sighted senators talked publicly of possible nationalization.

and November, by agreement of the big oil companies—from ¥14,000 to ¥27,000 per kiloliter for gas; naptha from ¥8,200 to ¥13,500 per kiloliter; heating oil (doubtless out of deference to consumers) from ¥11,500 to ¥13,500. It was, in the words of the FTPC chairman Takashi Toshihide, "an antisocial and modern economic crime."

In 1972 scandals had first broken over huge profiteering among the large trading companies. Disclosures about this continued. The further evidence of oil company excesses added up to a display of corporate greed too prominent for even the business-oriented Liberal Democratic government to ignore. Early in 1974 Premier Tanaka called almost one hundred representatives of industry to a meeting at the prime minister's residence to warn against profiteering and speculation, while the government, behind the scenes, began to pressure the companies to adjust their pricing. After scandals in land sales and speculations, the public had begun to wonder. Now their concern was turning into anger and protest.

The problem was not that big business was simply seeking profits. The price raising was in a sense, defensive. The theory was simple: Inflation is coming, so let's hike the prices and protect ourselves, i.e., this company. Although there was a certain feeling of economic responsibility to a particular industry (the word *cartel* was coming back into currency), a sense of public responsibility was not much in evidence. Once more, a basic defect of the vertical society showed itself. The average Japanese manager does things the safe way. To safeguard your particular company's position is simple prudence; and management and labor joined hands to do so. To help save the national economy—well, that was a job for the government.

The crisis of 1973–74 laid bare every special interest in Japan. The job of the government, of the establishment leaders in business, labor, the parties, the intelligentsia was to keep the various elements together, part of a functioning whole. It was not easy, with company presidents admitting crookedness in public and neighbors saying that the Communists were the only honest party on the block. The trust that kept the island society of Japan together, half-shattered in 1945 and rebuilt with care and will ever since, had taken a buffeting.

Both government and opposition parties showed themselves astonishingly ill-prepared to handle the energy crisis. (Exploiting it, as the Communists did, was another matter.) Initially the principal official reaction to the gross inflation which seized the country was to hope that it would go away or to discount its existence.[3] There was no lack of sound, sophisticated planning in the ministries and among Japan's excellent economic experts and academicians for trimming the growth rate and adjusting the nation's course. But one had a feeling that the planners had been kept too long in the back room. In their political dealings, notably in their communication with Japanese voters, the parties had themselves too long supported a policy of undiminished growth; or given it only token opposition. Their reasons were good ones, if considered from a narrowly partisan point of view.

The Liberal Democrats, having developed the economic growth strategy, were not about to throw it out. Political scandals could come and go, foreign policy could bend and know-tow, and public-sector improvements could be indefinitely postponed; but as long as the growth rates soared, and wages and bonuses with them, the voters would not eliminate Santa Claus. So the party's big business supporters reasoned. They saw expansion as a necessary condition of life. At conferences in the business meeting rooms of the shiny new Keidanren Kaikan, in the golf games at Karuizawa or the Three Hundred Club or over the world's most expensive drinks and snacks with the venerable geisha at the Akasaka houses, they made their concerns known to their classmates in the government.

The Socialist party was partial to the growth policy for a parallel reason. Over the years, instead of broadening its base, its support had narrowed to that provided by the Sohyo federation of left-wing unions. As long as the growth continued, unions could demand and receive ever-fatter base-up raises and bonuses. Expansion meant more employees in government departments and the government businesses, such as the railroads and the tobacco monopoly (where much of Sohyo's strength lies), as well as in private industry. Until 1973, at least, the Socialists' public denuncations of the growth policy had little conviction. It was far easier for them to concentrate on safe issues like American imperialism or the right to strike, which did not interfere with the unions' pay packages.

The Komeito and the smaller Democratic Socialist party had similar problems. Most of their constituents lived in cities or urban

---

3. Euphemism knows no bounds. For a while Premier Tanaka chose to call this phenomenon not inflation, but "a consistently sharp rise in commodity prices," for which there is a tidy expression (*kōjō-teki bukka tōki*) in Japanese.

areas and were themselves caught up in the prosperity, apparently. The Komeito, in particular, drew much of its support from workers in marginal service industries, cabarets and restaurants included, which were riding on the crest of the big corporate growth rates and escalating expense accounts.

Even the Communist party came on the antigrowth issue late in the game. The Communists, anxious to build a new image as a popular party, were not much more eager to be caught taking a shot at Santa Claus than the Liberal Democrats. Although they began to capitalize on the neglect of public housing and other improvements in the sixties, they made few criticisms of the growth policy itself. It was the distribution of growth's rewards that they complained about; and, ultimately, the exposé of big business profiteering gave them some powerful points.

It could hardly be said that citizens rallied to their government at the time of the inflation crisis; no more than it could be said that Americans rallied to theirs during similar problems in the United States. Indeed, the Japanese public's silence in this regard was deafening. After thirty years, the Japanese democracy had failed by the mid-seventies to develop much of a sense of public spirit. Patriotism had revived, that was true. But it was a patriotism directed against foreign pressures, more of a negative reaction than a positive emotion.

The various special interests in Japan had done very well. Big and small business, linked to the Liberal Democrats, had gotten their profits and their growth policy. Sohyo had gotten its huge base-up raises. So had the other unions, who cheerfully cooperated with business in its growth policy, as long as they could get a slice of the pie. Japanese parties had continued, largely speaking, to act the way Marx and the prewar Japanese fascists thought parties should act: The bourgeois party represented big business, the proletarian parties represented the workers. In Japan's emerging postindustrial society, this left the majority of the people voiceless, although increasingly angry. One recalled the typical students of a Japanese university. By and large, normal, average people who wanted simply to go on the campus and attend classes, they could have their campus closed down for days or even weeks by open warfare between two tiny rival groups of "activists."

Economically, socially, and educationally, Japan had attained a plateau of liberal capitalist welfare democracy, in the pattern of the United States, which Marx had never dreamed of. It was a triumphant consumer society. Unfortunately, everyone was represented but the consumer. In thirty years Japan had become the world's second biggest middle-class society. If there was any representative Japanese, it was no longer the farmer or the industrial worker, but the salaryman or the small tradesman—call them white-collar, knowledge industry, service industry, or whatever. The "salary-man," so-called, might be the pride of the economy, but he had become the forgotten man of Japanese politics.

He was in for more trouble, too. On the heels of the energy crisis came conclusive evidence that officials of his government, at the highest levels, were guilty of the most flagrant kind of political cor-ruption. The Tanaka-Lockheed incident of 1974–1976 resulted in the resignation, then the indictment of a Prime Minister for criminal behavior, the first case of this sort in Japan's post-war history. This would be followed by a continuing recession inside Japan and more exposure to the hard facts of world macro-economics, as the run-away surge of Japan's favorable trade balance brought pressures against Japanese exports from the United States and Europe. In 1978, five years after the 'oil shock,' there were still a great many pipers waiting to be paid.

The High Cost of Kick-backs

Japanese are fond of saying that there is a front-door and a back-door approach to almost anything. Going through the front door is to submit your formal application, make the deposition, take the test. Having done this, the conventional wisdom holds, you then insure a successful result by a variety of round-about methods, ranging from innocent after-hours talks with friends of friends through gifts and favors to the most obvious bribes and pay-offs. Given the imposing dimensions of front doors in Japan's legalistic society, where examinations, formal applications, and detailed bureaucratic procedures are the order of the day, a great deal of back-door action takes place. Its darker side ranges from business kick-backs and pay-offs to politicians and the Mafia-type connections of the *yakuza* gangs with parts of the entertainment world to the sums of money regularly handed out to university registrars (and as regularly exposed) as a back-door school entrance (*uraguchi nyugaku*) by desperate parents of the not-so-bright.

All societies have their graft, their bribes, and their fixers. Indeed, because of the personal honesty of the average Japanese and the close-linked groups he belongs to, the incidence of out-and-out thievery in Japan is relatively low; such everyday American problems as police corruption are minimal. In business and politics, however, the kick-back and the pay-off have a long history. And back-door approaches are made easier in Japan because of a native fondness for disguising or confusing the sources of authority. Some of the best remembered characters in Japanese history, and also some of the most rascally, are leaders like the twelfth-century Emperor Go-Shirakawa who formally retired, but continued to control their fiefs from the cover of the temple, the tea-house, or the monastery cloister. In modern times we have the analogy of the company chairman who ostentatiously resigns, but continues to order things from the sidelines. Or there are the retired politicians, like the *genro* ("elder statesmen") hold-overs from the Meiji era, whose word to loyal subordinates in power still has the force of law.

The shady political *kuromaku* (literally 'black curtain' men) take advantage of this tradition. A *kuromaku* is a behind-the-scenes fixer who operates on a big scale. His is a realm of "back money"—the undeclared political contribution, the barely disguised bribe, the money that corporations dole out to get contracts and pay off influential friends who can assist them. He may have a legitimate business base for his operations—Sasakawa Ryoichi, one of the biggest, enjoys a large income through his control of the legalized gambling on the motorboat races in Japan. Or he may have gotten his start, like the notorious Kodama Yoshio, running a vast off-shore procurement business for the Japanese Navy in World War II. He invariably has access to a network of highly placed politicians, almost all of them members of the majority Liberal-Democratic Party, and is apt to play a key role in supplementing the contributions of the politicians' various local support associations.

Most of the surviving *kuromaku*, like Kodama, have long histories of involvement with the semi-gangster super-patriots of the thirties. Kodama, in fact, began life as something of a populist. In 1929, aged eighteen, he rushed up to the Emperor's coach with a petition for the relief of farmers stricken by famine in northern Japan. For that he got six months in jail. Three years later he was arrested after taking part in an unsuccessful coup d'etat. (Some hand grenades he had carelessly hidden in his lodgings exploded.) He received another six months in jail and a suspended sentence for three years. Nine years later, on the eve of World War II, he planned to blow up Prime Minister Konoye's train, in the event that an improbable, but widely discussed, conference between Konoye and President Roosevelt were to materialize.

With such strong 'nationalist' credentials, it was only natural that Kodama find himself in the middle of the intrigues that were on the fringes of the Japanese occupation forces in China. His work for the Navy was so satisfactory that his employers—at the time of the 1945 surrender—allowed him to keep large sums of money and a supply of scarce raw materials, platinum, gold, and diamonds among them. In the shadowy half-world of *Yakuza* gangs and black-market traders which grew up behind the military rule of Occupation Japan, Kodama parlayed his wartime gains into a considerable fortune. According to some accounts, he contributed ¥150 million, a huge sum at the time, to the late Hatoyama Ichiro, the old-time politician who founded the Liberal Party. The credit gained from such enterprise he used liberally throughout the post-war period. For the old-fashioned *kuromaku*, besides their extra-legal activities, have acted as covert lobbyists. Since Japanese businessmen as well as politicians shun open favor-seeking as much as they avoid public controversy, the *kuromaku* often perform for Tokyo's big business world something of the same services which Washington law firms offer, if more overtly, in the United States. Kodama never hesitated to use henchmen from the *yakuza* in working out his deals. "When a thing is beyond the power of ordinary men," he once told a Japanese interviewer, "that is when I move into action."

The one reason an anachronism like Kodama could be tolerated in Japan's complex modern society, apart from the services he performed, was the low profile he kept; as did others like him. It was left for Tanaka Kakuei, with the material assistance of the Lockheed Aircraft Corporation, to bring Kodama and the whole shadowy network of back money he represented to public view and public justice, in a series of scandals which first broke on Japan late in 1974 and have continued to reverberate since. The "Lockheed scandal" has been compared to Watergate in the United States. For political scientists interested in measuring the fever charts of democracies, the parallels are often close and instructive. Yet the king-sized pay-offs of the American airframe manufacturer were only part of it. Lockheed was just one symptom of an endemic Japanese political problem, bearing only slightly more significance to the whole than the original Plumbers' break-in bore to the fabric of political corruption behind it.

Tanaka Kakuei, the self-made "computerized bulldozer" from Niigata Prefecture, had had at least one spectacular run-in with the law long before he became Prime Minister in 1972. In 1948, only thirty but already a post-war Diet veteran, he was named Vice Minister of Justice. Shortly afterward he was indicted for bribery, in a case involving coal mining interests in Kyushu. The conviction did not stick, however. Three years later he successfully appealed it. Tanaka went on to build up his political power in the party, over the years, with liberal gifts both to party influentials and his own supporters. Much of the money for this came from his own contracting business, which Japan's loose conflict of interest laws allowed him to retain while a Diet member.

His election to the presidency of the Liberal Democrats was facilitated, so the rumors went, by huge disbursements to a variety of people, from members of his own Diet faction to friendly newspapers. The whispers about vote-buying and "money politics" continued after he assumed the Prime Ministership. They grew more intense after the real estate speculation surrounding his prematurely announced plan for rebuilding the Japanese Archipelago.

If the times had continued good, the gossip about Tanaka's payoffs might have remained just that. The lack of an effective opposition, as previously noted, was in itself an incitement to loose practices among the Liberal Democrats. More important was a conspicuous lack of voter concern. As long as the high-growth period continued, the average voter, prospering himself, rarely grew indignant over alleged corruption among the businessmen and politicians responsible for the prosperity. Tanaka himself came to power on a surge of popularity which his aggressive reconstruction plans at first seemed to warrant. It was Tanaka's bad luck, however, to be the last of the big political spenders—and the most flagrant—at a time when economic shock waves told even the most trusting Japanese voter that the Good Old Days were going.

In a sense Tanaka's frankness was his undoing. A self-made individualist, he lacked the establishmentarian gloss on which most of his fellow party leaders set such store. With "Kaku-san" it was simply a matter of 'money talks.' He is not a bad man, in fact an immensely able man, likeable and engaging. His experience was formidable. He was elected to the Diet while in his twenties. He had served as chief party secretary for the Liberal Democrats, Finance Minister, and Minister of International Trade and Industry. During his rise to the political top, he had continued to build up his network of paying and receiving friends in business as well as among bureaucrats and politicians. His business friends had excellent establishment credentials. For the first time since the High-Growth Era began, however, big business was itself under attack.

At this delicate moment, *Bungei Shunju*, a widely read and respected monthly, began a well-researched series of articles on the Tanaka money connection, in its November, 1974, issue. For the first time there was documented the network of dummy corporations which Tanaka had used to channel contributions from business supporters to members of his Party faction. By the magazine's reckoning, the "back-money" running through this network was at least $250 million. Of this, it was suggested that more than $15 million had been used in intra-party pay-offs to secure Tanaka's presidency of the Liberal Democrats—and the premiership that went with it—in 1972. More than $200 million had been spent in the subsequent election for the House of Councilors. Tanaka was revealed to be the owner of Tokyo property worth $8 million and resort property worth $2 million, with other land worth $50 million if government restrictions on its sale were to be lifted. His dealings had a long history. The amount of land sold off to private interests during Tanaka's past term as finance minister, for example, was almost triple that sold by successors or predecessors.

Japan's huge daily newspapers, in the tradition of the Mandarinate, refused to touch the story for almost a week. Finally, however, its exposure in the foreign press forced them to cover it. Once off the leash, the reporters turned out in full cry and "money politics" went on page one. Barely two weeks after the *Bungei Shunju* story appeared, Tanaka resigned.

So far the scandal, although bad, was in the tradition of containable "incidents." Both the Party and press expressed satisfaction when the ex-Premier promised to adjust his income tax returns to account for some of the back money that had been exposed. He was succeeded in December, 1974, by Miki Takeo, the "Mr. Clean" of Japanese politics and the one faction leader of the LDP who was obviously above reproach. Although Miki tried to start unravelling the Tanaka money connection, he was given scant support by the Party elders.

Then disaster came from an unexpected quarter. In February, 1976, witnesses before a U.S. Senate subcommittee on multinational corporations testified that the Lockheed Corporation had spent some $12.5 million over a period of years in pay-offs to Japanese politicians and businessmen. The revelations came as part of a general confessional session in Washington which included mentions of bribes given to highly placed people in Europe. In Japan, where the public was still angry over the prospect of a Prime Minister with sticky hands, Tanaka's name quickly surfaced again. Lockheed went on page one and stayed there for more than a year.

The bare facts of the Lockheed scandal were simple enough: pay-offs and kick-backs to influential Japanese for the sale of the TriStar airbus to All Japan Airways and the new Lockheed patrol plane to the Self Defense Forces. The problem was how Prime Minister Miki and the Japanese prosecutors could get the names and numbers out of Washington for investigation and indictment. In Japan press, public, and opposition, not to mention Miki, were clamoring for the facts. But because disclosure would allegedly embarrass a friendly government, the U.S. State Department inexplicably supported Lockheed's recourse to the courts to prevent disclosure, thereby delaying for months the transmittal of actionable information to Japan.

As it turned out, there were several channels through which Lockheed paid out its bribes: Kodama Yoshio the old time *kuromaku;* Tanaka's unofficial business partner Osano Kenji; the Marubeni Corporation, the huge trading company which acted as Lockheed's agent; and various executives of All Japan Airways. In the end, however, all roads led back to Tanaka. Two years after his resignation he was indicted, arrested, and thrown into a cell during questioning—a far more ignominious fate than the Americans visited on Richard Nixon. In January 1977, Tanaka went on trial, along with his secretary and three executives of the Marubeni Corporation, Lockheed's official agents in Japan. Kodama was also put on trial, as was Osano Kenji, the principal financial backer of Tanaka's "money politics." Four other Diet members were publicly named as "gray-area officials" who received bribes but could not be prosecuted because of legal difficulties. There is clear evidence against nine more. Along with the former bosses of Marubeni, leading executives of All Japan Airways, the country's biggest privately run airline, were also put on trial.

Historically speaking, Tanaka-Lockheed was the worst scandal to hit Japan since the Siemens Incident of 1914. This interestingly enough also involved foreign business. At that time the German Siemens Schukert Company and the British Vickers Company, along with Mitsui Bussan, the largest of Japan's trading companies, were caught out in a straight case of bribery involving Navy procurement contracts. Tanaka's and Kodama's speculations were far bigger, peculiarly a product of the High-Growth Era. Their mosaic of business and political collusion represents the dark underside of Japan's rise to economic superpower status during the sixties, as fully as the Watergate crookedness was a by-product of the centralization of power in Washington.

There had been scandals in post-war Japan before Tanaka and Lockheed. A thin thread of corruption stretched from the Showa Denko bribery case in 1948 through the shipbuilding procurement scandal of 1954 down to the Japan Line stock manipulations of the sixties. In a sense, the increasingly large sums of money involved amounted to a rough index of Japan's rising prosperity. A major scandal like Showa Denko could bring down a Cabinet and change the political map drastically. More frequently one or two officials will give up their posts by way of apology—no one resigns with such grace as a veteran Japanese Cabinet Minister. But prosecutions were rare. The last time the public prosecutors tried to bring top party bosses to trial was in the shipbuilding subsidies case of 1954. They failed. Japanese prosecutors are extraordinarily cautious, indicting only where they can see overwhelming odds for conviction. And political cases are slippery.

The law is loose, to begin with. Although vote-buying in the Diet itself is strictly proscribed, there is nothing illegal about buying votes in an *intra-party* election, as Tanaka and so many others did. The very organization of the Liberal Democratic Party into factions, with heavy intra-party vote-switching and vote-buying, invites the kind of subsurface money politics which is difficult for either the law or the voters to detect. The fact, also, that the majority party's elected president automatically becomes prime minister is a central defect in this system, at least as long as party affairs remain unpoliced.

As Tachibana Takeshi, who directed the 1974 exposé, later wrote in *Bungei Shunju*: "The Lockheed incident is not just another 'made-in-America' development. It has shown itself to be merely the extension of a problem peculiar to Japan. . . . What distinguished Lockheed from other incidents and scandals of the post-war period is the fact that it was exposed and played out in full public view." Like the storied bumbling foreigners who barge into a teahouse with their shoes on, in their eagerness to get at the girls, the Lockheed people scattered their million-yen bribes with little regard even for the traditional niceties enjoyed in the *kuromaku's* world. Nor was there any friendly Japanese 'reporter's club' in Washington to modulate the shock effect of the Senate Committee's disclosures.[1]

---

1. Interestingly enough, a similar set of circumstances back in pre World War I years set off the Siemens affair. Then the tell-tale foreign investigation came from Germany. At a trial in Berlin, documents surfaced which proved that Japanese Navy officers and politicians had taken bribes in connection with ammunition procurement contracts. The news was telegraphed to Tokyo and appeared in the press there on January 23, 1914. The Diet began a series of investigations into Navy corruption. By February 10 there was a vote of no-confidence and Admiral Yamamoto Gombei, then Prime Minister, resigned with his entire cabinet.

Considering the seriousness of the problem, the immediate political results of the Tanaka-Lockheed scandals were unspectacular. Thanks yet again to the weakness of the Socialists, the Liberal Democrats managed to hang onto a thin majority in the subsequent elections of December, 1976. The biggest gains were made by an anti-establishment off-shoot of the majority conservatives, called the New Liberal Club, led by a younger Liberal Democratic politician named Kono Yohei. (His father, an old-fashioned Japanese politician, had incidentally been closely associated with Kodama.) Tanaka himself was re-elected by a large majority of loyal constituents in the northern snow country of Niigata, well-cared for after years of new highways, fancy development projects, and other favors from the central government. The principal casualty was Prime Minister Miki, who had pushed through the Tanaka investigation against strong opposition within his party. For two years Miki stood off the Liberal Democratic faction bosses by going over the heads of the party and appealing to the people through TV and the press. This was a novel practice in Japanese politics. His successor as Prime Minister, Fukuda Takeo, although an old political enemy of Tanaka's, was not overly interested in making the Tanaka investigations any wider than they already were.

In 1976, reflecting the heat from the Tanaka investigations, the National Personnel Authority severely limited the number of *amakudari* (descents from heaven), that is, senior bureaucrats who are allowed after retirement to take big jobs with private or semi-public companies, including those they had dealt with in the course of their government duties. The Lockheed scandal had brought to light the close ties, for example, between the Ministry of Transport and All Japan Airways (whose president and other executives were Ministry alumni), and showed the bad side of such good connections. In 1977, however, things swung back to normal. Fully 198 retired bureaucrats parachuted to safe landings as company directors. Former finance, construction, and tax bureaucrats led the descent.

The Tanaka trials continued, however. It seemed clear that, ultimately, most of the charges would stick. (In the spring of 1978, one of the accused testified that he had paid money direct to Tanaka.) The Japanese voter may not have had a big, reliable second party to fall back on. But he and his press could be counted on for more vigilance in the future about the conduct of either majority or coalition party leaders in government. It is doubtful whether the dark undergrowth of the *Kuromaku* and the "money politics" of the back room can flourish again as they did before.

Chapter XIV

Slow Growth and Fast Exports

"Exports and imports will never lose their equilibrium, as they circulate in accordance with the laws of nature, never ceasing recurrently to rise and fall . . ."

—Tsuda Mamichi, 1873

"If a man can build a better mouse-trap, though his house be in the middle of the woods, the world will find a well-beaten path to his door."

—Ralph Waldo Emerson

With their fondness for self-depreciation, Japanese often call their whole economy a "bicycle operation—*jitensha sogyo*." That is: as long as you keep pedaling furiously, the enterprise will go on; lose motion for a moment at your peril. This is not mere hyperbole. For all the power of post-war Japanese industry, its margins are often very thin. For all the brilliance of Japan's long-range economic planners, they are collectively stubborn about making quick, short-term adjustments. Rather than stop and look about or inquire directions from passersby, the bicycle rider takes refuge in his own industry and just keeps pedaling head down in the direction first pointed out to him, heedless of honking horns or flashing lights at the intersections. Or so it seems to others on the road.

The Japanese export crisis of 1977 and 1978 followed all too logically from the problems caused for the amazing growth economy by the energy crisis four years before. Seen in isolation, an export drive was the one obvious way for a recession-bound economy to keep itself on an even keel, which means to keep the new 10,000 ton per day steel furnaces active, to keep workers busy on assembly lines rolling out roughly eight million new cars each year, to keep the girls in neat smocks turning out the new Sonys, Sanyos, and

Panasonics in the noiseless, antiseptic electronic workshops, not to mention providing odd jobs for the thousands of sub-contractors clustering around the giant Japanese corporations, like the shops and rickety dwellings on the outskirts of a castle town.

Unfortunately, Japan's problem could not be solved in isolation. In a world whose macro-economic balances were already badly out of tune, Japan's "up-the-exports" response to the demand recession at home very quickly became an international problem. Extraordinarily, this was not foreseen in Tokyo. Despite past tremors of Nixon shock and oil embargo, the Japanese government busied itself with domestic concerns for more than a year when something might have been done to curb its soaring balance of payments. When the bureaucrats braced themselves to face the storm, it was almost too late.

By the end of 1975, with ample sets of statistics from the energy shock available to them, most of the Japanese establishment—businessmen, economists, and bureaucrats—had realized that the two-digit growth years were gone. The words "slow growth" dinned themselves into the national consciousness, at least insofar as the consciousness is formed by the continual antenna-rubbing of newspaper comment, government announcement, and unending TV round-table discussion of economic problems, backed up by frequent pronouncements of the business community. The perenially rising GNP, reported in the press as faithfully as the baseball scores, had gone down in 1974 for the first time in fifteen years. Inventories were piling up. Companies committed to full employment were resorting to early retirements and furloughs at less than full pay, in desperate efforts to avoid mass lay-offs.

The very successes of the fast-growth era helped deepen the recession. Technological breakthroughs and economies of scale were of little use, if markets did not expand. The lifetime employment system that so increased productivity when the labor force was young became more of a handicap as age levels (and pay packages) kept growing. Concentration on heavy development and marketing in selected products, so thoroughly applauded in the high-growth era, became a liability now, since these industries needed steadily rising sales to justify the large investments made in them. Ship-building was an extreme example. Built, pushed, and subsidized in tandem with the post-war steel industry, ship-building needed a constant flow of orders to meet its rising capacity and increasingly high technology. Not only were the orders from overseas now slackening; new competition was appearing, notably from the fast-growing Korean economy almost next door, which had set out early in the seventies —oil crisis or no—to surpass Japan's own growth record, with much the same methods.

At this point of economic standstill, the time had come, according to the Keynesian gospels, to stimulate consumer demand. An economy which had grown great through heavy capital investment should now ideally transfer its faith to a combination of heightened consumer demand and big government spending. Government spending—for new construction, parks, and other public works designed to improve the quality of life—should then stimulate ever more consumer spending in housing, leisure industries, cars, clothing, whatever. This at least was the tried and true American formula.

There were three obstacles, however, in the way of making this switch in Japan: the consumers, the government, and business. The faithful Japanese consumer, worried by oil shocks, political scandals, and the new possibility of lay-offs or unemployment, seemed to be tiring. He was cutting down his spending, especially on leisure or luxury goods. He did keep up his remarkable rate of saving. (Throughout the seventies, Japan as a nation continued to save roughly 30 per cent of its entire GNP; more than two-thirds of this represented private savings.) But his savings were going into banks which now had few heavy capital investors around to use the ready money.

The government, particularly since the collapse of Tanaka's ill-timed program for rebuilding the Japanese Archipelago, reaffirmed its traditionally negative view of heavy pump priming, whether in the form of massive public works, housing assistance, or expanded welfare and benefit plans. There are not many Keynesians in Tokyo's Ministry of Finance. It is in fact one of the world's last strongholds of militant monetarists, who believe that government's principal, if not its only, effective control-lever over the economy is regulation of the money supply.

The future problems of investment in Japan's high growth industries had indeed been anticipated. Strategists at the Economic Planning Agency and elsewhere had prepared marvelously detailed blueprints for switching Japanese development from labor-intensive products to "knowledge-intensive" and service industries. But the practical problems involved were not easy ones to solve. For yet another by-product of the high-growth prosperity was postponement of the plans for the so-called "post-industrial" era. The managers of a still flourishing textile plant, for example, on being told that they are part of a dying, uneconomic industry, will be pardoned if they do not obediently shut off production at once and attempt to turn their assembly line gnomes into clever engineers working out home information storage systems.

Both big business and labor were nervous about doing anything drastic. Japanese managers, whose companies get the bulk of their financing from banks, do not have to worry about demands for bigger profits from the shareholders. If a company can ride out the recession by selling at reduced prices and wait for better days, the banks will rarely object. The managers had an added problem in addition to keeping their own companies producing. Each large Japanese corporation has its host of related subcontractors. Ties are strong between them and their big foster-parents. They need to be sustained, whenever possible, despite their often un-economic and inefficient ways of doing business. With slow growth, their margins of profit—even of survival—were becoming thin.

When Fukuda Takeo, by training, instinct, and background a Finance Ministry bureaucrat, became Prime Minister at the end of 1976, his principal concern was how to cut down inflation. This goal was achieved in 1977. Consumer price increases were kept down and the wholesale price index became stable again. The rate of wage raises decreased. Jobs grew scarce, also. The unemployment exceeded 2 per cent of the work force in 1977, double what it had been in the high-growth days. The number of bankruptcies in 1977, as more and more subcontractors went out of business, was a record 18,471. They left $12 billion in outstanding debt.

When the government finally did propose a certain amount of public works pump priming, the highest in fifteen years, it was certainly not of a magnitude to move any recession away in a hurry. For businessmen who had to pay their people, more exports still seemed like the only answer.

Exports, as we noted, rarely exceed 12 per cent of Japan's total GNP—a far smaller percentage, for example, than Germany's current 27 per cent. But they had become more and more significant for being concentrated in a few key manufacturing industries: automobiles, electronic products, machinery, steel. Like good military strategists deploying overwhelming force at several key points, Japanese business, with government assistance, had built up tremendous competitive strength in these selected industries. And the more that was invested, built, and sold, the better and more competitive they grew. Their favorite export targets, apart from Southeast Asia, were Europe and, most important of all, the United States, the country on whose business they patterned theirs and with whose markets they are most familiar. Let us cite one striking example. In 1970 Japan produced a total of 5.3 million motor vehicles; in 1975, 6.9 million; in 1977, 8.5 million. Of the 1977 total, 4.35 million cars, trucks, and busses were exported. Fully 1.74 million of these went to eager, prospering Toyota, Datsun, Mazda, and Honda dealers in the United States.

Led by the small car, the color TV set, and high-quality steel, selected high-technology exports poured out of Japan. The imbalance between exports and imports from the European Economic Community steadily worsened, from $4.4 billion in 1976 to $5.2 billion in 1977. With the United States the figures were more spectacular. Export totals jumped from $11.3 billion in 1975 to $15.2 billion in 1976 to $18 billion in 1977! Imports into Japan from the United States, however, remained stagnant at a level of roughly $10 billion annually for the three years.

The reason was easy to see. The bulk of Japan's imports from America are foodstuffs and industrial raw materials. Food kept coming in, but Japanese business was in no mood to increase its imports of raw materials until there were some signs that the recession would break. Thanks to import restrictions by the Japanese and past failures of American companies to take advantage of export opportunities, there was no corresponding consumer demand from Japan for American manufactures, to balance the clamor for Hondas, Nikons, Yamahas, and new pocket calculators in the United States. The resultant escalation in trade imbalances led to howls of protest from American businesses and labor unions, which were losing markets and employment. Steel invoked anti-dumping laws. Zenith and Ford sued. Individual Congressmen reacted quickly to complaints about "unfair Japanese competition" from constituents in trouble. What had begun in 1976 as scattered protests by mid-1977 had escalated into an economic confrontation with the Japanese that dwarfed previous such exercises. Democratic spokesmen proved that they could thump the table at least as loudly as Republicans, in the series of inter-government confrontations that followed.

When systems of reference and 'trigger' prices for Japanese products, anti-dumping laws, and orderly marketing agreements proved inadequate to stem the tide, the Carter administration encouraged a precipitate slide of the dollar as an international exchange instrument, by conspicuously not supporting it. While this may have been a deserved reaction to years of Bank of Japan support for an under-valued yen, it happened too quickly for anyone's comfort. Speculators profited, as the dollar dropped from a level of roughly 290 yen in December, 1976, to the 185 level by September, 1978. Although designed to make American exports more competitive and imports more expensive—and to an extent effective—the end consequences of this simplistic strategy are awesome to contemplate.

In justice to the Carter administration, it must be said that strong efforts were made by the executive branch to turn back the growing demand in Congress for protectionist legislation, in favor of consultation. New administrative devices, such as the Trade Facilitation Committee, were set up in 1977 to negotiate problems of real or apparent inequities or barriers to free trade between the two countries. Nonetheless, rarely in recent history have so many leaders of one country publicly told the leaders of another sovereign power how to order their economy in detail, down to specific instructions as to the desired percentage of the Japanese domestic growth rate (on the assumption that an increased growth rate would enable Japan to import more and export less).

We should not pity the Japanese overmuch in this confrontation. A rapid increase in exports may have seemed an ideal immediate solution to the recession problems of Japanese business, but Japanese statesmen and businessmen should have been aware that the economic problems of the world's most successful trading nation cannot be solved in isolation. Yet until well into 1978, when the first big import-buying delegation of businessmen was rushed off to the United States, there was not even a serious effort made to tell the Japanese side of the problem.

Credibility also became an issue, as before in the Nixon-Sato negotiations of 1971. Early in 1977 members of the Fukuda cabinet told Vice President Mondale, their first official caller from Washington, that their goal was to transfer Japan's lopsided surplus to a $700 million deficit, virtually within a year. As it turned out, the projections at the beginning of 1978 showed a $10 billion overage developing on the Japanese side. Here was yet another instance of Americans taking a pious expression of Japanese hope as the next thing to a commitment, but one which the Japanese would not readily be allowed to forget.

Nor did Japan's appeals for free trade in 1978 fit well with a history of methodical protection of key home industries, which was continued with great skill in the post-war period. It is true that the Japanese government, under pressure, has recently moved away from protectionism and, since the sixties, 'liberalized' one section of its economy after another. But many barriers have stayed in place. Among them are high tariffs on such 'sensitive' items as computers and imported color film, and rigidly restricted quotas on agricultural products. In addition, a kind of *de facto* tariff continues to be enforced by the Customs Bureau, whose bureaucratic restrictions on the entry, payment terms, and inspection of non-Japanese products would arouse the envy of a late Ching dynasty mandarin.

Timing of course is critical to the successful entry of a new business. Thus, a company like Eastman Kodak might see some irony in its being allowed finally to invest in Japan, in 1976, after its local rival Fuji Films had been allowed to monopolize the Japanese market for two decades. Other cases of industry protectionism included automobiles and electronics. While such protectionism might conceivably be justified on political grounds, as in the American interest—for example, 'wouldn't you rather have a healthy, sound Japan than a depression-ridden society of malcontents about to go Communist?'—it has no place in an economic argument. In a 1968 interview, ten years before the new "exports" crisis, Henry Ford II had expressed the resentment of American businessmen who had so long been frozen out of a good market by Japan's protectionism. "I haven't got anything against open competition," he said, "if they can build a better car and sell it for less money, let them do it. But what burns me up is that I can't go into Japan. We can't build, we can't sell, we can't service, so we can't do a damn thing over there. Ford Motor Company still owns land in Japan, and we still have a building there that was put up before the war. I understand it leaks like a sieve but it's still there, built with our money, and we can't use it. I'd be in there tomorrow if the Japanese would let me. I'd be manufacturing cars and I'd give the Japanese a run for their money. But they won't let me in and that's why the whole thing is unfair." [1]

The Americans for their part continue to play what might be called "semi-domestic" politics with Japan. In this, they have taken advantage of Japan's next to total dependence on the United States for much of its raw material supply, as well as the vital role of the American consumer to the Japanese export economy. (In 1978, as in past years, 25 per cent of all Japanese exports went to the United States.) In a more subtle way, and perhaps without wholly realizing it, the American business negotiators and statesmen who publicly lecture the Japanese are relying on (and spending) a large reserve of good will and receptivity to American moods, ideas, and suggestions among a people whose leaders have been content to follow the lead of the Americans, still the People-to-Learn-From, in their postwar political and economic policies. Examine the comments of American public spokesmen to the Germans, with whom the United States has similar economic problems. One finds a far more subdued tone, diffident by comparison. But because the Japanese listen (and seem to listen even more than they do), the loud lecturing on the American side is of a sort that we would never inflict on anyone else.

After a while, in the finest tradition of court-room advocacy, it is easy to confuse a public negotiating posture with perceived reality. This is especially true when your selected adversary does not talk back. So American spokesmen, prodded by their special interest groups behind them, and aware of their press coverage, tend to become ever more adversarial in talking to the Japanese. Many Japanese statesmen, to a fault, welcome this public high-pressuring. It allows a less-than-strong factional government to say to its own special-interest groups, "What can we do? You see how violent they are. We must make concessions." Meanwhile, behind the facade of accommodation, a collective feeling is growing among the Japanese that the Americans, whom they have notably trusted, are using them as pawns to correct problems which are partly, if not largely, of the Americans' own making.

Indignation at Japan's export and import policies would indeed be more effective, if the American exporter were doing his bit to sell

1. It must be added, however, that Ford had previously turned down an opportunity to enter the Japanese market, in 1950, when the president of Toyota went to the United States, to try to negotiate a tie-up with Ford in Japan.

hard to Japan. With some exceptions, he is not. Where the weak point of the Japanese argument is Japan's historic protectionism against foreign imports, the American weak point is the export effort to Japan. American exports to Japan have not kept pace with appreciation of exports, for example, to the European Economic Community. Considering the closeness of the American-Japanese relationship and the opportunities available for selling goods and services, the achievement is poor.

Partly this shows the pessimism of American exporters, thrice shy of the Japanese market after rebuffs and reverses in the past. Over the years American businessmen have been reluctant to enter the Japanese market, almost as a matter of principle—unless in obvious cases of ready acceptance and quick profits. It is difficult. Japanese are at once too obviously worldly and too parochial. They are worldly in the sense that they offer this planet's readiest market for foreign goods, especially if they are made in U.S.A. But this almost built-in customer receptivity is obscured and outweighed, for most American businessmen, by the problems of doing business in a country which is also culturally and economically solid enough to insist on having products packaged and delivered in its own way, distributed through Japanese-type distributors, and made subservient to local Japanese writ, whether it is the formal Bank of Japan currency remittance regulations, the informal "window guidance" of MITI, or the subtle pressure of the *gyokai*, the Japanese trade associations who regulate competition in their own peculiar ways. Doing business successfully in Japan requires a large up-front commitment, resolution to take one's lumps over a period in the expectation of long-term gain and an adaptability for doing business the Japanese way, as far as possible, despite one's own habits and preconceptions. Some American businesses with an extra ration of tenacity have been richly rewarded for their pains. IBM, NCR, Caterpillar, Caltex, Dow, and others have made large and consistent profits. But about fifty per cent of American business enterprises in Japan have either ended in failure or been characterized by a desire for short-term profits which has given the American businessman in Japan an undeserved reputation as a fast-buck "landing" operator.

The weakness of American manufactured exports to Japan, however, is merely the extreme symptom of a deep-seated attitude that, for the richest country in the world, exporting manufactures is not really necessary. Over the post-war years, American business showed its preference for investing in plant overseas instead of sending American products out. This was particularly true of the multinationals, whose managements, in their hot pursuit of big bottom-line profits, showed little interest in where the numbers came from. Until recently, also, investment terms overseas were generally better and labor was generally far cheaper than in the United States.

Through the seventies, however, American labor costs have come down, relative to those in other developed countries, so that investment in off-shore plant is nowhere near the good buy it used to be. While the American economy is aging, foreign competitors have been building ever better mouse-traps in their own factories, with their own funds. And thanks to the new permanent energy crisis and heavy American consumption, the United States must pay more than $40 billion annually for what was once cheap foreign oil.

Testifying before a Senate subcommittee in 1978, Assistant Secretary of the Treasury Fred Bergsten stated the problem: "In times past, exports didn't matter much—either to the individual American firm or the country as a whole. . . .[now] the U.S. must become much more export oriented . . . our future prosperity at home is closely tied to our success at boosting sales abroad . . . a healthy, expanding export sector is essential for the long-run stability of our external accounts and thus the dollar . . . we must avoid protectionist trade policies and provide full government support for the export efforts of American firms."

Alone among the major powers the United States has given only minimal help to its exporters. The amount of money the American government spends on export promotion averages about half that of the other developed nations. The Export-Import Bank, originally thought of as an aid to business, has until lately at least been neglected. It is still the prisoner of restrictions on its lending; for example, Congressional approval is needed for each loan in excess of $60 million. In 1971 Congress set up Domestic International Sales Corporations to stimulate U.S. exports by tax advantages. But instead of expanding the utility of these DISC, which ultimately create more tax money through the profits they bring it, legislation was lately introduced to cut them back.

Similarly, tax reform proposals prepared for 1977 and 1978 virtually did away with income tax advantages for Americans living abroad, made vital living allowances taxable, and threatened many with ruinous double taxation. American companies find it increasngly more expensive to station American nationals overseas. Some of the restrictions imposed on American exporters are exhaustingly complex. Proposed environmental protection legislation, for example, makes government loans to U.S. exporters conditional on lengthy examinations of what their product would do to the environment in foreign countries! From anti-trust laws to the Federal Drug Administration, the U.S. government hobbles its export businesses with restrictions designed for American market conditions, which often have no relevance to their overseas markets. The result is to give more business to foreign competitors. If the government in Japan is seen as the serious partner and banker of Japan Inc.'s export business, the role of the United States government vis-a-vis American exporters is more like that of an unfriendly prosecuting attorney.

There was a larger problem of attitudes behind the American-Japanese "export" confrontation of 1977-78. For the automatic cry of "cheap Japanese imports" reflects an unspoken feeling of frustration that the Japanese are doing so well. It is like the inexplicable sense of irritation the American tourist gets when he looks at his first made-in-Japan bullet train in the Kyoto station. Have we been beaten at our own game? Or, to be more precise, have we forgotten our own game?

The United States, for a variety of reasons, has been for some time a slow-growth economy. Perhaps that is what most Americans want. It is certainly just that more money should be spent for improving the environment, crusading for human rights, offering equal opportunity to all citizens, increasing welfare services. Indeed, in a "post-industrial" economy, so the rubric goes, we need worry less about our industrial production, as more people switch over to service and knowledge industries. Nonetheless, the American economy still needs growth. For growth it needs more capital investment. And this it is not getting.

The authoritative Hudson Research Europe Ltd. in a 1978 report, had this to say: ". . . the U.S. economy has tended to grow more slowly than any other western industrialized economy apart from the U.K. . . . Productivity has actually grown more slowly than [in] any other western industrialized nation. . . . The poor growth record is matched by an equally poor trade performance, in spite of the steady appreciation of the currency of many competitor nations against the dollar. It is not widely realized that the U.S. regularly tends to run a persistent trade deficit and that the soaring deficits of the last two years are not a novelty for the nation. . . .

". . . We suggest that there is reason to expect this deficit to continue, with the possibility that the present size of the deficit—$30 billion approximately in 1977—will not in retrospect look so extraordinary. Capital spending is by international standards low; the unemployment rate has consistently been above that of its main competitors; even the inflation performance which is broadly considered to be good is, on examination, a more deep-seated trend than U.S. political leaders like to admit. . . . Nor is it generally understood that the U.S. technological lead is disappearing as the research and demand effort declines. . . ."

Now consider Japan. Its economy, fiscal policies, and labor laws drastically revised by Americans during the occupation, the Japanese set out in their economic planning to build a large capitalistic growth economy on the American model. They borrowed heavily from American technology, which American business was quick to sell them. They used it to develop quality products, which they sold throughout the world based on sound market research, the principles of which they had learned from Americans. But they made some significant changes of their own. They shunned deficit financing, kept government expenditures tight, sacrificed creature comfort to capital investment, kept government down—Japan's 20 per cent total government share of the GNP is far less than the 33 per cent in the United States. They preferred their own ideas about planning, productivity, and the 'adversary system' of running a business.

Without repeating statistics previously given about sheer GNP growth, let us review a few comparative statistics of Japan's progress. Between 1960 and 1976 industrial production showed an absolute increase of 340 per cent, as against 97 per cent for the U.S. Productivity over the same period rose from 52.6 (1967 = 100) to 204.6—as contrasted to the American 78.8 to 123.6. Over five recent years (1970–1975) fixed capital formation in Japan was 35.5 per cent of gross domestic product, contrasted with 17.4 in the United States. As for technology, the average development time between the invention of a new product and its actual employment has been 6.4 years in the U.S. In Japan it takes 3.6 years.[2]

2. The statistics are taken from the Hudson Research Europe's 1978 special report, "The American Economy: A Reappraisal."

The steel industry offers the most vivid and sobering contrast between post-war Japanese ideas about developing technology and American, not to mention differing points of view on labor-handling and capital formation. Over the past quarter century four basic technological improvements have been available to steel-makers everywhere: (1) the basic oxygen furnace, vastly superior to the outmoded open hearth; (2) the use of continuous casting equipment; (3) economies of larger blast furnaces; and (4) control of production by computer systems programming. The Japanese were quick to put them all to work. As a result, the sheer increase in Japan's steel production from 41 million metric tons in 1965 to 105 million metric tons a decade later (while the United States stayed generally static in the 120 million metric ton range) tells only part of the story. Far more important is the fact that Japanese steel is produced in new plants, continuously improved at a cost about 20 per cent lower than American. Current manpower costs of 5.5 hours per net ton shipment in the best Japanese plants are slightly more than half the American costs. And there is potential for reducing that 5.5 figure significantly. These are the reasons, simply put, why so many American customers have been buying their steel from Japan. Shipping costs included, Japanese steel can be delivered in the U.S. cheaper than the American competition; and its quality is of the best.[3]

After World War II the American steel companies did not modernize themselves as they should have. Whatever the reasons adduced for this, and there are some good ones—post-war reconstruction needs, inability to finance plant improvement from earnings, government regulations, serious labor disputes—their problems are internal ones, on whose solution, incidentally, the future economic health of the United States depends. They have little to do with Japan. Yet the steel companies, the steel unions, and their representatives in Congress for years have been making demands for govern-

3. Ironically, one of the great incentives for the Meiji reformers of Japan to modernize quickly, was the low cost of British iron, as compared to the products of Japan's import industry. As Tsuda Mamichi wrote in the *Meirokuzasshi* in 1873: "While a heavy commodity like iron is costly to transport, the price of iron brought several thousand miles distant by sea from England is actually cheaper than our domestic iron. This may be attributed entirely to the technological backwardness of our iron mining and production. There is an old saying that wisdom and foolishness are 3,000 miles apart. It is not alone in the iron industry that our country and Europe are 3,000 miles apart."

ment action against "unfair Japanese competition," "dumping," "price-cutting," "cheap labor," etc. These demands became more intense after the 1975 recession and the aftermath of the energy crisis put a quick end to the short-lived American steel boom of 1974.[4]

The Japanese steel-makers are hard, often rough competitors. But they have tended to regulate their exports to the United States, more sensitive than other Japanese businessmen to the threat of American official retaliation. They have invested heavily in anti-pollution equipment to cope with Japan's stringent domestic standards. Their own labor force is extremely well paid. Although they have frequently cut prices when the market is bad, as have the American producers, export prices only rarely fall below their own domestic levels. It is the European Economic Community which has done most of the "dumping" in recent years. At least, in the words of the Council on Wage and Price Stability, the Europeans have used "extremely aggressive discounting."

The problem of competitive steel-making was summarized in a July 1977 research report of Merrill, Lynch, Pierce, Fenner & Smith:

> Under conditions of free trade, we do not believe that the U.S. steel industry is economically sound over the long term. Japanese steel companies are, in our opinion, more efficient than their United States competitors and that advantage seems to be increasing. The OPEC oil price increase, combined with the two United States dollar devaluations in the 1970s caused the Japanese steel production cost gap to narrow against the United States, but the gap is now widening at a rapid rate and has returned to the almost 30 per cent Japanese cost advantage of 1972. Direct labor productivity in Japanese steel mills is about 50 per cent greater than the productivity in the U.S. on a tons/man/yield basis; but newer United States mills, such as the Burns Harbor facility of Bethlehem Steel compare well with newer Japanese mills in terms of productivity. Unfortunately only one new medium-sized mill has been built in the United States in the last sixteen years, compared with eight giant mills in Japan. . . .
>
> To become competitive, the U.S. steel companies need a massive building and modernization program, which they cannot afford, primarily, we believe, because of fifteen years of *de facto* price control and, consequently, inadequate capital formation. Furthermore, we believe that steel imports into the U.S. from Japan are not the problem

4. For most of the information here noted, I am indebted to an excellent forthcoming study on the American steel industry, *In Steel We Trust* by Luc Kiers, which puts this problem into perspective.

for the domestic steel industry. . . . In our opinion, imports are a symptom of the problem, not the problem itself.

By 1978 the Japanese themselves were working hard to restore order in the lopsided trade balance. Prime Minister Fukuda, when he visited the United States in May of that year, told President Carter and influential members of Congress of his immediate plans: (1) let the trigger-price system reduce steel exports by about 10 per cent; (2) cut the exports of color TVs to the United States by about 30 per cent; and (3) restrict automobile exports as well. This will take time to show results. Thanks to the phenomenon of leads and lags, the sliding dollar caused a short-term rush of Japanese exports, while it would take time, planning, and some good luck before American exports could carve out a greater niche in the Japanese market at their new lower prices. Meanwhile Fukuda promised to buy more enriched uranium, airplanes, and similar big-ticket items from the U.S.

No doubt similar solutions could be worked out for the Europeans, whose bargaining counters were not so great—other than their ultimate weapon of cutting off Japanese imports, which might prove self-damaging in the long run. The amount of Japan's trade with Eastern Europe and the Soviet Union has not been significant; and given the one-sidedness of Soviet terms, a major Japanese development effort in Siberia is unlikely. Trade with Southeast Asia continued to expand, however, and the prospect of really substantive trade with China has become a real one. It may make up for cutbacks in steel exports to the U.S. The Chinese have been anxious to secure Japanese heavy industrial products and even large-scale development, in return for coal and oil. Judging by recent commitments, the $2 billion worth of machinery and manufactures going to China from Japan in 1977 may be doubled by 1979, the Chinese to pay for this with increased shipments of crude oil. The $10 billion long-term development agreement signed by Mitsubishi Industries in Peking early in 1978 augured more of the same.

The basic problem for Japan, however, lay behind trade balances and the swollen current accounts surplus. They were only symptoms. How could the Japanese restructure their economy, cut down on its "bicycle operation" aspects, and steer their country into an era of more moderate growth, but with the growth coming in different places? Theoretically speaking, the solution was easy. Japan need only do what the American economic missionaries, preaching *ex cathedra* from Washington, suggested and many Japanese economists themselves advocated: increase government spending on public works and welfare programs; divert money from capital investment into improving the "quality of life," while stimulating consumer demand; offer more parks, public libraries, and cheaper housing at home, while investing more in under-developed countries; meanwhile increasing investment in well-developed trading partners, like the United States.

While much Establishment thinking in Japan runs in this direction, the effects of so drastic a reordering of priorities would transcend the merely economic. They would invite social and perhaps political changes which few could foresee. When the growth rate is running along nicely at 12 per cent, it is no problem to slice up the ever-expanding pie. But when growth is down to five or six per cent, there will be winners and losers, at each pie-cutting. In Japan, where the corporation largely fills the social welfare role of government in the United States, people cannot easily be fired. A Japanese Zenith Corporation cannot complain about "unfair competition," shut down its plants in Illinois and Indiana and start up again, happily cost-cutting, in Taiwan or Mexico. It has a social obligation as well as a business relationship with its employees. If domestic prices in Japan must go down to stimulate demand, what can be done about the cumbersome system of subcontractors, small distributors, and middlemen—the heart of the "bicycle operation"—all of whom have employees and families to support?

The Japanese government cannot order its business by fiat, as so many foreigners suppose. The lobbies and special-interest groups, particularly in agriculture, are fully as strong as they are in the United States. The Ministry of International Trade and Industry has its greatest leverage on companies which need something from it, whether subsidy, regulation, or indirectly offered bank support. A company like Toyota, for example, needs little such support. And Japan, after all, for all the business-government cooperation and "window guidance" systems, is a free enterprise economy and a democracy. Its people vote on its policies. It is a free vote.

The Japanese economy has proved extraordinarily strong, through dollar shocks, oil shocks, and trade-imbalance shocks. One of the ironies of the 1977–1978 American "export" crisis, in fact, is that the Japanese domestic economy, after several years of recession, was just at this time beginning an upturn, which ultimately would produce purchasing power and consumer demand enough to restore part of the trade balance. Like the United States, however, Japan is becoming a country of older people. While the percentage of senior citizens increases, the floods of young high-school graduates seeking employment are subsiding. Older industries, too, like textiles, are being rendered obsolete by competition from new 'economic miracles' like Korea and to a lesser degree Taiwan and Hong Kong. Korean economic planners, in fact, openly express impatience that Japan has not transferred more of its energies into post-industrial "knowledge" industries or service industries to make way for Korean steelmakers and shipbuilders. New Korean shipyards, full of orders, have set the managers of companies like Ishikawajima Harima frantically diversifying.

Meanwhile, the special interest groups in Japan continued to campaign quite justly for purer air, better housing, and more regulation of industry. And while jobs were shrinking, the universities kept up their assembly lines. By 1977 the percentage of college graduates going into the labor market each year had gone up to 27. Soon there might be more managers than workers.

There was, in short, a great deal of work for the coming generation of planners and managers to do.

"Which of the four power blocs in the world will Japan end up joining?" asked the American oil tycoon of the Japanese banker. "There will be four and a half," was the answer.

In September 1973, Japan's sacred Grand Shrine of Ise was rebuilt, for about the fiftieth time. After two months of ceremonies, beginning with the White-Stone Festival (*shiraishi matsuri*) in August when people bring pebbles from all over the country to surface the temple enclosures, the fresh cypress timbers of Shinto's Mecca were all in place, forming simple Polynesianlike structures with high thatched roofs, beautifully neatened huts, set on stilted platforms. Their enclosure was next to the old shrine buildings, about to be torn down, in accordance with the wish of the Empress Jito, twenty years after their erection. The empress had expressed her wishes in 689 A.D. Ever since, the shrine to the sun goddess *Amaterasu Omikami* has been systematically torn down and rebuilt completely at roughly twenty-year intervals, with the exception of a century of neglect during the civil wars of the 1500s and a few years after World War II, when the United States occupation was having spiritual doubts about the advisability of helping the demigods of Japan's ancestral faith find a new home.

It says something about the Japanese psyche that the rebuilding custom has been so faithfully kept; the twenty-year intervals between the ceremonies represent a convenient span for stock-taking, semicolons in history. The Japan that rebuilt Ise in 1953 with private contributions—the first time in history that the shrine was not state-supported—was just starting its extraordinary period of growth and recovery. The Japan of 1973 could afford to spend far more lavishly on a semipublic celebration, standing at the height of its postwar wealth and attainment.

What kind of society will send its representatives to Ise in 1993 to bring the fresh white pebbles, cut the timbers, and prepare the new replicas of its ancestors' treasures to furnish the shrines? How changed a country? How changed the people? Or will they merely continue at the shrine as always, changing only the physical, touchable things—the beams, the thatch, the vestments—as if to mock change, in the midst of engineering it?

In discussing some possibilities and alternatives for Japan in the coming years before Ise is rebuilt, it is wise to avoid metaphors involving crossroads, turning points, or western-style compass resettings. The Japanese are not rendered logically unhappy by sudden shifts of direction or by pursuing several varieties of culture and politics at the same time. That, indeed, is their peculiar genius.

This ability is most strikingly demonstrated in their art. In the introduction of his book, *Form, Style, Tradition*, Kato Shuichi, one of Japan's few original modern commentators, has written, "Different types of art, generated in different periods, did not supplant each other, but co-existed and remained more or less creative from the time of their first appearance up to our time. Buddhist statues, a major genre of artistic expression in the period from the sixth to the ninth century, continued to evolve in style during the following eras, even when the picture scroll opened new possibilities for the visual representation of the world in the Heian period. Brush works with India ink flourished during the Muromachi period, but one school of artists remained faithful to the techniques and style of the picture scroll. Under the Tokugawa regime a new style of painting and decorative art was established by Sotatsu and Korin, the technique of the woodcut print, and was elaborated to perfection. Yet artists never ceased to carve Buddhist statues or engage with great passion in brush-work painting. No major art form was actually replaced by another. Practically no style ever died. In other words, the history of Japanese art is not one of succession but of superposition."

Superposition is the operative word in Japanese politics, business, and society, as well as art. Like the Buddhist and Shinto altars in the same room, the culture of the West has come to live in Japan beside its ancient arts; the logic of the huge factory exists with the illogical dependence feelings of its workers and the clutter of the family machine shops and small store-front subcontractors that supply it. The face-downs of parliamentary politics are worked out within the society of consensus. Rough edges are trimmed only, so that they all fit under the same roof.

It is with this caution that I hazard a few speculations about the immediate Japanese future.

STRONG BUREAUCRATS, WEAK POLITICIANS: Japan is hardly alone among the world's societies in suffering from the most elemental of political problems: the basis of confidence between government and the governed has been severely shaken. In Japan, America, and the Western European countries the problem bears little relation to the simpler issues of freedom versus oppression. Broadly speaking, it is a matter of social ecology. In the same way that technology has suffocated cities and blighted countrysides to affront nature's balance, its attendant problems of mass urbanization, escalating industries, and enveloping communications have proved to be more formidable ones than modern governments were designed to cope with, or than modern citizenries can be expected to vote upon with intelligence.

Japan's social ecology has not been disturbed at the local level nearly so much as that of societies in the West. As we have noted, the tendency of the Japanese to cut their society into vertical segments has absorbed the onrush of modern invention, and adapted itself to it. The segmentation, we have seen, cuts through all types of society—business, local government, the schools, the professions, labor, and what is left of the farm. Within each segment people have carried on their local hierarchies and remained surprisingly true to their old feelings for group loyalties and commitments.

In recent years, however, Japan's mass communication society has worked on the one hand to weaken old loyalties and securities, yet failed to develop a strong sense of confidence, not to say participation, in the national government. Consensus has been slower in coming. As Japan's affluent society rediscovers its sense of limitation, disagreements which were repressed or papered over during the breathless drive to expand GNP continue to surface. Such political conflicts generally stem from local pressures or economic needs. They involve local or sometimes national anti-pollution or anti-noise movements, consumer activity, pros and cons of nuclear power development, criticism of big business or demands for greater expenditure for public welfare, education, housing, or civic development— all areas which were scanted by the emphasis on heavy capital investment in the high-growth period.

Ideology is rarely a major factor. The storms on the left over the Security Treaty with the United States have died down since the sixties (see below). Looking toward the eighties one can hardly help agreeing with the late Eda Saburo's comment that "the age of ideology is dead," in a country where more than 90 per cent of the citizenry regards itself (when polled) as "middle-class." Yet the effect of the ideological struggles of the sixties lingers on. For by insisting on their own brand of unrealistic "revolutionary" politics, the leaders of the left opposition hindered the development of responsible partisan argument within the framework of a democratic party system. There is sadly little of this today in Japan.

On the contrary, both the government Liberal Democrats and the opposition parties are jostling each other in their rush to appear safe, sane, and centrist. The Japanese voter would seem to want things this way. Lockheed scandals and export crises notwithstanding, he is still basically satisfied with what he has got and is looking for more of the same—rising affluence, greater comforts, the new sense of national prestige that comes from sheer industrial achievement. He has been encouraged in this attitude by the long succession of Liberal Democratic governments, which have broadly speaking delivered on their promises of greater affluence. There is little doubt, as we noted previously, that the Liberal Democratic majorities will continue to shrink, ultimately to be cut down by sheer voter ennui, if nothing else. They will yield either to an out-and-out coalition government or a new re-grouping of the parties, with the majority strength still in the center.

Recent history of the major Japanese parties shows a steady trend in this direction. Fate has not generally been kind to people trying to steer their parties in the direction of extreme or even sharply distinctive policies, whether right or left. Even temporarily popular American extremes like the Goldwater or Reagan right or the McCarthy or McGovern left have had no parallels in recent years in Japan.

Two off-shoots did sprout from the tangle of Liberal Democratic Party factions. The Seirankai ("Cool Wind Society") was formed in 1973 by a group of vocally right-wing Diet members, who advocated more active nationalist politics, to the point of considering nuclear armament, and threatened for a while to be developing an ideology of their own. They have been relatively inactive in recent years, except to speak out vociferously for Taiwan or the Republic of Korea. On the other end of LDP's factional spectrum, the New

Liberal Club for a time projected an image of some vigor (principally because of the relative youth of many of its candidates) and its 'clean government' slogans attracted many conservative voters who were repelled by the Liberal Democratic leadership's heavy association with the Tanaka-Lockheed scandals. As of 1978, however, there seemed increasingly little to set the New Liberals apart from the old ones.

The Socialists had a similar split when Eda Saburo, frustrated by the doctrinaire Marxist line of the majority, founded a new Socialist Citizens Alliance (*Shakai Shimin Rengo*) in 1977. (Another veteran Socialist, Den Hideo, later took a second group outside the Party.) Eda died soon afterwards, however, and most Socialist Diet members, in any case, were too beholden to their tight groups of supporters in the left-wing union bureaucracy to put forth any realistic new platform proposals. It is doubtful whether the Socialists will ever return to their role of leadership in the opposition. The Communists, who suffered heavily in the 1976 elections, have had their own widely publicized scandals to hurt them. Comparisons with the active Communist parties in France or Italy are largely irrelevant.

In a sense, this urge toward the center is a happy augur for the future. Japanese politicians on the stump can say the same soothing things about the good sense of the voter that Americans do. Similarly the distrust of government apparent in so many aggressive local rights movements is no more abnormal than in the United States. In his basic sense of democratic rights, if not duties, the Japanese citizen is far closer to the American than the latter realizes. No American, however, has to bear the blight of a quarter century of one-party rule, with the leading party of the opposition abdicating its role in the democratic debate, and with it its chances for ever attaining power through the electoral process.

In its governance of its country the Liberal Democratic Party took the easy path—ruffle as few feathers as possible, get a consensus by avoiding the big problems, stay in power at all costs—tactfully rotating its factions to share the political wealth. No American or European political party could have gotten away with this niggardly solution to its country's problems; but the Liberal Democrats were blessed with one of Japan's few natural resources, a developed bureaucracy—intelligent, dedicated, hard-working to the point where its arrogance seems almost justifiable. For almost two generations—some would say for two centuries—the bureaucracy has ruled Japan, in a rather Confucian solution to a modern problem.

The trouble is that a bureaucracy, even the world's best, is rarely self-programming. It needs some strong political direction, ideas, policies. It is not in itself a breathing political animal. Only the combination of a confident, competent bureaucracy, inexorably spinning its wheels, and a rather lackadaisical political leadership could produce a running disaster like the Narita airport fiasco, where an ill-advised and ill-planned airport, its very site a political football, has cost more than $3 billion dollars, while having its opening postponed for the greater part of a decade by a group of local farmers and a few remnants of the rough radical student groups of the sixties. The problem could have been solved by a combination of strong political direction, a sensibility to local concerns, and good staff planning. Yet successive cabinets merely ratified an originally bad decision, fortified by a bureaucracy naturally reluctant to interfere with any project on which a generation of past superiors had put its stamp.

Politicians and bureaucrats need each other. In the United States, the politician contributes far more than he does in Japan or even Europe to the governing process, if for no other reason than that the American bureaucracy is inferior to most. In Japan, where the best and the brightest enter the bureaucracy as a matter of honor, the shoe is on the other foot. In America, Japan, or any working democracy, government works poorly when the healthy tension sags between elected politicians who make decisions and the bureaucrats who implement (and almost make) them. Years of Japan's one-party regime, in which most premiers have been former bureaucrats, have thrown off the balance in favor of the bureaucracy—and conferred on the bureaucrats in the ministries an immunity and a semi-decision-making role which they should not have. Thus an asset becomes a liability. In a world where decisions must be made fast—and accountably—this overweighted bureaucracy has become a problem in spite of its tactical efficiency. The bureaucrats do their jobs too well. The politicians do their jobs not enough.

The dangers to any democracy in this disproportion should be apparent. The years between now and 1993 will see more political emergencies, demanding more quick decisions and sensible public communication than any bureaucracy can be expected to provide. If the current system cannot perform, if the planners are unable to guide Japan's transition to a "post-industrial" economy with greater social services and more consumer demand, if the government is unable to negotiate out of its troubles with foreigners inundated by the Japanese export floods, if the engines of new production do not turn over sufficiently to put the idle and semi-idle back to work,[1] then the electorate may in time grow restless and look for people with solutions outside of the center. It is my conclusion that such people or movements would come from the right, rather than the fractionalized and intellectually bankrupt left.

INDUSTRY AND CREATIVITY: The stability of Japan's future politics, insofar as a dangerous polarization can be averted, depends at least in the short run on the health of Japan's economy. It is my own judgment, after both watching this economy and living in it for more than a decade, that it has: (1) the virility to continue its steady growth, but at rates more like 5 to 6 per cent annually than the 10 to 12 per cent ascensions of the past; and (2) the adaptiveness to make the shifts necessary to insure continued health and growth. The word "adaptiveness" I use in its widest sense and it is an important condition. For Japan's future economic success depends not merely on shifts in industries and trade patterns, but on more basic changes of outlook and interest in the whole society— changes that go beyond mere economic planners' capacity to effect.

It is increasingly evident that the pattern of industry must change faster. With energy and labor both in short supply, it behooves Japan to hasten its shift toward the knowledge-intensive industries, continuing the upward spiral of technology that one hopes is assumed to culminate in nations of enlightened executives working four-day weeks, tending their machines and computers, with most of the blue collars worn by college kids at examination time. In this regard ex-Premier Tanaka's blueprint, from his book, *Building a New Japan*, is worth quoting: "We should develop R&D—intensive industries that utilize knowledge, technology, and ideas (computers, aircraft, electric automobiles, industrial robots, and marine development), sophisticated assembly industries (communications equipment, business machines, antipollution devices, and educational equipment), fashion industries (sophisticated clothing, furniture, and household utensils), and knowledge industries producing and marketing knowledge and information (information-processing services, video industries, and systems en-

---

1. It is estimated that in 1978 about 1.5-2.0 million Japanese workers on company payrolls would already have been laid off or fired in other capitalist economies.

gineering), and at the same time have other general manufacturing industries become more knowledge-intensive through process sophistication and product improvement. Advancing such a course of development requires that we develop new techniques and train personnel suited to the needs of the new age. In the knowledge-intensive industries, the quality and capability of people is the key factor in development."

But knowledge-intensive industries, including the kind of sophisticated decision making and planning required of modern governments, demand more than quantities of good people. They require the cooperation of the intelligensia, the official bureaucracy, and business. The three compartmented segments of Japanese life must work together, for there can be little slack in planning for the eighties and scant margin for error. Insofar as the intelligensia hesitates to cooperate with government and business, Japan is heading into the twenty-first century with one hand tied behind its back. Nor have business and the government, until now, distinguished themselves in wooing the nation's intellectuals. As we have noted, in no major country of the world is this compartmentalization so sharp, the lack of cooperation among units of the vertical society so glaring and unproductive.

Urging the intelligentsia's cooperation with government need not imply a repetition of the continuing American pattern, where presidents and other prominent politicians try to collect Harvard professors the way people in the twenties used to keep race horses; nor am I suggesting that Japanese professors should start carrying an ambassador's baton in their bookbags. The strains of the Vietnam War snapped many ties between government and professoriat. But for all its abuses, real and fancied, the cooperation between American academics and their government has been, on the whole, of great benefit to the United States, and the world.

Man does not live by engineering alone, or by other people's patents, either. The lack of effective communication between government, business, and the academy is not merely a social failing, it is one of Japan's most serious potential limitations as a modern industrial great power. In politics as well as business, Japan suffers badly from a lack of creative idea power partly because of this massive communications failure.

This half-sundered world of compartments and private principalities is, after all, the vertical society's outstanding defect, just as the loyalties, devotion to tasks, intragroup cooperation, and sense of solidarity it engenders are its obvious strengths. In March 1974 the entire country joined in welcoming the long sought-for Lieutenant Onoda, the solitary Japanese soldier who had held out for almost twenty-nine years in the jungles of Lubang Island in the Philippines (and had killed numbers of Filipino villagers in the process) because he had not been directly ordered to surrender by his superior in 1945. Although Onoda had, in recent years, found out something of the world through radio and chance encounters with Japanese tourists, he remained faithful to his original orders not to give up even if an official surrender announcement were made from Tokyo. In the end, his former wartime superior flew down to the Philippines, with a party of Japanese officials, and showed him surrender documents. None the worse for wear, he returned to a triumphal welcome by a country mostly curious but partly fascinated by this case of the Japanese *giri* religion carried to its extreme. The emperor could hardly have picked a better man for the job, at the time.

Yet the memory of the war calls back the other side of the coin, as well. Not merely the cruelties, conscious and heedless, inflicted by Japanese officers and soldiers on presumed lesser breeds outside the law, but the gaping divisions between segments of Japanese society itself. War brings out the extremes in any national character. The same society that instilled the obedience of an Onoda produced a war machine that was hopelessly out of gear because of interministry and intercompany squabbling, a war in which the lack of cooperation between the army and the navy reached ridiculous extremes. Probably the *reductio ad absurdum* was the decision in at least one battle, Okinawa, for the army and land-based navy elements to fight their last-ditch battles separately, without reinforcement or communication between them.

It was the survival of this compartmentalization of society that was so cogently attacked by the Nobel Prize-winning physicist Ezaki Leona, who coincidentally arrived in Japan not long before Onoda, on a visit from his home in the United States. "Wouldn't you say," he asked the novelist and commentator Shiba Ryutarō, in the course of a press discussion, "that very few dialogues or meaningful conversations take place in Japan? Japanese are not really happy with dialogues; with them they tend rather to become quarrels. Yet it is in the dialogue, the thrust of thesis and antithesis, that creativity is nurtured. Because there is no real ethical receptivity for the dialogue in Japan, creativity itself is hampered. The lack of meaningful dialogues is one major reason why Japanese scholarship has only developed in spots and segments. When something new in scholarship appears in America, American scholars like to hear about it and discuss it, in preference to reading the details in printed matter. They like to grasp its significance as a living, moving thing, so to speak. . . .

"Of course, Japanese physicists can conduct dialogues with other physicists. But as far as a scientist conducting a dialogue with a novelist or a company director, Japanese don't relish that sort of encounter. . . . Yet when the same kind of people are always gathered together on the same kind of occasion, the talk really comes to nothing."

Ezaki has written and said more detailed things on this subject, but his comments above underline the problem. Unless Japanese scholars learn to handle interdisciplinary research and problems—and can join the other segments of their society in solving them—Japan's hopes for developing its knowledge-intensive industry will never become a reality. Nor will Japan take the place in international scholarship and cultural exchanges that the sum of its talents, individually considered, deserve.

INVESTMENTS, NOT JUST EXPORTS: A brand-new Honda plant is going up in Ohio; a Kawasaki motorcycle plant in Lincoln, Nebraska; YKK zippers are manufactured in Georgia; Sony is already building television sets in San Diego and turning out tapes in Alabama; while Matsushita's acquisition of Motorola's television business in 1974 makes it the third largest seller of television sets in the United States domestic market. Mitsubishi, Teijin, and Nippon Electric in Brazil, Nissan in Mexico and Chile, Bridgestone Tire in Singapore, and Sanyo Electric in Indonesia all foreshadow the new thrust of Japanese industry outward, toward multinational operations. By the end of 1976, Japan's total direct investment in foreign countries approached $20 billion. The best MITI estimates project an aggregate of over $30 billion by the end of 1980 and $80 billion by 1985. The $80 billion figure would come to something over half of *current* American investment abroad, but the percentage of gain is impressive. Comparing investment balances for 1966 and 1975, for example, Japan's grew 16 times, compared to the American growth of 2.4, while Germany, the next growth contender in investments, increased by a magnitude of 6.4.

The impulse to expand production overseas, as well as selling, had been germinating in Japanese business planning long before the energy crisis of 1973. It was made inevitable by rising production costs inside Japan, as well as by the tightening labor shortage. American business' dramatic, if sometimes nearsighted, successes in multinationalizing were long studied. The problem not only of raw material scarcities, but of substantial tariff resistance to Japanese exports in the United States and Europe hastened the trend. The rising protests against Japanese economic imperialism, centering in Southeast Asia and Korea, added another dimension to the multi-national business effort. It has become apparent to the Japanese, just as it was drummed into a whole generation of would-be interna-tionalized Americans, that one cannot indefinitely use overseas plants and branches as crude profit-making operations, but one must, on the contrary, be prepared to give to the host country as well as take from it.

About the market acceptability of Japanese products there is no question. Japan's share of Southeast Asia's total trade [3] by the end of 1974 was estimated at 30 per cent; by 1980 it is projected to pass the 40 per cent mark. Its imports, it is true, are rising by 8 per cent yearly. Indonesia supplies 15 per cent of Japan's oil supplies and 24 per cent of its bauxite. The Philippines provides Japan with more than 40 per cent of its copper. But the trade balance remains strongly in favor of Japan. Thanks to the huge in-pouring of manu-factured goods over the past fifteen years, countries like Thailand and Indonesia are virtually in hock to the Japanese.

The aid and loan assistance given by Japan to Southeast Asia, in particular, is considerable, but hardly commensurate with the size of the investment. Most of it is in the form of services and manu-factures which must come from Japan. ("This is not aid, it's trade promotion," a prominent Malaysian statesman once remarked.) Plans for a $20 billion Japanese Marshall Plan for the area advanced by the economist Okita Saburo were vetoed as too ambitious. Al-though aid and credits continue to flow, they were administered until recently in a selective way, with little pretense at developing the area. Most Japanese publications make much of the statistical just under 1 per cent of its GNP which Japan allots for overseas aid,

3. As used here, the geographical designation includes Indonesia, Ma-laysia, Singapore, the Philippines, Burma, Thailand, Taiwan, Vietnam, Cambodia, Laos, and Hong Kong.

but go light on specifics like high interest rates and preclusive buying arrangements to which much of the aid is tied.

Nowhere so much as in their overseas trading in Asia (as distinct from the United States and Europe, where they pay considerable attention to local ground rules), have the Japanese so obviously shown themselves frustrated politicals. They are territory seekers, in a sense, in merchants's clothing. In the course of their quest, their tendency to live and work in enclaves apart from the local populations has brought criticism almost everywhere. Their exclusiveness was no small factor in provoking the riots which greeted Premier Tanaka on his visit to Southeast Asia early in 1974, the pent-up results of Asian resentment at what some describe as "economic colonialism."

This, again, is reminiscent of World War II, when the Japanese brought armies into countries which at first looked on them as liberators, but then proceeded to alienate the natives by regarding them as just that. There was no nonsense about equal treatment then. The classic injunction of Premier Tojo Hideki to "respect the opinions of the natives and to take a true fatherly attitude towards them" is redolent of British India at its most condescending. The accompanying cruelties of the Japanese army managed to alienate Thai, Burmese, and Indonesians, among others, who had been disposed to accept the original propaganda of Asia for the Asians at its face value.

The same mistakes have been made by the next generation after World War II. For the accounts of the clannishness of the Japanese, their indifference, if not contempt for the local populations, and their exploitation of local markets rival the worst criticisms of American business coca-colonisation. Some of them are amusingly similar.[4] But the indignation that spectacularly surfaced in the Thai boycott of Japanese goods in 1972—and had similar echoes in Indonesia, Korea, and the Philippines before and after—was no laughing matter. Prime Minister Fukuda's visit to the five ASEAN countries—Indonesia, Malaysia, the Philippines, Singapore, and Thailand—in 1977 was a major step by Japan in improving its relations, as was the announced $1-billion loan for ASEAN development proj-

---

4. Like the storied American overseas—if anything worse—the Japanese, according to a 1973 questionnaire in Thailand, are denounced as loudmouthed, clannish, lecherous, unwilling to trust local people, boastful, and chauvinistic, with the additional un-American attributes of being stingy with change and overly fond of conferences.

ects. The so-called Fukuda doctrine has given belated recognition to the fact that an "equal partnership needs to be stated and implemented between Japan and these countries. As with the increase in investment in the United States, however, the effort in the Southeast must be made wholeheartedly, in the literal sense of the word, or it will result only in more bitterness. The aid development programs and increasing investment of management and capital overseas represent Japanese business' great area of expansion for the future decades, if not also a guarantee to its survival as a vigorous factor in the world's economy.

AN INTERNATIONAL ROLE: Let us assume that Japanese business harnesses creative resources, continues its multinational direction, changes its spots, and fields a corps of dedicated profit-sharing internationalists—hardly an easy transition to envisage. Yet even such a drastic economic redirection will be of little use, if it is not accompanied by a new activism in Japan's foreign relations. This does not mean a new diplomatic attitude, nor necessarily a change in present alliances and relationships, so much as it means a national awareness that Japan has a political mission in the world, which is as much a part of its great-power status as the business of producing, buying, and selling.

No one is discharging this mission now. Such a mission is made necessary by a world that has turned far more than most of us realize in the direction of one roughly unified international society.

The world power blocs are cracking and weakening. However tempting it may be to base one's diplomacy on the big-power balance idea, the time may soon come when the current balance of power will lose its relevance. The gap between the underdeveloped countries and the developed—the North-South problem so beloved of Japanese symposium participants, is growing, not decreasing. Force is an increasingly precarious arbiter, as Israelis, Americans, Indians, and even the Soviet security police have begun to discover. People with resources in their land are finally realizing their value, as well as the blackmail potential inherent in their possession. The sheer pursuit of affluence in countries like the United States and Japan, where this is a problem, is becoming wearying to the peoples involved, especially their younger generations.

In one sense, alignments are becoming more fluid and countries are more responsive to economic pressures. In another, wealth is becoming hostage to a politics that is played in many different ways. New arbiters as well as new spokesmen are needed.

Since Japan has become for the time a business society, perhaps the easiest way to present the political challenge is to draw up a job description:

WANTED

A nation of long history and proven competence, whose achievements command some respect, to represent the non-big powers of the earth. Since the spokesman's principal role is to advocate peaceful solutions of problems, it should be unarmed, or possessed of only a small armed force. It should be committed to free institutions, although not insistent on the way other societies order theirs. It should not be European, since Europeans are more or less permanently type cast in the mind of Asia and Africa, as exploiters rather than civilizers. Yet it should be a developed, not an underdeveloped, nation, secure enough in its own achievement to invite some constructive imitation, yet sympathetic with the problems of rapid growth and desired change. Although committed to basic individual freedoms, it should ideally be a collectivist society, instinctively sympathetic to the collective and group solutions which seem in keeping with growing scarcities.

The prospective candidate should be at least relatively free from the past commitments of the Cold War, not by any means because the Cold War's problems were illusion, but because the battle lines of that war have become increasingly less relevant to a great proportion of the world's inhabitants. They are more occupied with problems of subsistence.

Candidate's people should be industrious, well-disciplined, and highly intelligent, able to understand long-term goals as well as short-term advantages. They must be dedicated to permanent world peace and free intercourse among nations, not as merely desirable, but as essential things. They must be well educated and disposed to understand the problems of people other than their own. They must be, above all, convinced that the job is secure, necessary, and satisfying as well as challenging. They must then be prepared to unite behind their representatives, not only in fulfilling their missions, but in convincing the rest of the world they have a right to it.

Compensation: Open. Good chance for advancement for the right country; although no bonuses from the stockholders may be expected.

The mission of spokesman for the small, the unarmed, and the underdeveloped has been taken over by different countries, at varying times, with relative ease. For some years postwar Jawaharlal Nehru assumed it for India, principally on the basis of his own commanding personality, despite the fact that his country was disunited and badly administered, enjoys an almost hopeless gulf between a small ruling elite and millions of the poor and unrepresented, and has never shrunk from military solutions to problems (e.g., Hyderabad, Goa, Pakistan). President Tito of Yugoslavia at one time pre-empted the role quite successfully, although as the head of a European semipolice state, his actual affinities with those he claimed to represent were small. Gamal Abdul Nasser of Egypt and Sukarno of Indonesia wrapped themselves in the Third World mantle largely for domestic purposes, but with considerable success. Most recently, there is China. The China of Chairman Hua has emerged in the late seventies as a far different force than it was before. Nonetheless the sight of China, with its growing nuclear arsenal assuming in the United Nations the role of spokesman for the weak and unarmed is irony of some degree.

Japan is another matter. As an international spokesman, the very weakness of the fragile superpower can be an asset. Japan may be only a half-power, as so many Japanese are fond of stating. But this is true only in terms of obvious military force. Add Japan's human and technical resources to any of the existing great powers, or power blocs, however, and that power's strength, for war or for peace, would double, even by current standards.

As a permanent member of the United Nations Security Council, and Japan should so become, it would be the only one without nuclear weapons. In a world of nuclear superstition and nuclear stalemate this is not an inconsiderable asset. Japan could speak for many; and there are many who would welcome such a champion.

The Japanese, to put it mildly, are not generally thought of as a nation of "one-worlders." In view of their demonstrated insularity, particularly, such a forthright stance sounds unlikely or preposterous. It would be far more preposterous, however and dangerous to continue Japan's postwar role into the future. Between 1953 and 1973 Japan's diplomacy was almost totally subordinated to its policy of economic growth. If abasement or compromise was necessary to secure commercial concession or avoid public confrontation, it was duly performed. From Japan's passive me-tooism, following America's lead at the United Nations, to the tons of humble pie eaten by Japanese suppliants, commercial and diplomatic, visiting Peking, to the embarrassingly soft handling of the dictatorial South Korean government in the Kim Dae Jung case [5] the foreign policy of Japan resembled nothing so much as the labor relations policy of its big companies. "If the union makes demands, give them what they want in a hurry, and let's get back to production." This is not to say that Japanese diplomacy is not skillful or highly professional. In fact, it achieved its objectives for the growth period. The trade lanes were kept open. Japanese goods kept flowing out to the world and imports kept coming in, while the affluent society could entertain itself with the loud rock choruses of abundance within its bustling islands.

This determined policy of low-posture diplomacy, while doubtless helpful in the short run to Japanese business interests, won Japan few friends. Japan's general diplomatic diffidence served to diminish the influence even of worthy plans or proposals. The national preference for group action, plus constant shifts in foreign ministers, have given the world almost no actors to dramatize Japan's presence on the world stage. There have been no Chou En Lais, Anwar Sadats, Willy Brandts, Henry Kissingers, or Harold Wilsons—not even an Andrei Gromyko. Never in modern history has a country with such wealth and breadth of relationships, a major power by any consideration, projected such a squeaky international voice.

This policy is now dead, partly because it was too successful. The short-term aid programs, the huge preclusive buying, the grabbiness of Japanese market seekers have registered on the rest of the world. They are now resisted. Especially as resources dwindle, economics, including the give and take of trade, has been hopelessly infiltrated by politics. Not only in the Communist countries has economics been reduced to what Clausewitz called "a political instrument." The cherished Japanese doctrine of *seikei bunri*, "keep politics and economics separate," has been a dead letter for some time. The petroleum crisis provided its epitaph.

5. In 1973, Korean government hoodlums kidnapped Kim Dae Jung, a former Korean presidential candidate and opposition spokesman from his hotel in Tokyo and smuggled him to Seoul, where he was put under house arrest and refused permission to leave Korea. Japan did not push its demands to return Kim to Tokyo, doubtless with an eye to the big Japanese business investment in Korea—although the whole incident was a flagrant violation of international law and human decency. Later, in 1974, two Japanese were arrested and tried for subversive activities in Seoul, under similarly dubious circumstances. Again, Japanese protests were weak and ineffectual.

Looking toward the future, from the vantage point of the late seventies, one can see signs that the Japanese are changing. The "political side" of economics has become an often used phrase in Japan, particularly during the export crises of 1977–78. Not only the Fukuda Doctrine in Southeast Asia, but the overtures Fukuda made to the United States, during his 1978 visit, for mutual development of nuclear fusion and other energy projects, are bold statements, if we compare them to the low-posture Japanese policy of former days. So is the resolute defense of Japan's and other countries' right to their own brand of nuclear development in the Tokai Mura confrontation with the United States in 1977. Japan's ambitious funding of the United Nations University is another heartening signal. It is a matter of regret that the United States and other developed countries have been so niggardly in adding their own contributions. Whether the eighties will see the emergence of a more articulate Japan, capable of explaining its power as well as directing it, is an open question. But the successive international currents of anti-Japanese antagonism in the late seventies demonstrate the vulnerability of a nation content merely to react to pressures and movements around it.

So, in fact, the Japanese people may have to answer that political want ad, or one like it. In any case, their old job has been disappearing from under them, one of the few cases of plant obsolescence they have had to cope with.

JAPANESE, AMERICANS, AND SECURITY: The protection afforded by the U.S.-Japan Security Treaty of 1960 is the basis of Japan's defense and security. All hopes for Japan's unique international position as the one unarmed superpower, are premised on the Security Treaty's guarantees. They have been accepted by the Japanese government and the great majority of its people. Despite a past history of riots, protest marches, and steady opposition from the Left through the sixties, the treaty has come to be thought of as part of the national furniture—a bit unsightly in spots, but better to have than not. By the mid-seventies opposition parties, including both Socialists and Communists, had virtually ceased using it as a campaign issue. The American bases remain, if on a reduced scale. Agitation had died down even over the presence of two-thirds of an American Marine division on the over-crowded island of Okinawa. Certainly few people, writing about Japan in the mid-seventies, would have thought much comment necessary either about the continuance of U.S. bases in Japan or the validity of the American security guarantee.

This is no longer so. In a few years the perspective has changed. As they look at the decades ahead, Japanese who once worried that the U.S. military presence might drag their country into a big-power Asian war are now troubled that a continued weakening of that presence will invite either aggression or irrestible pressures.

The ultimate source of the threat is obvious. Since the mid-sixties, Soviet ground and naval forces in the Far East have nearly doubled; air strength has significantly increased. The Sea of Japan, by the admission of U.S. Navy commanders, has become a virtual Soviet lake. While the land build-up could be understood in terms of the Sino-Soviet border confrontations, it is hard to extend this reasoning to a power display in which Soviet naval and air forces are continually on patrol, east of Japan as well as to the west. To match the military increases, Soviet diplomacy has grown continually harsher, particularly since the Japanese failure to advance sizeable investments in Siberia. The Soviet behavior in the fishing negotiations, so vital to Japan, seem more and more like studied exercises in humiliation. By the evidence of their eyes and ears, the Japanese find little detente in Northeast Asia, where they happen to live.

By contrast, American military power continues to shrink. The number of U.S. aircraft in the Japan area is little better than half what it was ten years ago. The U.S. Navy's strength in the area is now only half that of the Soviet Union. In 1977, the Japanese Defense Agency published a rather unusual White Paper which among other things noted the fact that "a change is occurring in the military balance between the United States and the Soviet Union." In the history of the Defense Agency, surely the most docile of the world's war ministries, this was the first time such a concern had been voiced.

Although the White Paper pointed out the growing Soviet superiority in nuclear weapons and the need to modernize and improve the American counterparts, the real problem lay not so much in ultimate nuclear capabilities or the amount of conventional armament the U.S. is able to deploy. A great many Japanese are concerned not so much about questions of tonnages, aircraft, and general military capabilities as they are about the political intentions of the United States, the American national will, and how it relates to *them*.

President Carter's decision to withdraw U.S. troops from Korea, announced in 1977 at the very outset of his administration, was the catalyst that turned vague concerns into real worries. The Japanese were not consulted about this decision; they were informed. They made no public protest, although they managed to convey a sense of uneasiness—at least to those Americans who cared about real reactions, as opposed to official acquiescence. When the pull-back, which apparently resulted from a piece of campaign strategy, was officially slowed down—more in response to bad reactions in the United States than in Japan—the Japanese duly expressed relief. But the almost casual ease with which a thirty-year-old American policy was changed, has caused some soul-searching.

The Japanese had counted on an American presence in Korea not merely to check the real danger of aggression from the opportunistic Stalinoid regime in North Korea, but equally as much as a check on the moody opportunism of South Korea's President Park Chung Hee, whose arbitrariness has hardly mellowed with time. Japan has an immense stake in the peace of the Korean peninsula, not only with its economic investment in the south, but because any political trouble across the Tsushima Straits is too close to home for comfort.

The total strength of Japan's Self Defense Forces is minuscule compared to its neighbors'—an army of 153,000 men, a Navy of 200,000 tons, an airforce of some 830 planes. The defense budget is kept within 1 per cent of the total GNP and less than 7 per cent of the government's general accounts budget. Thusfar any temptation to think of nuclear weaponry has been resisted; the American nuclear umbrella remains Japan's ultimate security guarantee. But there is an undercurrent of worry over Japan's deficiencies in conventional armament. If there were trouble in Korea and North Korea were backed by the Soviet Union or China, would the United States intervene? Now? In five years? In ten? Would even a negotiated peace, short of all-out war, result in the closing of the Tsushima Straits. Would an American Navy pull-back follow the announced pull-back of ground forces? If so, would Japan's life-lines of trade—oil and other raw materials included—be hostage to the slightest Soviet pressure?

These may be groundless fears when contemplated from the safe vantage points of Chicago or New York. Viewed from Tokyo, or Sapporo, or Shimonoseki they may seem more like possibilities. To the extent that Japanese start worrying about such matters, they begin to work their way into diplomacy and national policy.

The Chinese have their own fears about Soviet aggrandizement and they have been communicating them to Japan. Since 1976 a succession of Japanese military personnel and advisors has been invited to Peking, by the same Communist bureaucrats who until recently criticized the existence of the Self Defense Forces as evidence of dangerous revanchist tendencies. The Chinese have invited visits of Japanese training ships and asked for Japanese opinions on Soviet Naval build-ups. "It is a good thing for an independent nation to have its own self-defense forces," one Peking official was reported as telling a visiting Japanese mission. "It is understandable that Japan advance her foreign policy with the U.S.-Japan Security Treaty as its basis. Yet, isn't it a bad thing that Japan is so completely dependent on the United States?"

The dependency preys on the mind of many Japanese, especially when American discontent over matters of economics, trade, or nuclear policy, for example, Japan's alleged flouting of President Carter's cherished non-proliferation ideas, is expressed in such polemical terms. How reliable is the American commitment? is a question which has not been asked in Japan for many years. It will be asked in the future, more often.

A NEW GENERATION: The net effect of what has been written in these pages has conveyed, I would hope, the picture of an extraordinarily gifted people, of tremendous energy, with a rare facility for harmony and constructive, almost spontaneous, collective effort. No country in history has adapted itself so brilliantly to changed situation and new invention and lost so little of its own identity in the process. The economic success story of the sixties and seventies was a fitting climax to the Meiji century of learning, advancement, and, on occasion, aggression that made Japan a factor in the world.

The Japanese will never be like Americans, no more than the French, the Mexicans, or the Russians. Their attributes, as people and as a nation, are unique. They should be studied in our schools, read in our libraries, and entertained among us, not as random curiosities, but as members of our world society, who have developed a distinctive relationship with the United States. They too have become People-to-Learn-From.

Nonetheless, they face formidable problems in the near future. Japan is possibly the most goal-oriented country on earth. As Robert Bellah put it, "[it is a society where] performance or achievement becomes a primary value." [6] If the kind of performance changes, will the value change with it?

It was easy for the Japanese as a nation to work together during the days of spectacular economic growth. The whole country developed—and indeed maintains—a kind of athlete's pride in the sheer physical achievement. The momentum of past successes, their own growing efficiency, and a peculiarly Japanese capacity to absorb any amount of shock, as long as their direction is unchanged—these attributes all helped the Japanese surmount the crises that capped the high-growth era, when they came. Come energy shock, Nixon shock, Soviet threats, or export problems, the society shook itself and rallied. Plans were revised, new positions taken, and the pursuit of happiness in the archipelago continued.

But as Japan faced the eighties decade there was for the first time a sense of staleness about the political and business governance of the society. The generation which had come to early maturity in time to run the post-war economic miracle was visibly tiring. The reactions were slower. The step was heavier. There was a sense of marking time, in the cultural life of the country as well as in its work.

What of the next generation? How close was the youth of Japan to coming into its own? How much sense of identity and continuity did it share with its elders?

On the face of things the signs were disturbing. In the summer of 1973, for example, the Prime Minister's office had commissioned a "World Youth Consciousness Survey," with the help of various international research organizations, that compared the current attitudes of Japan's younger generation (aged eighteen to twenty-four) with the youth of ten other countries.[7]

A society where production is given top priority and individual happiness ignored; a national concentration on what is good for the country as a whole, to the exclusion of individual hopes and aspirations; a political system that runs counter to people's hopes; a materialistic way of life that thinks money can solve any problem; a society that treats its aged shamefully—these were the major dis-

---

6. In his notable book, *Tokugawa Religion*, Bellah described the system of values riveted on Japan during the 250 years of the Tokugawa Shogunate, which have remained surprisingly intact to this day.

7. The United States, Brazil, the United Kingdom, France, Sweden, India, the Philippines, Switzerland, West Germany, and Yugoslavia.

contents of the young Japanese. Although the youths of other nationalities polled had similar complaints, their percentages were comparatively low. Only in Japan did the pessimism run so heavy. Among Americans, West Germans, and Swedes, the percentage of "dissatisfaction with your way of life" ran about 45 per cent. Japan's was 73.5 per cent, by far the highest of the group.

Japan's youth were particularly cold toward their government and their own part in it. When asked if they would take active steps about their dissatisfactions with government, fully 73 per cent of the Japanese young people answered, "The problems are beyond the reach of individuals." This feeling was reflected in other government polls, taken on more specific levels. Although almost half of the young people in the Home Affairs Ministry's sampling were dissatisfied with politics, only 15.5 per cent of them supported a particular party. No particular drift either to the left or right was discernible.

Many of the answers were peculiarly Japanese. The worry about the aged was hardly unexpected in a country where filial piety is still strong, but where urban welfare care has yet to replace adequately the village family living that once gave the old such security.

Although most Japanese youths complained about low wages, they were least worried of all about competition or limitations on their advancement in their job—once again reflecting the stability of the Japanese permanent employment system and the relatively high degree of social mobility the meritocracy offers.

The pessimism was underscored in the second 'youth consciousness' survey, the results of which were announced in July, 1978. Although some of the extreme statistics had improved somewhat—e.g., the percentage of 'dissatisfied' had dropped from 73.5 to 57—Japan's youth still ranked among the world's darkest pessimists. If the young Japanese in 1978, as in 1973, remained proud of their cultural heritage, their measure of "dedication to one's own country" was strikingly lower than that in any other countries polled. As in 1973, a heavy majority of the youth in the world's second most productive working democracy justified political passivity by stating the problems of their society to be "beyond the reach of any individual."

What is bothering them? "Senso o shiranai kodomotachi," as a popular song of the late sixties ran, "Children who have not known war," the generation growing up at the time of the Ise Shrine's renewal in 1973 is the first in Japan's modern history without experience of war. The first audiovisually nurtured generation, they are the first to think of meat and bread on the same level as fish and rice, the first to regard the car as a necessary utility-cum-status symbol rather than a remote luxury. The most highly educated generation in Japan's history, they are the least politically committed.

Educated under a democracy quite different from anything their fathers or grandfathers knew, they are far more individualist than their elders. The Japanese equivalent of "What's in it for me?" is heard rather frequently in the seventies. Objects of possession are familiarly used with the English pronoun attached, for instance, a house of one's own is called *mai homu* ("my home") or *mai ka* ("my car"). They have an almost obsessive itch to travel overseas. They come closer to having an internationalist point of view than any generation since the small group of Meiji reformers started reading German military manuals and dragged their complaining wives to Western-style dances at the Rokumeikan ballroom in the 1880s. And they were just a single, if influential, stratum. This generation is internationally oriented in the mass. In their popular art, culture, and entertainment, the sports they play—the youth of Japan are probably closer to their American contemporaries, in particular, than the children of any other large nation.

Indeed, their frustrations and concerns are very similar to those of American young people. They inherited a world of opulence which they did not make, but one whose problems are probably more obviously insoluble than at any time in history. They have been overindulged. The word of the overdependent society—*amae*—keeps recurring in Japanese analyses of the youth polls. They are conspicuously underled. Inspiring leaders are even harder to find in Japan than in the United States. The dreams of an affluent Japan are largely realized. But neither the Socialist vision of revolution nor the capitalist ideal of an even bigger, richer future holds much meaning for them.

But if they are underled, Japan's young people are certainly overdirected. The changes in the physical externals of life in Japan in the leap toward affluence, unlike the United States, were not echoed by a corresponding semirevolution in social attitudes among the movers and shakers of the society despite the frequent demonstrations. There have been few lasting protest movements, and no basic re-examinations of society like that following the racial revolution in the United States. When the happy hedonist comes back from his skin diving, driving his sportscar breakneck around the curves, or drag racing through the harried streets of small towns, he reports on Monday to much the same kind of office in Tokyo that his father worked in. Indeed, his father or his uncle or his grandfather is still running it.

That is perhaps a basic source of Japanese youth's pessimism and indifference, although it is not much talked about. Thanks to the Japanese manner of decision making, it is true, young men participate in important planning and decisions probably ahead of many of their contemporaries in the United States and certainly in Europe. In the actual possession of authority, however, they lag far behind. Going into the eighties, Japan is still a society run by old men at the top.

There are many signs that this is passing. Tanaka Kakuei's becoming premier at fifty-four was certainly among them. But the average government ministry, the average company, although administered by men in their fifties, is ultimately directed by ministers, company presidents, and chairmen whose ages range from the early sixties to the late seventies, with eighty- or eighty-five-year-old chairmen by no means exceptional. A man like Averell Harriman, who managed to be astonishingly active in government and diplomacy until his eighties, was an oddity in the youth-emphasis society of the United States. In Japan he would have been merely a normal member of the recognized establishment. The dynamic thirty-five- to forty-year-old president is a fixture in American business, but it is extremely difficult to get a thirty-nine-year-old director approved by a Japanese company.

The young people of Japan are aware that the world around them is different from theirs. They are surrounded, in a sense, by a youth culture that sets the same premium on being cool, with it, in it, and above all YOUNG, in its ads, its songs, its television, as the United States. But they still live in the land of seniority. The Ise Shrine may get new timbers every twenty years, but it and its country continue to be run by the same presidents, the same ministers, the same high priests.

When Japanese youth's discontent surfaces, it generally explodes in violence—the steel-pipe beatings of rival student activists in the universities, the bizarrely cruel activities of the so-called Red Army, their activists spraying innocent passengers with machine guns at an Israeli airport, or young senior high school students exploding bombs at police boxes. These are the acts only of tiny minorities; but they are a warning sign. Japan had a similar warning, in different situations, in the young officer's revolts of the thirties. There was the same confused idealism, the same search for an easy solution to confusing problems. "You can see only the alliance of plutocracies and political parties. . . . Japan is decaying in chaos and corruption. Stand up, fellow Japanese, for revolution." Those words were written not in the 1970's by a student activist of the time, but in 1932 by Lieutenant Minami Taku, one of the leaders of the May 15 military revolt. While the officers of the thirties revolted against economic scarcities, the modern protestors are vaguely denouncing affluence. It is at least a backhanded tribute to the new Japan's internationalism that the Red Army activists of the seventies are trying to liberate not just Japan but the world.

Mostly, however, the discontent of Japanese youth is submerged and infrequently expressed. One can hardly blame them for being confused and pessimistic. The new generation can readily see the deficiencies in its suddenly expanded schools, its government, and the thousands of offices where people who have been brought up to respect individualism bow respectfully to the assistant department head and listen solemnly to the managing director's formal quotations from Ninomiya Sontoku [8] to work harder and take pride in what you do. For the first time in many years, a Japanese generation is looking up from its workbench and asking, just a bit, where the effort is supposed to end and what it is for.

There are no revolutions on the horizon. Despite the pessimistic poll answers, the youth of Japan have remained remarkably stable, through a period of extreme changes in their environment. And given the talent of Japanese society for superposition, there is no reason to suppose that the drag racing individualist of the weekend will not continue to work hard at the old-fashioned virtues from Monday to Friday. He remains a conformist in many ways. The sex habits of Japan's young people, for one thing, are almost astonishingly conservative, in contrast to the revolution in sex attitudes taking place in Europe and the United States. In the American sense, the Japanese youth is far from an individualist, anyway. Everything

8. Ninomiya (1787–1856), a storied business-man-sage of the late Tokugawa Era, has been held up to Japanese schoolchildren ever since as a model of piety, thrift, and industry. As the well-known children's verse runs, "gathering up firewood, weaving the straw to make our sandals, helping our parents, taking care of the young children, close to our brothers and sisters, we observe filial piety—our model is Ninomiya."

the young Japanese do, they tend to do, as always, in circles, committees, parties, groups.

Yet no generation has ever needed an additional challenge so badly. No generation has ever so badly needed people to inspire it, as well as to direct the traffic. The last of the men who were born in the Meiji era will soon be retiring, perforce. To maintain its position, not to say expand it, Japan needs a whole new Meiji generation who are as adventurous in meeting the outside world as the Meiji reformers were; but who will, in addition, take on the responsibility of living in the world and contributing to it, as well as using its inventions.

The changes that need to be made among the Japanese are most drastic. They are drastic not so much in the strain they put on Japan's physical capabilities, which are immense; but they are drastic in the changes they demand in Japanese attitudes. To play a role in the world other than that of the traveling salesman, Japan needs people who can talk to foreigners, live with foreigners, relax in the presence of foreigners, argue with them, and arouse their enthusiasm. This, in turn, means discarding the almost professional diffidence the Japanese have been exhibiting toward the rest of the world, as well as its corollary, the inner arrogance of Japanism which the diffidence often barely conceals.

To do this, the new generation need not abandon its heritage. On the contrary, it needs to rediscover part of that heritage. Besides giving Japanese society peace and harmony, the Tokugawa Shogunate imposed on the Japanese people a rigid sense of hierarchy, an unnatural restraint, an excess of formalism which was quite at odds with the early, freer traditions of Japanese culture that flourished in the Heian period. A century after the Tokugawa exclusion was broken, Japan retains many of the marks of a closed country in its attitudes toward the rest of the world. Such aloofness the Japanese can no longer afford.

The new generation appears to have a certain self-confidence. If, in addition, this generation can discover, or rediscover, a talent for communicating with the rest of the world, the gain to Japan and to the world will far outweigh the temporary confusion caused in the country by changed values and standards. If the solidarity, the trust, the work enthusiasms, and the loyalties of Japanese society can ever translate themselves, if they can, in a word, both explain and project the unique values of that society to a world outside it, the world will be much the richer. And there will be a freer society within Japan itself on hand to scatter the white pebbles and hew out the cypress beams for the shrine at Ise in 1993.

Index

The MERIDIAN Quality Paperback Collection

℗

The Best in Fiction from PLUME

☐ **THE NEW YORK TIMES BOOK OF BACKGAMMON**
by James and Mary Zita Jacoby.   (#Z5248—$3.95)

☐ **JOY OF COOKING** by Irma S. Rombauer and
Marion Rombauer Becker   (#Z5259—$6.95)

☐ **UNDER THE VOLCANO** by Malcolm Lowry.   (#Z5103—$3.95)

☐ **CAREERS TOMORROW** by Gene LR Hawes.   (#Z5214—$4.95)

☐ **AMERICAN FILM NOW** by James Monaco   (#Z5212—$7.95)

☐ **PRICKSONGS AND DESCANTS** by Robert Coover.
(#Z5208—$3.95)

☐ **THE FOUR-GATED CITY** by Doris Lessing. (#Z5135—$5.95)*

☐ **LANDLOCKED** by Doris Lessing.   (#Z5138—$3.95)*

☐ **MARTHA QUEST** by Doris Lessing.   (#Z5095—$3.95)*

☐ **A RIPPLE FROM THE STORM** by Doris Lessing.
(#Z5137—$3.95)*

☐ **A PROPER MARRIAGE** by Doris Lessing.   (#Z5093—$3.95)*

☐ **THE DANGLING MAN** by Saul Bellow.   (#Z5251—$3.95)

☐ **THE VICTIM** by Saul Bellow.   (#Z5154—$3.95)

*Not available in Canada
In Canada, please add $1.00 to the price of each book.

Buy them at your local bookstore or use this convenient
coupon for ordering.

THE NEW AMERICAN LIBRARY, INC.
P.O. BOX 999, BERGENFIELD, NEW JERSEY 07621

Please send me the PLUME and MERIDIAN BOOKS I have checked
above. I am enclosing $_____(please add $1.50 to this
order to cover postage and handling). Send check or money order
—no cash or C.O.D.'s. Prices and numbers are subject to change
without notice.

Name_____

Address_____

City_____State_____Zip Code_____
Allow 4-6 weeks for delivery.
This offer is subject to withdrawal without notice.